Ulrich Knauer and Kolja Knauer

Algebraic Graph Theory

De Gruyter Studies in Mathematics

Volume 41

Ulrich Knauer and Kolja Knauer

Algebraic Graph Theory

Morphisms, Monoids and Matrices

2nd edition

DE GRUYTER

Mathematics Subject Classification 2010
05C05, 05C10, 05C12, 05C20, 05C25, 05C12, 05C38, 05C50, 05C62, 05C75, 05C90, 20M17, 20M19, 20M20, 20M30, 18B10, 18B40

Authors
Prof. Dr. Dr. h.c. Ulrich Knauer
Carl von Ossietzky Universität
Institut für Mathematik
Ammerländer Heerstr. 114–118
26129 Oldenburg
Germany
ulrich.knauer@uni-oldenburg.de

Dr. Kolja Knauer
Université Aix-Marseille
Laboratoire d'Informatique
et des Systèmes, LIS UMR 7020
163 avenue de Luminy
13288 Marseille Cedex 9
France
kolja.knauer@lis-lab.fr

ISBN 978-3-11-061612-5
e-ISBN (PDF) 978-3-11-061736-8
e-ISBN (EPUB) 978-3-11-061628-6
ISSN 0179-0986

Library of Congress Control Number: 2019946001

Bibliographic information published by the Deutsche Nationalbibliothek
The Deutsche Nationalbibliothek lists this publication in the Deutsche Nationalbibliografie; detailed bibliographic data are available on the Internet at http://dnb.dnb.de.

Typesetting: VTeX UAB, Lithuania
Printing and binding: CPI books GmbH, Leck

www.degruyter.com

Preface

This book is a collection of the lectures I have given on algebraic graph theory. These lectures were designed for mathematics students in a Master's program, but they may also be of interest to undergraduates in the final year of a Bachelor's curriculum.

The lectures cover topics which can be used as starting points for a Master's or Bachelor's thesis. Some questions raised in the text could even be suitable as subjects of doctoral dissertations. The advantage afforded by the field of algebraic graph theory is that it allows many questions to be understood from a general mathematical background and tackled almost immediately.

In fact, my lectures have also been attended by graduate students in informatics with a minor in mathematics. In computer science and informatics, many of the concepts associated with graphs play an important role as structuring tools—they enable us to model a wide variety of different systems, such as the structure of physical networks (of roads, computers, telephones, etc.) as well as abstract data structures (e. g., lists, stacks, trees); functional and object oriented programming are also based on graphs as a means of describing discrete entities. In addition, category theory is gaining more and more importance in informatics; therefore, these lectures also include a basic and concrete introduction to categories, with numerous examples and applications.

I gave the lectures first at the University of Bielefeld and then, in various incarnations, at the Carl von Ossietzky Universität Oldenburg. They were sometimes presented in English and in several other countries, including Thailand and New Zealand.

Selection of topics

The choice of topics is in part standard, but it also reflects my personal preferences. Many students seem to have found the chosen topics engaging, as well as helpful and useful in getting started on thesis research at various levels.

To mark the possibilities for further research, I have inserted many "Questions," as well as "Exercises" that lead to illuminating examples. Theorems for which I do not give proofs are sometimes titled "Exerceorem," to stress their role in the development of the subject. I have also inserted some "Projects," which are designed as exercises to guide the reader in beginning their own research on the topic. I have not, however, lost any sleep over whether to call each result a theorem, proposition, exerceorem, or something else, so readers should neither deduce too much from the title given to a result nor be unduly disturbed by any inconsistencies they may discover—this beautiful English sentence I have adopted from the introduction of John Howie's *An Introduction to Semigroup Theory*, published by Academic Press in 1976.

https://doi.org/10.1515/9783110617368-201

Homomorphisms, especially endomorphisms, form a common thread throughout the book; you will meet this concept in almost all the chapters. Another focal point is the standard part of algebraic graph theory dealing with matrices and eigenvalues. In some parts of the book, the presentation will be rather formal; my experience is that this can be very helpful to students in a field where concepts are often presented in an informal verbal manner and with varying terminology.

Content of the chapters

We begin, in Chapter 1, with basic definitions, concepts, and results. This chapter is very important, as standard terminology is far from being established in graph theory. One reason for this is that graph models are so extremely useful in a great number of applications in diverse fields. Many of the modelers are not mathematicians and have developed their own terminology and results, without necessarily caring much about existing theory. Chapter 1 contains some new variants of results on graph homomorphisms and the relations among them, connecting them, in turn, to the combinatorial structure of the graph.

Chapter 2 makes connections to linear algebra by discussing the different matrices associated to graphs. We then proceed to the characteristic polynomial and eigenvalues, topics that will be encountered again in Chapters 5 and 8. There is no intention to be complete, and the content of this chapter is presented at a relatively elementary level.

In Chapter 3, we introduce some basic concepts from category theory, focusing on what will be helpful for a better understanding of graph concepts.

In Chapter 4, we look at graphs and their homomorphisms, in particular binary operations such as unions, amalgams, products, and tensor products; for the latter two operations I use the illustrative names cross product and box product. It turns out that, except for the lexicographic products and the corona, all of these operations have a category-theoretical meaning. Moreover, adjointness leads to so-called Mor constructions; some of the ones presented in this chapter are new, as far as I know, and I call them diamond and power products.

In Chapter 5, we focus on unary operations such as the total graph, the tree graph and, principally, line graphs. Line graphs are dealt with in some detail; in particular, their spectra are discussed. Possible functorial properties are left for further investigation.

In Chapter 6, the fruitful notion of duality, known from and used in linear algebra, is illustrated with the so-called cycle and cocycle spaces. We then apply the concepts to derive Kirchhoff's laws and to "square the rectangle." The chapter finishes with a short survey of applications to transportation networks.

Chapter 7 discusses several connections between graphs and groups and, more generally, semigroups or monoids. We start with Cayley graphs and Frucht-type results, which are also generalized to monoids. We give results relating the groups to combinatorial properties of the graph as well as to algebraic aspects of the graph.

In Chapter 8, we continue the investigation of eigenvalues and the characteristic polynomial begun in Chapters 2 and 5. Here, we present more of the standard results. Many of the proofs in this chapter are omitted, and sometimes we mention only the idea of the proof.

In Chapter 9, we present some results on endomorphism monoids of graphs. We study von Neumann regularity of endomorphisms of bipartite graphs, locally strong endomorphisms of paths, and strong monoids of arbitrary graphs. The chapter includes a fairly complete analysis of the strong monoid, with the help of lexicographic products on the graph side and wreath products on the monoid side.

In Chapter 10, we discuss unretractivities, i. e., under what conditions on the graph do its different endomorphism sets coincide? We also investigate questions such as how the monoids of composed graphs (e. g., product graphs) relate to algebraic compositions (e. g., products) of the monoids of the components. This type of question can be interpreted as follows: when is the formation of the monoid product-preserving?

In Chapter 11, we come back to the formation of Cayley graphs of a group or semigroup. This procedure can be considered as a functor. As a side line, we investigate (in Section 11.2) preservation and reflection properties of the Cayley functor. This is applied to Cayley graphs of right and left groups and is used to characterize Cayley graphs of certain completely regular semigroups and strong semilattices of semigroups.

In Chapter 12, we resume the investigation of transitivity questions from Chapter 8 for Cayley graphs of strong semilattices of semigroups, which may be groups or right or left groups. We start with Aut- and ColAut-vertex transitivities and finish with endomorphism vertex transitivity. Detailed examples are used to illustrate the results and open problems.

Chapter 13 considers a more topological question: what are planar semigroups? This concerns extending the notion of planarity from groups to semigroups. We choose semigroups that are close to groups, i. e., which are unions of groups with some additional properties. So we investigate right groups and Clifford semigroups, which were introduced in Chapter 9. We note that the more topological questions about planarity, embeddings on surfaces of higher genus, or colorings are touched on only briefly in this book. We use some of the results in certain places where they relate to algebraic analysis of graphs—the main instances are planarity in Section 6.4 and Chapter 13, and the chromatic number in Chapter 7 and some other places.

Each chapter ends with a “Comments” section, which mentions open problems and some ideas for further investigation at various levels of difficulty. I hope they will stimulate the reader’s interest.

How to use this book

The text is meant to provide a solid foundation for courses on algebraic graph theory. It is highly self-contained, and includes a brief introduction to categories and functors and even some aspects of semigroup theory.

Different courses can be taught based on this book. Some examples are listed below. In each case, the prerequisites are some basic knowledge of linear algebra.

- Chapters 1 through 8—a course covering mainly the matrix aspects of algebraic graph theory.
- Chapters 1, 3, 4, 7, and 9 through 13—a course focusing on the semigroup and monoid aspects.
- A course skipping everything on categories, namely Chapter 3, the theorems in Sections 4.1, 4.2, 4.3, and 4.6 (although the definitions and examples should be retained) and Sections 11.1 through 11.2.
- Complementary to the preceding option, it is also possible to use this text as a short and concrete introduction to categories and functors, with many (somewhat unusual) examples from graph theory, by selecting exactly those parts skipped above.

About the literature

The literature on graphs is enormous. In the bibliography at the end of the book, I give a list of reference books and monographs, almost all on graphs, ordered *chronologically* starting from 1936; it is by no means complete. As can be seen from the list, a growing number of books on graph theory are published each year. Works from this list are cited in the text by author name(s) and publication year enclosed in square brackets.

Here, I list some books, not all on graphs, which are particularly relevant to this text; some of them are quite similar in content and are cited frequently.

- N. Biggs, *Algebraic Graph Theory*, Cambridge University Press, Cambridge 1996.
- M. Behzad, G. Chartrand, L. Lesniak-Forster, *Graphs and Digraphs*, Prindle, Weber & Schmidt, Boston 1979. New (fifth) edition: G. Chartrand, L. Lesniak, P. Zhang, *Graphs and Digraphs*, Chapman and Hall, London 2010.
- D. Cvetković, M. Doob, H. Sachs, *Spectra of Graphs*, Academic Press, New York 1979.
- C. Godsil, G. Royle, *Algebraic Graph Theory*, Springer, New York 2001.
- G. Hahn, G. Sabidussi (eds.), *Graph Symmetry*, Kluwer, Dordrecht 1997.
- P. Hell, J. Nešetřil, *Graphs and Homomorphisms*, Oxford University Press, Oxford 2004.
- H. Herrlich, G. Strecker, *Category Theory*, Allyn and Bacon, Boston 1973.
- W. Imrich, S. Klavžar, *Product Graphs*, Wiley, New York 2000.

- R. Kaschek, U. Knauer (eds.), *Graph Asymmetries*, Discrete Mathematics 309 (special issue) (2009) 5349–5424.
- M. Kilp, U. Knauer, A. V. Mikhalev, *Monoids, Acts and Categories*, De Gruyter, Berlin 2000.
- M. Petrich, N. Reilly, *Completely Regular Semigroups*, Wiley, New York 1999.
- D. B. West, *Introduction to Graph Theory*, Prentice Hall, Upper Saddle River, NJ 2001.

Papers, theses, book chapters, and other references are given in the text where they are used.

Acknowledgments

This book was originally planned as a joint work with Lothar Budach. But the untimely death of Lothar, before we could really get started, cut short our collaboration.

First of all, I thank the De Gruyter publishing house and especially the Mathematics Editor Friederike Dittberner, as well as Anja Möbius, for their cooperation and help.

Thanks go to all mathematicians, living or dead, whose work I have cannibalized freely—starting with Horst Herrlich, from whose book *Axiom of Choice* (Springer, Heidelberg 2006) I have taken this sentence.

Moreover, I thank all the students who have contributed over the years to improving the course—especially Barbara Langfeld, who looked through a final version of the text. One of my first students was Roland Kaschek, who became a professor in New Zealand, and one of most recent was Xia Zhang, who is now a professor in China. I thank all the colleagues who have used parts of my notes for their own courses and lectures.

I thank Mati Kilp (Estonia) for his contributions to several parts of the book, and I am grateful to Kolja Knauer for his comments, suggestions and contributions specially to Chapter 13. I also thank Sayan Panma (Thailand) who contributed many results to Chapters 11 and 12 while he was a PhD student at Oldenburg.

I thank Theresia Meyer for doing much of the typesetting of earlier versions, drawing pictures and providing various other sorts of technical help.

Above all, I thank my wife, Jane Richter, for her many good ideas as a nonmathematician, her encouragement, and her patience with me during the intense periods of work.

Oldenburg and Berlin, April 2011 — Ulrich Knauer

Preface for the second edition

We made the authorship younger, without changing their names too much.

We corrected mistakes and misprints as they were pointed out to us or found by ourselves. The list of literature is no longer ordered by appearance date, but by author's names. And it is updated. Still we mention all books relevant to our subjects as we know or found them, whether cited or not in the List of books. In addition, we collect the works cited in the text, which do not go in the list of books in a new List of cited papers, theses etc. In both lists we noted the pages where the items are cited. We made some minor rearrangements. So, for example, we added a section on the genus of graphs to Chapter 1, thereby shortening later parts. We broadened the scope of the chapters on Cayley graphs of semigroups by several new hints. We thoroughly worked over the Section "Planar Clifford semigroups" and completed it with new results, and added a section on planar semilattices.

And, hopefully, we improved formulations, proofs and statements, and added more examples, figures, and pictures. Also Index and Symbol Index became longer.

We will publish mistakes and corrections found after appearance of the book on the following webpage:

http://www.degruyter.com/books/978-3-11-061612-5

We again thank De Gruyter and the Mathematics Editor, Dr. Apostolos Damialis, who initiated the second edition of this work, as well as Nadja Schedensack and, in particular, the Project Manager Ina Talandienė, for their cooperation and help. We also repeat our thanks to Jane Richter, who patiently ignored our frequent discussions about the book.

Note added in proof

We want to keep the memory of our friend and teacher Horst Herrlich (11.09.1937–13.03.2015).

Berlin, Marseille, Barcelona, 2019 Ulrich Knauer and Kolja Knauer

https://doi.org/10.1515/9783110617368-202

Contents

1 Directed and undirected graphs

In this chapter, we collect some important basic concepts. These concepts are essential for all mathematical modeling based on graphs. The language and visual representations of graphs are such powerful tools that graph models can be encountered almost everywhere in mathematics and informatics, as well as in many other fields.

The most obvious phenomena that can be modeled by graphs are binary relations. Moreover, graphs and relations between objects in a formal sense can be considered the same. The concepts of graph theory also play a key role in the language of category theory, where we consider objects and morphisms.

It is not necessary to read this chapter first. Anybody who is familiar with the basic notions may just refer back to this chapter as needed for a review of notation and concepts.

1.1 Formal description of graphs

We shall use the word "graph" to refer to both directed and undirected graphs. Only when discussing concepts or results that are specific to one of the two types of graph we will use the corresponding adjective explicitly.

Definition 1.1.1. A ***directed graph*** or ***digraph*** or also ***oriented graph*** is a triple $G = (V, E, p)$ where V and E are sets and

$$p : E \to V^2$$

is a mapping. We call V the set of ***vertices*** or ***points*** and E the set of ***edges*** or ***arcs*** of the graph. Sometimes we will write these sets as $V(G)$ and $E(G)$. The mapping p is called the ***incidence mapping***.

The mapping p defines two more mappings $o, t : E \to V$ by $(o(e), t(e)) := p(e)$; these are also called incidence mappings. We call $o(e) \in V$ the ***origin*** or ***source*** and $t(e) \in V$ the ***tail*** or ***end*** of $e \in E$.

As p defines the mappings o and t, these in turn define p by $p(e) := (o(e), t(e))$. We will mostly be using the first of the two alternatives

$$G = (V, E, p) \quad \text{or} \quad G = (V, E, o, t).$$

We say that the vertex x and the edge e are ***incident*** if x is the source or the tail of e. The edges e and e' are said to be ***incident*** if they have a common vertex.

An ***undirected graph*** is a triple $G = (V, E, p)$ such that

$$p : E \to \{V' \subseteq V \mid 1 \leq |V'| \leq 2\}.$$

An edge e with $o(e) = t(e)$ is called a ***loop***. The preimage $p^{-1}(x, y)$ is called a ***multiple edge*** if $|p^{-1}(x, y)| > 1$. In this case, elements of $p^{-1}(x, y)$ are said to be ***parallel***.

https://doi.org/10.1515/9783110617368-001

Let $G = (V, E, o, t)$ be a directed graph, let e be an edge, and let $x = o(e)$ and $y = t(e)$; then we also write $e : x \to y$. The vertices of graphs are drawn as points or circles; directed edges are arrows from one point to another, and undirected edges are lines, or sometimes two-sided arrows, joining two points. The name of the vertex or edge may be written in the circle or close to the point or edge. See Figure 1.1 for an illustration.

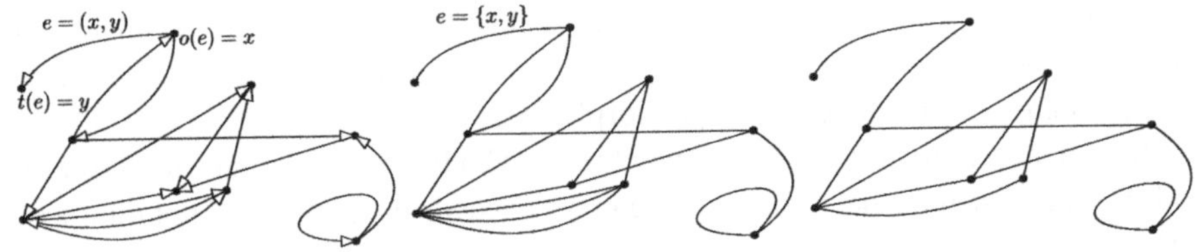

Figure 1.1: From left to right: A directed graph, its underlying undirected multigraph, and the simple graph obtained by forgetting multiedges.

Definition 1.1.2. Let $G = (V, E, p)$ be a graph. If G has no multiple edges, then we call G a ***simple graph***. Otherwise, we call G a ***multigraph*** or ***multiple graph***; sometimes the term ***pseudograph*** is used.

If $G = (V, E, p)$ is a simple graph, we can consider E as a subset of V^2, identifying $p(E)$ with E. We then write $G = (V, E)$ or $G = (V_G, E_G)$, and for the edge e with $p(e) = (x, y)$ we write (x, y).

Simple graphs can now be defined as follows: a ***simple directed graph*** is a pair $G = (V, E)$ with $E \subseteq V^2 = V \times V$. Then we again call V the set of ***vertices*** and E the set of ***edges***.

A ***simple undirected graph*** is a simple directed graph $G = (V, E)$ such that

$$(x, y) \in E \Leftrightarrow (y, x) \in E.$$

The edge (x, y) may also be written as $\{x, y\}$ or xy.

This undirected graph is denoted by $\underline{G}$ and is called the completion of G or the underlying (undirected) graph of G.

Mappings $w : E \to W$ or $w : V \to W$ are called ***weight functions***. Here, W is any set, called the ***set of weights***, and $w(x)$ is called the ***weight*** of the edge x or of the vertex x.

Definition 1.1.3. A ***path*** a from x to y or an $\boldsymbol{x, y}$ ***path*** in a graph G is a sequence $a = (e_1, e_2, \ldots, e_n)$ of edges with $o(e_1) = x$, $t(e_n) = y$ and $t(e_{i-1}) = o(e_i)$ for $i = 2, \ldots, n$. We write $a : x \to y$ and call x the ***start (origin, source)*** and y the ***end (tail, sink)*** of the path a. The sequence $x_0, \ldots, x_n$ is called the ***trace*** of the path a. The set $\{x_0, \ldots, x_n\}$ of all vertices of the trace is called the ***support of the path a***, denoted by supp a.

A path is said to be ***simple*** if every vertex appears at most once on the path. A path is said to be ***closed***, or is called a ***cycle***, if the start and end of the path coincide. A simple closed path, i. e., a simple cycle, is called a ***circuit***. The words ***(simple) semi-path, semicycle***, or ***semicircuit*** will be used if, in the sequence of edges, the tail or origin of each edge equals the origin or tail of the next edge. This means that at least two consecutive edges have opposite directions. The notions of trace and support remain unchanged. In a simple graph, every (semi)path is uniquely determined by its trace. We can describe a path also by its vertices $x_0, \dots, x_n$ where $(x_0, x_1), \dots, (x_{n-1}, x_n)$ are edges of the path. For undirected graphs, the notions of path and semipath are identical.

For the sake of completeness, we also mention the following definition: the ***trivial*** x, x ***path*** is the path consisting only of the vertex x. It is also called a ***lazy path***.

The reader should be aware that, in the literature, the words "cycle" and "circuit" are often used in different ways by different authors.

Lemma 1.1.4. *For $x, y \in G$, every x, y path contains a simple x, y path. Every cycle in G is the union of circuits.*

Proof. Take $x, y \in G$. Start on an x, y path from x and proceed until one vertex z is met for the second time. If this does not happen, we already have a simple path; otherwise, we have also traversed a circuit. Remove this circuit, together with all its vertices but z, from the path. Continuing this procedure yields a simple x, y path. If we start with a cycle, we remove one edge $e = (y, x)$, and this gives an x, y path. Now collect the circuits as before. At the end, we have a simple x, y path, which together with e gives the last circuit. □

Definition 1.1.5. Let $G = (V, E)$, and let $a = (e_1, \dots, e_r)$ be a path with $e_i \in E$. Then $\ell(a) := r$ is called the ***length*** of a.

We denote by $F(x, y)$ the set of all x, y paths in G. Then $d(x, y) := \min\{\ell(a) \mid a \in F(x, y)\}$ is called the ***distance*** from x to y.

We call $\operatorname{diam}(G) := \max_{x,y \in G} d(x, y)$ the ***diameter*** of G. The length of a shortest cycle of G is called the ***girth*** of G. In German the figurative word *Taillenweite*, meaning circumference of the waist, is used.

Remark 1.1.6. In connected, symmetric graphs the distance $d : V \times V \to \mathbb{R}_0^+$ is a metric, if we set $d(x, x) = 0$ for all $x \in V$. In this way, (V, d) becomes a metric space. If $\{\ell(a) \mid a \in F(x, y)\} = \emptyset$, then $d(x, y)$ is not defined. Often one sets $d(x, y) = \infty$ in this case.

Definition 1.1.7. For a vertex x of a graph G, the ***outset*** of x is the set

$$\operatorname{out}(x) := \operatorname{out}_G(x) := \{e \in E \mid o(e) = x\}.$$

The elements of

$$N^+(x) := N_G^+(x) := \{t(e) \mid e \in \mathrm{out}_G(x)\}$$

are called the ***successors*** of x in G. The ***outdegree*** of a vertex x is the number of successors of x; that is,

$$\overleftarrow{d}(x) = \mathrm{outdeg}(x) := |\mathrm{out}(x)|.$$

Definition 1.1.8. The graph $G^{\mathrm{op}} := (V, E, t, o)$ is called the ***opposite graph*** to G.

The ***inset*** of a vertex x is the outset of x in the opposite graph G^{op}, so

$$\mathrm{in}(x) = \mathrm{in}_G(x) := \mathrm{out}_{G^{\mathrm{op}}}(x) = \{e \in E \mid t(e) = x\}.$$

The elements of

$$N^-(x) := N_G^-(x) := N_{G^{\mathrm{op}}}^+(x) := \{o(e) \mid e \in \mathrm{in}_G(x)\}$$

are called ***predecessors*** of x in G. The ***indegree*** of a vertex x is the number of predecessors of x; that is,

$$\overrightarrow{d}(x) = \mathrm{indeg}(x) := |\mathrm{in}(x)|.$$

A vertex which is a successor or a predecessor of the vertex x is said to be ***adjacent to x***.

Definition 1.1.9. In an undirected graph G, a predecessor of a vertex x is at the same time a successor of x. Therefore, in this case, $\mathrm{in}(x) = \mathrm{out}(x)$ and $N(x) := N^+(x) = N^-(x)$. We call the elements of $N(x)$ the ***neighbors*** of x. Similarly, $\mathrm{indeg}(x) = \mathrm{outdeg}(x)$. The common value $d_G(x) = d(x) =: \deg(x)$ is called the ***degree*** of x in G.

An undirected graph is said to be ***regular*** or ***d-regular*** if all of its vertices have degree d. For directed graphs, similar notation can be defined.

1.2 Connectedness and equivalence relations

Here, we make precise some very natural concepts, in particular, how to reach certain points from other points.

Definition 1.2.1. A directed graph G is said to be:
- ***weakly connected*** if for all $x, y \in V$ there exists a semipath from x to y;
- ***one-sided connected*** if for all $x, y \in V$ there exists a path from x to y or from y to x;
- ***strongly connected*** if for all $x, y \in V$ there exists a path from x to y and from y to x.

For undirected graphs, the three concepts coincide. We then simply say that the graph is ***connected***; we shall also use this word as a common name for all three concepts.

If G satisfies none of the above three conditions, it is said to be ***unconnected*** or ***disconnected***.

Example 1.2.2. The following three graphs illustrate the three properties above, in the order given.

Definition 1.2.3. A connected graph is said to be ***n-vertex connected*** if at least n vertices must be removed to obtain an unconnected graph. Analogously, one can define ***n-edge connected*** graphs.

Remark 1.2.4. A ***binary relation*** on a set X is usually defined as a subset of the Cartesian product $X \times X$. This often bothers beginners, since it seems too simple a definition to cover all the complicated relations in the real world that one might wish to model. It is immediately clear, however, that every binary relation is a directed graph and vice versa. This is one reason that much of the literature on binary relations is actually about graphs. Arbitrary relations on a set can similarly be described by multigraphs.

An ***equivalence relation*** on a set X, i. e., a reflexive, symmetric and transitive binary relation in this setting, corresponds to a disjoint union of various graphs with loops at every vertex (reflexivity) which are undirected (symmetry), and such that any two vertices in each of the disjoint graphs are adjacent (transitivity). Note that the above mentioned disjoint union is due to the fact that an equivalence relation on a set X provides a partition of the set X into disjoint subsets and vice versa.

1.3 Some special graphs

We now define some standard graphs. These come up everywhere, in virtually any discussion about graphs, so will serve as useful examples and counterexamples.

Definition 1.3.1. In the ***complete graph*** $K_n^{(l)}$ with n vertices and l loops, where $0 \le l \le n$, any two vertices are adjacent and l of the vertices have a loop.

The ***totally disconnected*** or ***discrete graph*** $\overline{K}_n^{(l)}$ with n vertices and l loops has no edges between distinct vertices and has loops at l vertices. If $l = 0$, we write K_n or $\overline{K}_n$.

A simple, undirected path with n edges is denoted by P_n.

An undirected circuit with n edges is denoted by C_n.

An ***r-partite graph*** admits a partition of the vertex set V into r disjoint subsets $V_1, \ldots, V_r$ such that no two vertices in one subset are adjacent.

An r-partite graph is said to be ***complete r-partite*** if all pairs of vertices from different subsets are adjacent. The complete bipartite graph with $|V_1| = m$ and $|V_2| = n$ is denoted by $K_{m,n}$; similarly for complete r-partite graphs. A picture of the complete 3-partite graph $K_{1,2,3}$ is in Example 9.5.6.

Example 1.3.2 (Some special graphs).

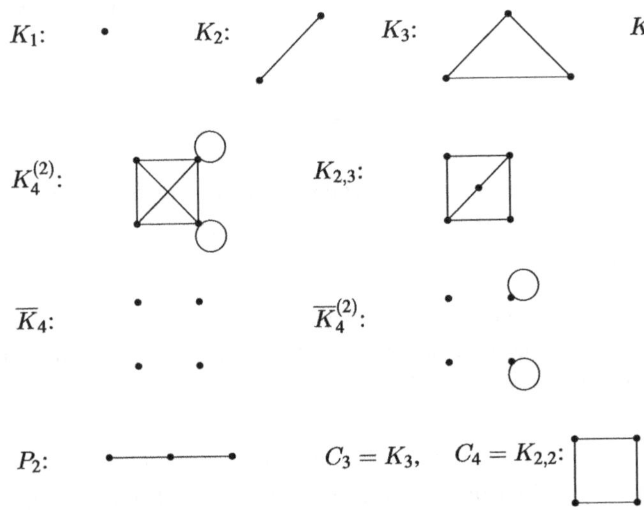

Definition 1.3.3. For $n \geq 0$, the ***n-cube*** Q_n has a vertex set the set $\{0,1\}^n$ of 0, 1-vectors of length n. We have $Q_0 = K_1$, $Q_1 = P_1$, and $Q_2 = C_4$. A drawing of Q_3 can be found in Figure 1.7 and a drawing of Q_4 is given in Figure 13.9.

Definition 1.3.4. For $n \geq 4$ and $\frac{n}{2} > k \geq 1$ the ***generalized Petersen graph*** $G(n,k)$ is the graph on $2n$ vertices $\{v_0, w_0, v_1, w_1, \ldots, v_{n-1}, w_{n-1}\}$ with edges $\{v_iv_j\}$ if $i-j$ is 1 modulo n, edges when indices are seen modulo $\{w_iw_j\}$ if $i-j$ is k modulo n, and edges $\{v_iw_i\}$ for all $0 \leq i \leq n-1$. See Figure 1.2 below, for $G(8,3)$. The ***Petersen graph*** itself is $G(5,2)$; see Figure 5.1, page 94. We have $Q_3 = G(4,1)$ and more generally one calls $G(n,1)$ the ***n-prism***; see Figure 13.3, page 263, for further examples. The ***Dodecahedron*** is the graph $G(10,2)$; see Figure 5.2, page 94.

Definition 1.3.5. A graph without (semi)circuits is called a ***forest***. A connected forest is called a ***tree*** of G. A connected graph G' with the same vertex set as G is called a ***spanning tree*** if it is a tree. If G is not connected, the union of spanning trees for the components of G is called a ***spanning forest***.

We give the following well-known result without proof.

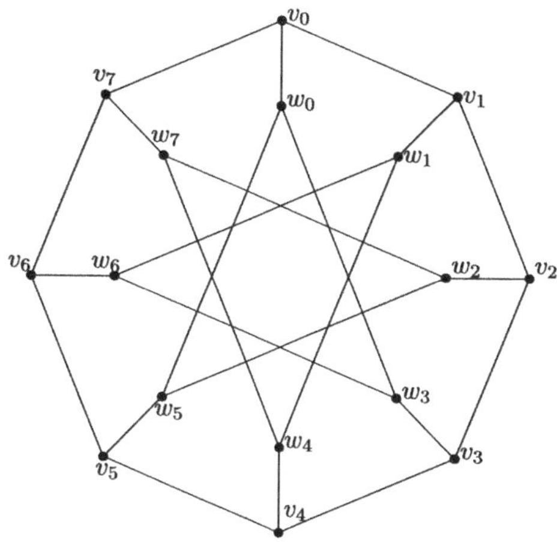

Figure 1.2: The ***Möbius–Kantor graph*** is the generalized Petersen graph $G(8,3)$.

Theorem 1.3.6. *Let G be a graph with n vertices. The following statements are equivalent:*

(i) *G is a tree.*
(ii) *G contains no semicircuits and has* $n - 1$ *edges.*
(iii) *G is weakly connected and has* $n - 1$ *edges.*
(iv) *Any two vertices of G are connected by exactly one semipath.*
(v) *Adding any edge to G produces exactly one semicircuit.*

Theorem 1.3.7. *A graph is bipartite if and only if it has no semicircuits with an odd number of edges.*

Proof. For "⇒," let $V = V_1 \bigcup V_2$. Since edges exist only between V_1 and V_2, all circuits must have an even number of edges.

For "⇐," let G be connected and take $x \in V$. Take V_1 to be the set of all vertices which can be reached from x along paths using an odd number of edges. Set $V_2 := V \setminus V_1$. If G is not connected, proceed in the same way with its connected parts. Isolated vertices can be assigned arbitrarily.

If there is an edge $\{y,z\}$ in one of the V_i, then take a path from x to y, then the edge $\{y,z\}$ and then a path from y to x, such that both paths have lengths of the same parity. The resulting semicycle is odd. Lemma 1.1.4 implies that this semicycle contains an odd semicircuit. □

We recall the following definition: a pair $(P, \le)$, where P is a set with a reflexive, antisymmetric, transitive binary relation $\le$, is called a ***partially ordered set*** or a ***poset***. For $x, y, z \in P$, we write $x < y$ if $x \le y$ and $x \ne y$. We say that y ***covers*** x, written $x \prec y$, if $x < y$ and if $x \le z < y$ implies $x = z$. See also Remark 1.2.4. We give several graphical representations of a poset.

Definition 1.3.8. The ***comparability graph*** of $(P, \leq)$ is an undirected simple graph with vertex set P and edge set $\{\{x,y\} \mid x < y\}$. The ***cover graph*** of a poset $(P, \leq)$ is the graph defined on the vertex set P and edge set $\{\{x,y\} \mid x \prec y\}$. The ***Hasse diagram*** $\boldsymbol{H_P}$ of $(P, \leq)$ is a graphical representation of the cover graph of the poset $(P, \leq)$, such that such that y is above x if $x \prec y$. Defining the edge set by $\{(y,x) \mid x \prec y\}$ gives a Hasse diagram where arcs are directed "down." See Figure 1.3 for an illustration.

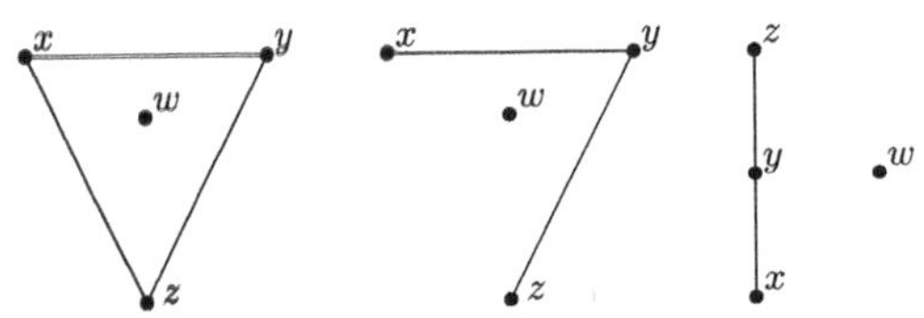

Figure 1.3: From left to right: comparability graph, cover graph, and Hasse diagram of the poset on $P = \{x, y, z, w\}$ with $x \leq y, x \leq z, y \leq z$.

Definition 1.3.9. A ***rooted tree*** is a triple $(T, \leq, r)$ such that:
- $(T, \leq)$ is a partially ordered set;
- H_T is a tree; and
- $r \in T$ is an element, the ***root of the tree***, where $x \leq r$ for all $x \in T$.

A ***marked rooted tree*** is a quadruple $(T, \leq, r, \lambda)$ such that $(T, \leq, r)$ is a rooted tree and $\lambda : T \to M$, with M being a set, is a mapping (weight function), which in this context is called the ***marking function***. We call $\lambda(x)$ a ***marking*** of x.

1.4 Homomorphisms

In mathematics, as in the real world, mappings produce images. In such images, certain aspects of the original may be suppressed, so that the image is in general simpler than the original. But some of the structures of the original, those which we want to study, should be preserved. Structure-preserving mappings are usually called homomorphisms. For graphs, it turns out that preservation of different levels of structure or different intensities of preservation quite naturally lead to different types of homomorphism.

First, we give a very general definition of homomorphisms. We will then introduce the so-called covering, which has some importance in the field of informatics. The general definition will then be specialized in various ways, and later we will use almost exclusively these variants. A reader who is not especially interested in the general aspects of homomorphisms may wish to start with Definition 1.4.3.

Definition 1.4.1. Let $G_1 = (V_1, E_1, o_1, t_1)$ and $G_2 = (V_2, E_2, o_2, t_2)$ be two directed graphs. A ***graph homomorphism*** $\theta : G_1 \to G_2$ is a pair $\theta = (\theta_V, \theta_E)$ of mappings

$$\begin{aligned} \theta_V &: V_1 \to V_2 \\ \theta_E &: E_1 \to E_2 \end{aligned}$$

such that $o_2(\theta_E(e)) = \theta_V(o_1(e))$ and $t_2(\theta_E(e)) = \theta_V(t_1(e))$ for all $e \in E_1$.

If $\theta : G_1 \to G_2$ is a graph homomorphism and v is a vertex of G_1, then

$$\theta_E(\mathrm{out}_{G_1}(v)) \subseteq \mathrm{out}_{G_2}(\theta_V(v)) \quad \text{and} \quad \theta_E(\mathrm{in}_{G_1}(v)) \subseteq \mathrm{in}_{G_2}(\theta_V(v)).$$

Definition 1.4.2. If $\theta_E|_{\mathrm{out}_{G_1}(v)}$ is bijective for all $v \in V$, we call θ a ***covering*** of G_2. If $\theta_E|_{\mathrm{out}_{G_1}(v)}$ is only injective for all $v \in V$, then it is called a ***precovering***.

For simple directed or undirected graphs, we will mostly be working with the following formulations and concepts rather than the preceding two definitions.

The main idea is that homomorphisms have to preserve edges. If, in the following, we replace "homo" by "ega," we have the possibility of identifying adjacent vertices as well. This could also be achieved with usual homomorphisms if we consider graphs that have a loop at every vertex. Illustrating pictures follow in Example 1.5.3.

Definition 1.4.3. Let $G = (V, E)$ and $G' = (V', E')$ be two graphs. A mapping $f : V \to V'$ is called a:

- ***graph homomorphism*** if $(x, y) \in E \Rightarrow (f(x), f(y)) \in E'$;
- ***graph egamorphism (weak homomorphism)*** if $(x, y) \in E$ and $f(x) \neq f(y) \Rightarrow (f(x), f(y)) \in E'$;
- ***graph comorphism (continuous graph mapping)*** if $(f(x), f(y)) \in E' \Rightarrow (x, y) \in E$;
- ***strong graph homomorphism*** if $(x, y) \in E \Leftrightarrow (f(x), f(y)) \in E'$;
- ***strong graph egamorphism*** if $(x, y) \in E$ and $f(x) \neq f(y) \Leftrightarrow (f(x), f(y)) \in E'$;
- ***graph isomorphism*** if f is a strong graph homomorphism and bijective or, equivalently, if f and f^{-1} are graph homomorphisms.

When $G = G'$, we use the prefixes "endo," "auto" instead of "homo," "iso."

We note that the term "continuous graph mapping" is borrowed from topology; there continuous mappings reflect open sets, whereas here they reflect edges.

Remark 1.4.4. Note that, in contrast to algebraic structures, bijective graph homomorphisms are not necessarily graph isomorphisms. This can be seen from Example 1.4.9; there the nonstrong subgraph can be mapped bijectively onto the graph G without being isomorphic to it.

Remark 1.4.5. Denote by $\mathrm{Hom}(G, G')$, $\mathrm{Com}(G, G')$, $\mathrm{EHom}(G, G')$, $\mathrm{SHom}(G, G')$, $\mathrm{SEHom}(G, G')$, and $\mathrm{Iso}(G, G')$ the homomorphism sets.

Analogously, let End(G), EEnd(G), Cnd(G), SEnd(G), SEEnd(G), and Aut(G) denote the respective sets when $G = G'$. These form monoids.

Indeed, End(G) and SEnd(G), as well as EEnd(G) and SEEnd(G), are monoids, i. e., sets with an associative multiplication (the composition of mappings) and an identity element (the identical mapping). Clearly, End(G) is closed. Also, SEnd(G) is closed, since for $f, g \in \text{SEnd}(G)$ we get

$$(fg(x), fg(y)) \in E \overset{f \text{ strong}}{\Longleftrightarrow} (g(x), g(y)) \in E \overset{g \text{ strong}}{\Longleftrightarrow} (x, y) \in E.$$

Definition 1.4.6. For $f_0 \in \text{EHom}(G, G')$, which identifies exactly two adjacent vertices, the graph $f_0(G)$ is also called an ***elementary contraction*** of G. The result of a series of elementary contractions $f_n(f_{n-1}(\dots(f_0(G))\dots))$ is called a ***contraction*** of G. It is denoted by G/E' if $E' \subseteq E$ is the set of contracted edges.

A graph H is called a ***(graph) minor*** of the graph G if H can be formed from G by deleting edges and vertices and by contracting edges.

Subdividing an edge $e = xy$ in a graph G means replacing e by a path xzy, where z is a new vertex. A graph H is a ***subdivision*** of G is H can be obtained from G by subdividing a sequence of edges. See Figure 1.4 for an illustration.

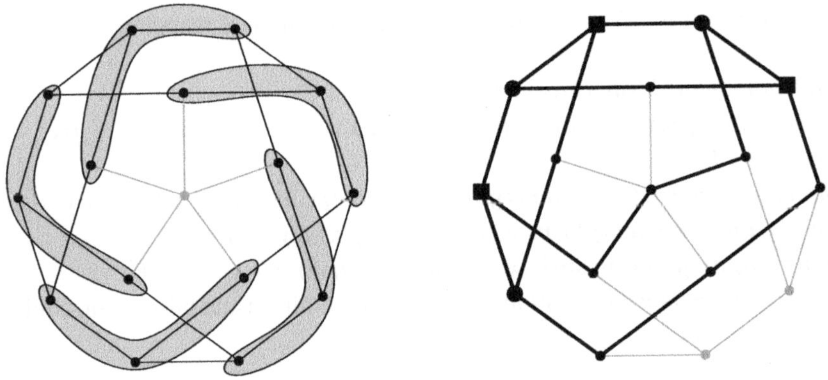

Figure 1.4: Left: Deleting gray edges and vertices and contracting gray bubbles yields a K_5 minor. Right: Deleting gray edges and vertices yields a subdivision of $K_{3,3}$.

Note that if H is a subdivision of G, then G can be obtained from H by contractions. In particular, G is a minor of its subdivisions but the converse does not hold. For instance, the graph in Figure 1.4 does not contain a subdivision of K_5 because it has maximum degree 3.

Proposition 1.4.7. *Let G and G' be graphs and $f : G \to G'$ a graph isomorphism. For $x \in G$, we have* $\text{indeg}(x) = \text{indeg}(f(x))$ *and* $\text{outdeg}(x) = \text{outdeg}(f(x))$.

Proof. We prove the statement for undirected graphs.

As f is injective, we get $|N_G(x)| = |f(N_G(x))|$.

As f is a homomorphism, we get $f(N_G(x)) \subseteq N_{G'}(f(x))$, i. e., $|f(N_G(x))| \leq |N_{G'}(f(x))|$.

As f is surjective, we have $N_{G'}(f(x)) \subseteq f(G)$; and, since f is strong, we get $|N_{G'}(f(x))| \leq |N_G(x)|$.

Putting the above together, using the statements consecutively, we obtain $|N_G(x)| = |N_{G'}(f(x))|$.

Now we use $\deg(x) = |N_G(x)|$ and $\deg(f(x)) = |N_{G'}(f(x))|$ to get the result. □

Subgraphs

The different sorts of homomorphisms lead to different sorts of subgraphs. First, let us explicitly define subgraphs and strong subgraphs.

Definition 1.4.8. Let $G = (V, E)$. A graph $G' = (V', E')$ is called a ***subgraph*** (or *partial subgraph*) of G if there exists an injective graph homomorphism $f : V' \to V$.

A graph G' is called a ***strong subgraph*** (or *induced subgraph* or *vertex induced subgraph*) if there exists an injective strong graph homomorphism $f : V' \to V$.

Example 1.4.9 (Subgraphs).

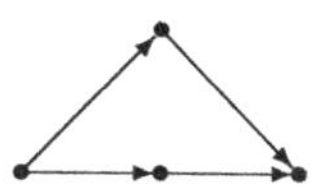

is a not strong subgraph while

is a strong subgraph of G:

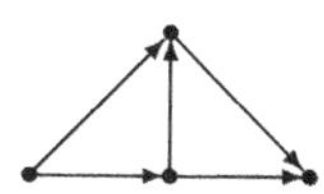

Remark 1.4.10. A (nontrivial) strong subgraph has fewer vertices than the original graph, but all edges of the original graph between these vertices are contained in the strong subgraph.

A subgraph in general contains fewer vertices or fewer edges than the original graph.

(Semi)paths, (semi)cycles and (semi)circuits are subgraphs.

Definition 1.4.11. A ***strong, one-sided,*** or ***weak component*** of a graph is, respectively, a maximal strongly, one-sided or weakly connected subgraph. Compare Example 1.2.2.

A complete subgraph is also called a ***clique*** of G. The number of vertices $\omega(G)$ of a largest clique of G is called the ***clique number*** of G.

Now we introduce the “edge dual” concept of a clique.

Definition 1.4.12. Two vertices $x, y \in V$ are called ***independent vertices*** if $(x,y) \notin E$ and $(y,x) \notin E$. The ***vertex independence number*** is defined as

$$\beta_0(G) := \max\{|U| : U \subseteq V,\ \text{independent}\}.$$

Analogously, two nonincident edges are called ***independent edges***, and we can define the ***edge independence number*** $\beta_1(G)$.

A set of pairwise independent edges of G is called a ***matching*** in G. A k-regular spanning subgraph is called a k***-factor*** of G; a 1-factor of G is called a ***perfect matching*** of G.

1.5 Half-, locally, quasi-strong, and metric homomorphisms

In addition to the usual homomorphisms, we introduce the following four sorts of homomorphisms. As always, homomorphisms are used to investigate the structure of objects. The large number of different homomorphisms of graphs shows how rich and variable the structure of a graph can be. In Section 1.9, we summarize which of these homomorphisms have appeared where and under which names; we also suggest how they might be used in modeling.

The motivation for these other homomorphisms comes from the concept of strong homomorphisms or, more precisely, the notion of comorphism, i. e., the continuous mapping. A continuous mapping "reflects" edges of graphs. The following types of homomorphism reduce the intensity of reflection. In other words, an ordinary homomorphism $f : G \to G'$ does not reflect edges at all. This means it could happen that $(f(x), f(y))$ is an edge in G' even though (x,y) is not an edge in G, and there may not even exist any preimage of $f(x)$ which is adjacent to any preimage of $f(y)$ in G. The following three concepts "improve" this situation step by step.

From the definitions, it will become clear that there exist intermediate steps that would refine the degree of reflection.

Definition 1.5.1. Let $G = (V,E)$ and $G' = (V',E')$ be graphs, and let $f \in \mathrm{Hom}(G,G')$. For $x, y \in V$, set

$$X := f^{-1}(f(x)),$$
$$Y := f^{-1}(f(y)).$$

Let $(f(x), f(y)) \in E'$. Then f is said to be:

- ***half-strong*** if there exists $\tilde{x} \in X$ and $\tilde{y} \in Y$ such that $(\tilde{x},\tilde{y}) \in E$;
- ***locally strong*** if $\begin{cases} \forall x \in X,\ \exists y_x \in Y \text{ such that } (x, y_x) \in E \text{ and} \\ \forall y \in Y,\ \exists x_y \in X \text{ such that } (x_y, y) \in E; \end{cases}$
- ***quasi-strong*** if $\begin{cases} \exists \tilde{x}_0 \in X \text{ such that } \forall \tilde{y} \in Y,\ (\tilde{x}_0, \tilde{y}) \in E \text{ and} \\ \exists \tilde{y}_0 \in Y \text{ such that } \forall \tilde{x} \in X,\ (\tilde{x}, \tilde{y}_0) \in E. \end{cases}$

We call $\tilde{x}_0$ and $\tilde{y}_0$ ***central vertices*** or, in the directed case, a ***central source*** and a ***central sink*** in X and in Y with respect to $(f(x), f(y))$.

Remark 1.5.2. With the obvious notation, one has

$$\begin{aligned}\mathrm{Hom}(G,G') &\supseteq \mathrm{HHom}(G,G') \supseteq \mathrm{LHom}(G,G') \supseteq \mathrm{QHom}(G,G')\\ &\supseteq \mathrm{SHom}(G,G') \supseteq \mathrm{Iso}(G,G'),\\ \mathrm{End}(G) &\supseteq \mathrm{HEnd}(G) \supseteq \mathrm{LEnd}(G) \supseteq \mathrm{QEnd}(G)\\ &\supseteq \mathrm{SEnd}(G) \supseteq \mathrm{Aut}(G) \supseteq \{\mathrm{id}_G\}.\end{aligned}$$

Note that apart from $\mathrm{SEnd}(G)$, $\mathrm{Aut}(G)$, and $\{\mathrm{id}_G\}$, the other subsets of $\mathrm{End}(G)$ are, in general, not submonoids of $\mathrm{End}(G)$. We will talk about the ***group*** and the ***strong monoid*** of a graph, and about the ***quasi-strong monoid, locally strong monoid,*** and ***half-strong monoid*** of a graph if these really are monoids.

Example 1.5.3 (Different homomorphisms). We give three of the four examples for undirected graphs. The example for the half-strong homomorphism in the directed case shows that the other concepts can also be transferred to directed graphs.

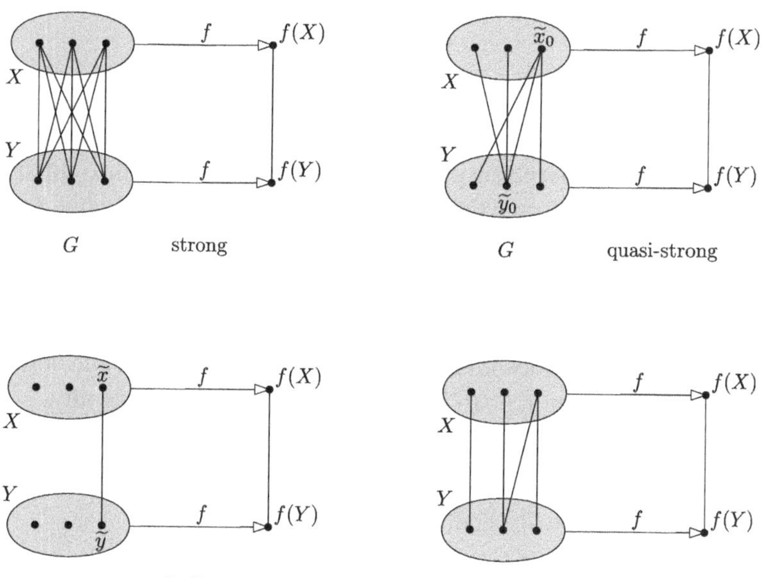

From the definitions, we immediately obtain the following theorem. To get an idea of the proof, one can refer to the graphs in Example 1.5.3.

Theorem 1.5.4. *Let $G \neq K_1$ be a bipartite graph with $V = V_1 \bigcup V_2$. Let (x_1, x_2) be an edge with $x_1 \in V_1$ and $x_2 \in V_2$. We define an endomorphism r of G by $r(V_1) = \{x_1\}$ and $r(V_2) = \{x_2\}$. Obviously, $r \in \mathrm{HEnd}(G)$. Moreover, the following hold:*

- $r \in \text{LEnd}(G)$ *if and only if G has no isolated vertices;*
- $r \in \text{QEnd}(G)$ *if and only if* V_1 *has a central vertex* $\tilde{x}_0$ *with* $N(\tilde{x}_0) = V_2$ *and correspondingly for* V_2*;*
- $r \in \text{SEnd}(G)$ *if and only if G is complete bipartite.*

Proposition 1.5.5. *A noninjective endomorphism f of G is strong if and only if for all* $x \in V$ *with* $f(x) = f(x')$ *one has* $N_G(x) = N_G(x')$.

Note that for adjacent vertices x and x', this is possible only if both have loops.

Proof. Necessity is clear from the definition. Now suppose that $N_G(x) = N_G(x')$ for $x, x' \in V(G)$. Construct f by setting $f(x) = x'$ and $f(y) = y$ for all $y \neq x, x'$. It is clear that $f \in \text{SEnd}(G)$. □

Definition 1.5.6. A homomorphism f from G to G' is said to be ***metric*** if for any vertices $x, y \in V(G)$ there exist $x' \in f^{-1}f(x)$ and $y' \in f^{-1}f(y)$ such that $d(f(x), f(y)) = d(x', y')$. Denote by $\text{MEnd}(G)$ the set of metric endomorphisms of G and by $\text{Idpt}(G)$ the set of ***idempotent*** endomorphisms, i. e., $f \in \text{End}(G)$ with $f^2 = f$, of G.

Corollary 1.5.7. *If* $\text{Aut}(G) \neq \text{SEnd}(G)$, *then* $\text{SEnd}(G) \setminus \text{Aut}(G)$ *contains at least two idempotents.*

As usual we make the following definition.

Definition 1.5.8. A homomorphism f from G to $f(G) \subseteq H$ is called a ***retraction*** if there exists an injective homomorphism g from $f(G)$ to G such that $fg = \text{id}_{f(G)}$. In this case, $f(G)$ is called a ***retract*** of G, and then G is called a ***coretract*** of $f(G)$ while g is called a ***coretraction***.

If H is an unretractive retract of G, i. e., if $\text{End}(H) = \text{Aut}(H)$, then H is also called a ***core*** of G.

Remark 1.5.9. It is straightforward to check that G' is a retract of G if and only if there exists $f \in \text{End}(G)$ such that $G' \cong f(G)$ and f restricted to $f(G)$ is the identity.

Remark 1.5.10. One has

$$\text{Idpt}(G), \text{LEnd}(G) \subseteq \text{MEnd}(G) \subseteq \text{HEnd}(G).$$

Example 1.5.11 (HEnd, LEnd, QEnd are not monoids). The sets HEnd, LEnd, QEnd are not closed with respect to composition of mappings. To see this, consider the following graph G:

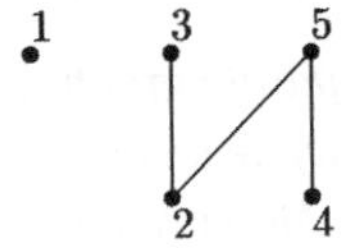

together with the mappings $f = \binom{1\,2\,3\,4\,5}{3\,4\,5\,4\,5}$ and $g = \binom{1\,2\,3\,4\,5}{1\,2\,3\,2\,5}$. Now $f \in \text{QEnd}(G)$ and $g \in \text{HEnd}(G)$ but $f^2 \in \text{HEnd}(G)\backslash\text{LEnd}(G)$ and $g \circ f \in \text{End}(G)\backslash\text{HEnd}(G)$. These properties are not changed if we add another vertex 0 to the graph which we make adjacent to every other vertex. The graph is then connected but no longer bipartite.

Question. Do Idpt and MEnd always form monoids? Can you describe graphs where this is the case?

1.6 The factor graph, congruences, and the homomorphism theorem

The study of factor graphs by graph congruences turns out to be fundamental for the general investigation of homomorphisms. The connection to arbitrary homomorphisms is established through the canonical epimorphisms, and this leads to the homomorphism theorem for graphs. We formulate the theorem only for ordinary graph homomorphisms.

Factor graphs

Definition 1.6.1. Let $\varrho \subseteq V \times V$ be an equivalence relation on the vertex set V of a graph $G = (V, E)$, and denote by x_ϱ the equivalence class of $x \in E$ with respect to ϱ. Then $G_\varrho = (V_\varrho, E_\varrho)$ is called the ***factor graph*** of G with respect to ϱ, where $V_\varrho = V/\varrho$ and $(x_\varrho, y_\varrho) \in E_\varrho$ if there exist $x' \in x_\varrho$ and $y' \in y_\varrho$ with $(x', y') \in E$, where ϱ is called a ***graph congruence***.

Example 1.6.2 (Congruence classes, factor graphs). In Figure 1.5, we exhibit some graphs together with congruence classes (encircled vertices) and the corresponding factor graphs:

Remark 1.6.3. By the definition of G_ϱ, the canonical epimorphism

$$\begin{aligned} \pi_\varrho : G &\to G_\varrho \\ x &\mapsto x_\varrho \end{aligned}$$

(which is always surjective) is a half-strong graph homomorphism.

Note that, in general, a graph congruence ϱ is just an equivalence relation. If we have a graph $G = (V, E)$ and a congruence $\varrho \subseteq V \times V$ such that there exist $x, y \in V$ with $(x, y) \in E$ and $x \varrho y$, then $(x_\varrho, x_\varrho) \in E_\varrho$, i. e., G_ϱ has loops.

If we want to use only loopless graphs, then $\pi_\varrho : G \to G_\varrho$ is a graph homomorphism only if

$$x \varrho y \Rightarrow (x, y) \notin E.$$

Therefore, we make the following definition.

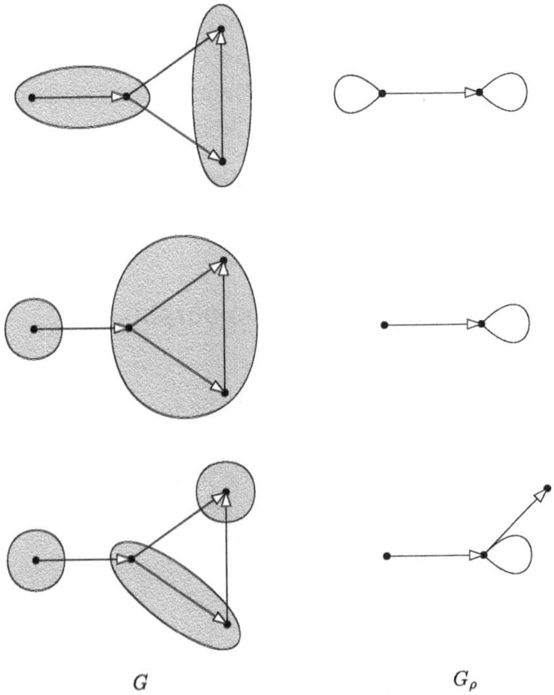

Figure 1.5: Three congruence relations in a graph and the corresponding factor graphs.

Definition 1.6.4. A ***(loop-free) graph congruence*** ϱ is an equivalence relation with the additional property that $x \varrho y \Rightarrow (x,y) \notin E$.

Definition 1.6.5. Let G_ϱ be the factor graph of G with respect to ϱ. If the canonical mapping $\pi_\varrho : G \to G_\varrho$ is a strong (respectively quasi-strong, locally strong, or metric) graph homomorphism, then the graph congruence ϱ is called a ***strong (resp., quasi-strong, locally strong, or metric) graph congruence***.

Example 1.6.6 (Connectedness relations). On $G = (V,E)$, with $x,y \in V$, consider the following relations:

$$x \varrho_1 y \Leftrightarrow \text{there exists an } x,y \text{ path and a } y,x \text{ path or } x = y;$$

$$x \varrho_2 y \Leftrightarrow \text{there exists an } x,y \text{ semipath or } x = y.$$

$$x \varrho_3 y \Leftrightarrow \text{there exists an } x,y \text{ path or a } y,x \text{ path.}$$

The relation ϱ_1 is an equivalence relation; the factor graph G_{ϱ_1} is called a ***condensation*** of G.

The relation ϱ_2 is an equivalence relation; the factor graph G_{ϱ_2} consists only of isolated vertices with loops.

The relation ϱ_3 is not transitive and, therefore, not an equivalence relation.

The homomorphism theorem

For convenience, we start with the so-called mapping theorem, i. e., the homomorphism theorem for sets, preceded by the usual result on mapping-induced congruence relations. Then, as for sets, we formulate the homomorphism theorem for graphs.

Proposition 1.6.7. *Let G and H be sets, and let $f : G \to H$ be a mapping. Using f we obtain an equivalence relation on G, the so-called* **induced congruence**, *if we define, for $x, y \in G$,*

$$x \varrho_f y \Leftrightarrow f(x) = f(y).$$

Moreover, by setting $\pi_{\varrho_f}(x) = x_{\varrho_f}$ for $x \in G$, we get a surjective mapping onto the factor set $G_{\varrho_f} = G/\varrho_f$. Here, x_{ϱ_f} denotes the equivalence class of x with respect to ϱ_f and G/ϱ_f the set of all these equivalence classes.

Proof. It is straightforward to check that ϱ_f is reflexive, symmetric, and transitive, i. e., it is an equivalence relation on G. Surjectivity of π_{ϱ_f} follows from the definition of the factor set. □

Proposition 1.6.8. *Let G and H be graphs, and let $f : G \to H$ be a graph homomorphism. Using f, we obtain a graph congruence by defining, for $x, y \in V(G)$,*

$$x \varrho_f y \Leftrightarrow f(x) = f(y).$$

Moreover, by setting $\pi_{\varrho_f}(x) = x_{\varrho_f}$ for $x \in G$, we get a surjective graph homomorphism onto the factor graph $G_{\varrho_f} = G/\varrho_f$. Here, x_{ϱ_f} denotes the congruence class of x with respect to ϱ_f and G_{ϱ_f} the factor graph formed by these congruence classes.

Proof. As for sets, we know that ϱ_f is an equivalence relation and π_{ϱ_f} is a surjective mapping by Proposition 1.6.7. Now use Remark 1.6.3. □

Proposition 1.6.9 (The homomorphism theorem for sets). *For every mapping $f : G \to H$ from a set G to a set H, there exists exactly one injective mapping $\overline{f} : G_{\varrho_f} \to H$, with $\overline{f}(x_{\varrho_f}) = f(x)$ for $x \in G$, such that the following diagram is commutative, i. e., $f = \overline{f} \circ \pi_{\varrho_f}$:*

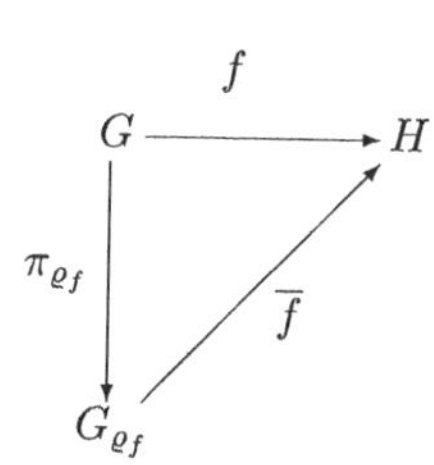

Moreover, the following statements hold:

(a) *If f is surjective, then $\overline{f}$ is surjective.*
(b) *If we replace ϱ_f by an equivalence relation $\varrho \subseteq \varrho_f$, then $\overline{f} : G_\varrho \to H$ is defined in the same way, but is injective only if $\varrho = \varrho_f$.*

Proof. Define $\overline{f}$ as indicated. We shall show that $\overline{f}$ is well-defined. Suppose that $x_{\varrho_f} = x'_{\varrho_f}$ in G_{ϱ_f}; then $x\varrho_f x'$, and thus $\overline{f}(x_{\varrho_f}) = f(x) = f(x') = \overline{f}(x'_{\varrho_f})$.

It is clear that $\overline{f}$ makes the diagram commutative and is the uniquely determined mapping with these properties. Indeed, if a mapping $\overline{f'}$ has the same properties, then $\overline{f'}(x_{\varrho_f}) = \overline{f'}\pi_{\varrho_f}(x) = f(x) = \overline{f}\pi_{\varrho_f}(x) = \overline{f}(x_{\varrho_f})$ for all $x_{\varrho_f} \in G_{\varrho_f}$.

It is also clear that the two additional properties are valid. In particular, the inclusion $\varrho \subseteq \varrho_f$ ensures that $\overline{f}$ is well-defined also in this case. □

Theorem 1.6.10 (The homomorphism theorem for graphs). *For every half-strong graph homomorphism $f : G \to H$, there exists exactly one injective strong graph homomorphism $\overline{f} : G_{\varrho_f} \to H$, with $\overline{f}(x_{\varrho_f}) = f(x)$ for $x \in G$, such that we have the following commutative diagram, i. e., $f = \overline{f} \circ \pi_{\varrho_f}$:*

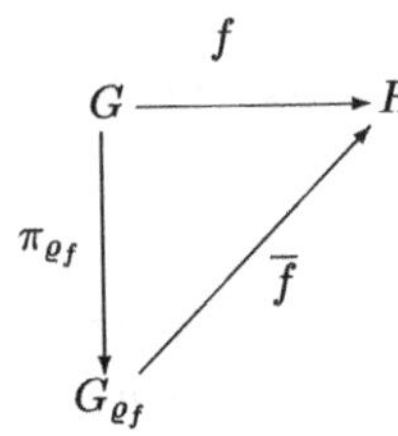

Moreover, the following statements hold:

(a) *If f is surjective, then $\overline{f}$ is surjective.*
(b) *If we replace ϱ_f by a graph congruence $\varrho \subseteq \varrho_f$, then $\overline{f} : G_\varrho \to H$ is defined in the same way, but is injective and strong only if $\varrho = \varrho_f$.*
(c) *If (a) and (b) are fulfilled, then $G_{\varrho_f} \cong H$.*

Proof. Define $\overline{f}$ as indicated, just as we did for sets in Proposition 1.6.9. Then $\overline{f}$ is well-defined, is unique and makes the diagram commutative.

We only have to show that π_{ϱ_f} and $\overline{f}$ are graph homomorphisms. For π_{ϱ_f} this comes from Proposition 1.6.8. Take $(x_{\varrho_f}, y_{\varrho_f}) \in E(G_{\varrho_f})$ and consider $(\overline{f}(x_{\varrho_f}), \overline{f}(y_{\varrho_f})) = (f(x), f(y))$. As f is half-strong, there exists a preimage $(x', y') \in E(G)$ of $(x_{\varrho_f}, y_{\varrho_f}) \in E(G_{\varrho_f})$, which implies $(f(x), f(y)) \in E(H)$.

The two additional properties are the same as for sets, so nothing further needs to be proved. □

Remark 1.6.11. In the language of category theory, the essence of the homomorphism theorem is that every homomorphism has an ***epi-mono factorization*** in the given category. Note that in the graph categories considered, epimorphisms (epis) are surjective and monomorphism (monos) are injective. The monomorphism is called an ***embedding*** of the factor graph into the image graph. The embedding is strong if f is at least half-strong.

Corollary 1.6.12. *Surjective endomorphisms and injective endomorphisms of a finite graph (set) are already automorphisms.*

Example 1.6.13. We consider again the homomorphism f from Example 1.5.11. Here, the congruence classes are $\{1\}$, $\{2,4\}$, and $\{3,5\}$, so π_{ϱ_f} maps every vertex to its congruence class, and $\bar{f}$ is the embedding which takes 1_{ϱ_f} to 3, 2_{ϱ_f} to 4 and 3_{ϱ_f} to 5. The result of this procedure can be visualized as follows:

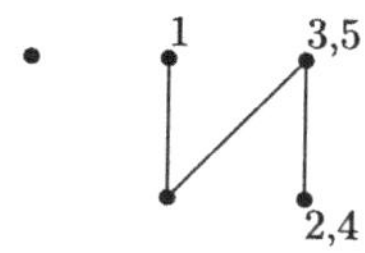

Application 1.6.14. As an application, we observe that the homomorphism theorem can be used to determine all homomorphisms from G to H as follows. We first determine all congruences on G, giving all possible natural surjections π. Then, for each congruence relation ϱ which is given by its congruence classes, i. e., for every π_ϱ, we determine all possible embeddings of G_{π_ϱ} into H. Each of these embeddings corresponds to some f, all of which are different but induce the same congruence.

In Example 1.6.13, we have $G = H$ and obtain all embeddings as follows. The class $\{1\}$ can be mapped onto any vertex of G, and after that the classes $\{2,4\}$ and $\{3,5\}$ forming an edge in G_{π_ϱ} can be mapped onto every edge of G which does not contain the image of $\{1\}$ in the actual embedding. In particular, if we map $\{1\}$ onto 1 we have six possible embeddings, and they all give quasi-strong endomorphisms. If we map $\{1\}$ onto 3 or 4, we have four possible embeddings in each case, two of which give quasi-strong, which are not strong, and the other two give not even half-strong endomorphisms. If we map $\{1\}$ onto 2 or 5, we have two possible embeddings, which in each case give not half-strong endomorphisms. So, overall, this congruence relation gives ten quasi-strong and eight not half-strong endomorphisms.

The same method for groups is formulated in Project 9.1.8.

1.7 The endomorphism type of a graph

For a more systematic treatment of different endomorphisms, we define the endomorphism spectrum and the endomorphism type of a graph.

Definition 1.7.1. For the graph X, consider the following sequence from Remark 1.5.2 (brackets around G are omitted for simplicity):

$$\operatorname{End} G \supseteq \operatorname{HEnd} G \supseteq \operatorname{LEnd} G \supseteq \operatorname{QEnd} G \supseteq \operatorname{SEnd} G \supseteq \operatorname{Aut} G.$$

With this sequence, we associate the sequence of respective cardinalities,

$$\operatorname{Endospec} G = (|\operatorname{End} G|, |\operatorname{HEnd} G|, |\operatorname{LEnd} G|, |\operatorname{QEnd} G|, |\operatorname{SEnd} G|, |\operatorname{Aut} G|),$$

and we call this 6-tuple the ***endospectrum*** or ***endomorphism spectrum*** of G. Next, associate with the above sequence a 5-tuple $(s_1, s_2, s_3, s_4, s_5)$ with

$$s_i \in \{0,1\} \quad \text{for } i = 1, \dots, 5,$$
$$\text{where } s_i = 1 \text{ stands for } \neq \text{ and } s_i = 0 \text{ stands for } =,$$

such that $s_1 = 1$ means that $|\mathrm{End}\, G| \neq |\mathrm{HEnd}\, G|$, $s_2 = 0$ means that $|\mathrm{HEnd}\, G| = |\mathrm{LEnd}\, G|$, etc. We use decadic coding and call the integer $\sum_{i=1}^{5} s_i 2^{i-1}$ the ***endotype*** or ***endomorphism type*** of G and denote it by *endotype* G.

If $\mathrm{End}\, G = \mathrm{Aut}\, G$, we call the graph G ***unretractive*** or E-A ***unretractive***; if $\mathrm{End}\, G = 1$, we call the graph ***rigid***; and if $\mathrm{Aut}\, G = 1$, we call the graph ***asymmetric***. More generally, if $XG = X'G$ for $X, X' \in \{\mathrm{End}, \mathrm{HEnd}, \mathrm{LEnd}, \mathrm{QEnd}, \mathrm{Aut}\}$, we call the graph X-X' ***unretractive***.

In principle, there are 32 possibilities, i. e., endotype 0 up to endotype 31.
We will now prove that graphs of endotypes 1 and 17 do not exist.

Proposition 1.7.2. *Let G be a finite graph such that* $\mathrm{End}\, G \neq \mathrm{HEnd}\, G$. *Then* $\mathrm{HEnd}\, G \neq \mathrm{SEnd}\, G$.

Proof. Take $f \in \mathrm{End}\, G \setminus \mathrm{HEnd}\, G$. Then there exists $(f(x), f(x')) \in E(G)$ but for all $\overline{x}, \overline{x}'$ with $f(\overline{x}) = f(x)$ and $f(\overline{x}') = f(x')$ one has $(\overline{x}, \overline{x}') \notin E(G)$. From finiteness of $\mathrm{End}\, G$, we get an idempotent power f^i of f, i. e., $(f^i)^2 = f^i$, and thus $f^i \in \mathrm{HEnd}\, G$; see Remark 1.5.10. In particular, since $(f^i(x), f^i(x')) \in E(G)$, we have that $f^i(x)$ and $f^i(x')$ are fixed under f^i, and thus they are adjacent preimages. Moreover, $f^i \notin \mathrm{SEnd}\, G$ since not all preimages are adjacent, in particular $(x, x') \notin E(G)$. □

Before analyzing the endotypes of graphs in more detail, we consider all endotypes with regard to whether or not $\mathrm{Aut}\, G = 1$.

Proposition 1.7.3. $|\mathrm{Aut}\, G| = 1$ *implies* $|\mathrm{SEnd}\, G| = 1$.

Proof. Take $f \in \mathrm{SEnd}\, G \setminus \mathrm{Aut}\, G$. Then there exist $x, x' \in V(G)$, $x \neq x'$, with $f(x) = f(x')$ and $N(x) = N(x')$ by Proposition 1.5.5. Then the mapping which permutes exactly x and x' is a nontrivial automorphism of G. □

The preceding result shows that for endotypes 16 up to 31 we always have $\mathrm{Aut}\, G \neq 1$, since $\mathrm{SEnd}\, G \neq \mathrm{Aut}\, G$ in these cases. So ***endotypes 0 to 15 get an additional a if one has asymmetry, i. e., if*** $\mathbf{Aut} G = 1$.

We can say that endotype 0 describes unretractive graphs and endotype $0a$ describes rigid graphs. Endotypes 0 up to 15 describe S-A unretractive graphs, and endotypes $0a, 2a, \dots, 15a$ describe asymmetric graphs. Endotype 16 describes E-S unretractive graphs which are not unretractive. Endotype 31 describes graphs for which all six sets are different.

Theorem 1.7.4. *There exist simple graphs without loops of endotype* $0, 0a, 2, 2a, 3, 3a, \ldots, 15, 15a, 16, 18, 19, \ldots, 31$.

Proof. See M. Böttcher and U. Knauer [11, 12] □

The following result is an approach to the question of to what extent trees are determined by their endospectrum. It also shows that the endospectrum in general does not determine graphs up to isomorphism.

We will use some new notation. We call $K_{1,n}$ a ***star***. Take two stars $K_{1,n}, K_{1,m}$ with $n \geq 3$, $m \geq 2$ or vice versa and identify one edge of one star with one edge of the other star. The result is called a ***double star***. A definition in terms of the lexicographic product can be found in the table of Theorem 1.7.5.

Theorem 1.7.5. *Let G be a tree, with $G \neq K_2$. The following table characterizes G with respect to endotypes, which are given by their decadic coding in the first column and explicitly in the second column. Classes of endomorphisms are abbreviated by their first letters, and $v_G \neq \Delta$ indicates the existence of two different vertices in G which have exactly the same neighbors; cf. Definitions 9.5.1 and 10.2.2. For the notation in the last column, see the generalized lexicographic product in Section 4.4. Note that P_n has n edges and $n+1$ vertices.*

N^0	*Endotype*	v_G	diam	*Examples or complete descriptions*
6	$E = H \neq L \neq Q = S = A$	$= \Delta$	≥ 4	*P_4 is the smallest*
10	$E = H \neq L = Q \neq S = A$	$= \Delta$	$= 3$	*P_3 is the only one*
16	$E = H = L = Q = S \neq A$	$\neq \Delta$	$= 2$	*Exactly the stars, i. e., $K_{1,n}$ for $n \geq 2$*
22	$E = H \neq L \neq Q = S \neq A$	$\neq \Delta$	≥ 4	*P_4 with one end-vertex doubled, e. g., $P_4[\overline{K}_2, K_1, K_1, K_1, K_1]$, which is the smallest*
26	$E = H \neq L = Q \neq S \neq A$	$\neq \Delta$	$= 3$	*Exactly the "double stars," i. e., $P_3[\overline{K}_n, K_1, K_1, \overline{K}_m]$ with $n \geq 2$ or $m \geq 2$*

Asymmetric trees G, i. e., G such that $|\mathrm{Aut}\, G| = 1$, are possible only with endotype 6; in other words, they have endotype 6a. The smallest is the path of length 5, with one pending vertex at the third vertex, i. e., a vertex of degree 1.

A proof follows after Proposition 1.7.14.

Lemma 1.7.6. *Let G be a graph such that $N(x) \subsetneq N(x')$ for some $x, x' \in G$ with $(x, x') \notin E(G)$. Then* $\mathrm{HEnd}\, G \neq \mathrm{LEnd}\, G$.

Proof. Define $f(x) = f(x') = x'$ and $f(y) = y$ for all $y \neq x \in G$. Then obviously $f \in \mathrm{HEnd}\, G$. But $f \notin \mathrm{LEnd}\, G$, because for $x'' \in N(x') \setminus N(x)$ one has $(f(x''), f(x)) = (x'', x') \in E(G)$, $f^{-1}(x'') = \{x''\}$, $f^{-1}(x') = \{x, x'\}$ but $(x, x'') \notin E(G)$, i. e., not every preimage of x' is adjacent to some preimage of x''. □

The following two lemmas are clear.

Lemma 1.7.7. *Suppose G is a tree with $x, x' \in G$ such that $N(x) \subsetneq N(x')$. Then* $\mathrm{diam}(G) \geq 3$.

Lemma 1.7.8. *Let G be a tree with* $\mathrm{diam}(G) \geq 3$. *Take $x, x', x'' \in G$ with $\{x'\} = N(x) \subsetneq N(x') \subseteq \{x, x''\}$. Then, by defining $f(x) = x''$ and $f(y) = y$ for $y \neq x$, we get $f \in$* $\mathrm{HEnd}\, G \setminus \mathrm{LEnd}\, G$.

Lemma 1.7.9. *Let G be a double star. Then* $\mathrm{QEnd}\, G \neq \mathrm{SEnd}\, G$.

Proof. Take $\{x_0', x_0, x_1, x_1'\}$, a longest simple path in G. Define $f(N(x_0)) = \{x_1\}$ and $f(N(x_1)) = \{x_0\}$. Then $f \in \mathrm{QEnd}\, G$, since $x_1 \in f^{-1}(x_1)$ is adjacent to every vertex in $N(x_1) = f^{-1}(x_0)$ and $x_0 \in f^{-1}(x_0)$ is adjacent to every vertex in $N(x_0) = f^{-1}(x_1)$. But $f \notin \mathrm{SEnd}\, G$ as $(x_0', x_1') \notin E(G)$. □

Proposition 1.7.10. *Let G be a tree with* $\mathrm{diam}(G) \geq 4$. *Then* $\mathrm{QEnd}\, G = \mathrm{SEnd}\, G$.

Proof. Take $f \in \mathrm{QEnd}\, G$. Then there exists $(x, x') \in E(G)$ such that $(f(x), f(x')) \in E(G)$, and we may assume that x and x' are central with respect to $(f(x), f(x'))$. Then $U := f^{-1}(f(x)) \subseteq N(x')$ and $U' := f^{-1}(f(x')) \subseteq N(x)$. As $\mathrm{diam}(G) \geq 4$, there exists $y \in N(U')$ such that $(y, \overline{x'}) \in E(G)$ for some $\overline{x'} \in U'$. Then $(f(y), f(\overline{x'})) = (f(y), f(x')) \in E(G)$, and since $f \in \mathrm{QEnd}\, G$ we get that y, say, is adjacent to all vertices in U', and hence to x' in particular. But then $|U'| = 1$, because otherwise there would be a cycle $\{y, x', x, \overline{x'}, y)$ in G, which is impossible since G is a tree. Moreover, every vertex in U has degree 1 with the common neighbor x'. Together with Proposition 1.5.5, this implies that $f \in \mathrm{SEnd}\, G$. □

Proposition 1.7.11. *If G is a tree with* $\mathrm{diam}(G) \geq 4$, *then* $\mathrm{LEnd}\, G \neq \mathrm{QEnd}\, G$.

Proof. As $\mathrm{diam}(G) \geq 4$, the tree contains $P_4 = \{x_0, x_1, x_2, x_3, x_4\}$. Define $f : G \to G$ as follows: all vertices with even distance from x_2 are mapped onto x_2; all other vertices are mapped onto x_1.

Then $f \in \mathrm{LEnd}\, G$, since every preimage of x_2 is adjacent to some preimage of x_1 and vice versa. But $f \notin \mathrm{QEnd}\, G$ because no vertex exists in the preimage of x_1 which is adjacent to x_0 and to x_4, as G has no cycles. □

Lemma 1.7.12. *For stars $G = K_{1,n}$, one has* $\mathrm{End}\, G = \mathrm{SEnd}\, G$.

Proof. We may assume that $n > 1$. If $|f(G)| > 2$ for an endomorphism f, the central vertex of the star is fixed and, therefore, f is strong. If $|f(G)| = 2$, i. e., $f(G) = K_2$, then f is also strong. □

Proposition 1.7.13. *Let $G \neq K_2$ be a tree with* $\mathrm{diam}(G) \leq 3$. *Then* $\mathrm{LEnd}\, G = \mathrm{QEnd}\, G$.

Proof. If $G \neq K_2$ is a tree with $\mathrm{diam}(G) \leq 3$, then G is a star or a double star. In the first case, the statement is contained in Lemma 1.7.12. So let G be a double star, i. e., suppose that there exist $x_0, x_1 \in G$ with $V(G) = N(x_0) \bigcup N(x_1)$ and $(x_0, x_1) \in E(G)$. Take $f \in \mathrm{LEnd}\, G$. Then it is impossible to have $f(y) = x_1$ and $f(y') \neq x_1$ for $y, y' \in N(x_0) \setminus \{x_1\}$. So f identifies only vertices inside $N(x_0) \setminus \{x_1\}$ or inside $N(x_1) \setminus \{x_0\}$, possibly followed by an automorphism of the resulting graph, and we have $f \in \mathrm{SEnd}\, G$. □

Proposition 1.7.14. *For any graph G, one has* SEnd G = Aut G *if and only if* $R_G = \Delta$, *i. e.,* $N(x) \neq N(x')$ *for all* $x \neq x' \in G$.

Proof. If the vertices $x \neq x'$ have the same neighbors, then $f(x) = x'$ is a non-bijective strong endomorphism, provided all other vertices are fixed. □

Proof of Theorem 1.7.5. It is clear that the third column of the table covers all possible trees.

The first column of equalities $E = H$ is obvious for all trees.

In the second column, the inequalities $H \neq L$ are furnished by Lemmas 1.7.8 and 1.7.6. The equality $H = L$ for type 16 is taken care of by Lemma 1.7.12.

The inequalities $L \neq Q$ are provided by Proposition 1.7.11, and the equalities $L = Q$ are given by Proposition 1.7.13.

The equalities $Q = S$ are taken care of by Proposition 1.7.10 and for type 16 again by Lemma 1.7.12. The inequalities are given by Lemma 1.7.9, noting that P_3 is also a double star.

The relations between S and A come from Proposition 1.7.14.

Now consider the "examples" and "complete descriptions" in the last column of Theorem 1.7.5. The statements about endotypes 6, 10, and 22 follow, by inspection, from what was said about v_G and diam. The statement about endotype 16 follows from Lemma 1.7.12 together with the fact that $v_G \neq \Delta$ and $\mathrm{diam}(G) = 2$. The statement about endotype 26 comes from he statement about endotype 10, if we double at least one end vertex, since then $v_G \neq \Delta$.

The last assertion about asymmetric trees is also implied by 4.13 in R. Novakovski and I. Rival [67]. Indeed, $|\mathrm{Aut}\, G| = 1$ is possible only if SEnd G = Aut G (see Proposition 1.7.3), i. e., only for endotypes smaller than 16; so in our situation only endotype 6 remains.

The statement concerning the smallest examples follows by inspection. □

In the following table, we use the (disjoint) union of graphs in a naive way. A formal definition (as coproduct) will follow in Chapter 3.

Endotype	Graph	Endotype	Graph
0	K_2	16	$\overline{K}_n, K_{1,n}, n \geq 2$
2	$K_1 \bigcup K_2$	18	$\overline{K}_n \bigcup K_2, n \geq 2$
4	$\bigcup_{n\geq 2} K_2$	19	$\overline{K}_n \bigcup (\bigcup_{n\geq 2} K_n), \overline{K}_m \bigcup K_{1,n}, n \geq 2, m \geq 1$
6		22	
7		23	
10	P_3	26	double stars
11	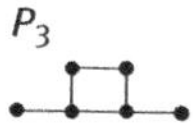	27	
15		31	

Theorem 1.7.15. *Bipartite graphs are exactly of the following endotypes, where the graphs or their common structures are given where possible.*

Proof. See U. Knauer [53].

Note that adding an isolated vertex to a connected graph which is not of endotype 0 or 16 adds 1 to the value of the endotype. This gives examples of graphs of endotypes 7, 11, 23, and 27 when starting with suitable trees from Theorem 1.7.5. The procedure yields graphs with endotype 2 or 18 when starting with graphs of endotype 0 or 16. □

Question. For which of the trees in Theorem 1.7.5 do the sets which are not monoids in general form monoids? The question makes sense for LEnd and endotypes 6, 10, 22, and 26.

It seems possible that trees are determined by their endomorphism spectrum up to isomorphism. Obviously, this is not the case for the endotype. Would this be a worthwhile question to investigate? Some more information about this can be found in U. Knauer [52].

1.8 On the genus of a graph

Final to this introductory chapter we introduce some of the well-known topological descriptions of graphs.

A graph is said to be *(2-cell)* ***embedded*** in a surface M if it is "drawn" in M in such a way that edges intersect only at their common vertices and, moreover, the surface decomposes into open discs after removal of vertices and edges of the graph. These discs form the ***faces*** or ***regions*** of the embedded graph.

A graph is said to be ***planar*** if it can be embedded in the plane or equivalently, on the sphere. We say that this embedding is a ***plane*** graph. By the ***genus*** of a graph G, we mean the minimum genus among all surfaces in which G can be embedded. So if G is planar, then the genus of G is zero. A graph is said to be ***outer planar*** if it has an embedding in the plane such that one face is incident to every vertex.

It is clear that we have the following.

Lemma 1.8.1. *The genus of a subgraph or of a contraction of G is not greater than the genus of G.*

First, we recall the following well-known results on the genus of graphs.

Theorem 1.8.2 (Kuratowski, Wagner). *For a finite graph G, the following are equivalent:*
(i) *G is planar,*
(ii) *G does not contain a subgraph that is a subdivision of K_5 or $K_{3,3}$ (Kuratowski),*
(iii) *G does not contain K_5 or $K_{3,3}$ as a minor (Wagner).*

So basically these two graphs are the prototypes of nonplanar graphs. However, note that the graph in Figure 1.4 shows that both statements in Theorem 1.8.2 are qualitatively different. A similar result, that is much easier to prove holds for outer planar graphs.

Theorem 1.8.3 (Chartrand, Harary). *For a finite graph G, the following are equivalent:*
(i) *G is outer planar,*
(ii) *G does not contain a subgraph that is a subdivision of K_4 or $K_{2,3}$,*
(iii) *G does not contain K_4 or $K_{2,3}$ as a minor.*

For the next result, we have to recall that a surface is called ***orientable*** if it has a "consistently oriented triangulation." This is the case for the sphere, the cylinder, and the torus, but not for the projective plane, the Möbius strip or the Klein bottle. In other words, a surface is nonorientable if it contains a Möbius strip.

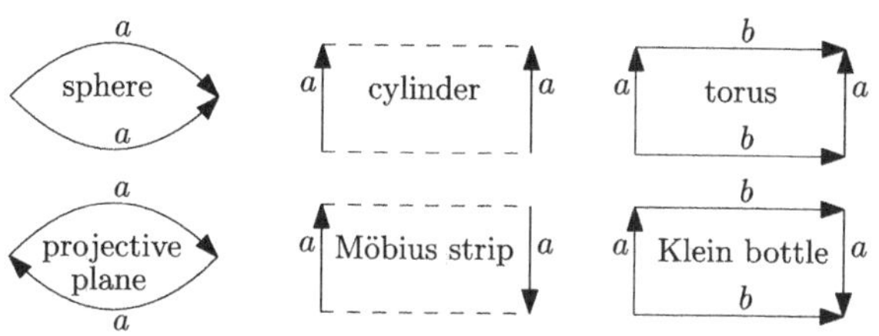

Figure 1.6: The above mentioned surfaces represented as polygons where appropriate arcs have to be by identified.

Theorem 1.8.4 (Euler (1758), Poincaré (1895)). *A finite graph that has n vertices and m edges and is 2-cell embedded on an orientable surface M of genus g with f faces satisfies the Euler–Poincaré formula; i. e.,* $n - m + f = 2 - 2g$.

This is specialized as follows.

Theorem 1.8.5 (Euler's formula). *Every simple connected plane graph G with vertex set V, edge set E and face set F fulfills* $|V| - |E| + |F| = 2$. *In particular, G has at most* $3|V| - 6$ *edges, i. e.,* $|E| \leq 3|V| - 6$ *and at most* $2|V| - 4$ *edges if the embedding has no triangular faces, i. e.,* $|E| \leq 2|V| - 4$.

Note that this formula can be proved quite easily by induction on the number of edges $|E|$ of G.

Remark 1.8.6 (Geometric dual). The ***geometric dual*** G^* of a plane graph G is the plane graph which has the faces of the original graph G as vertices; so it has a new vertex set, and two vertices in G^* are adjacent if and only if the two faces in G have a common edge.

This procedure can be generalizes to graphs embedded on a surface of genus greater than zero. Note that already in the plane different embeddings may be possible and, therefore, different geometric duals will exist. See Figure 1.7 for an illustration of dual graphs.

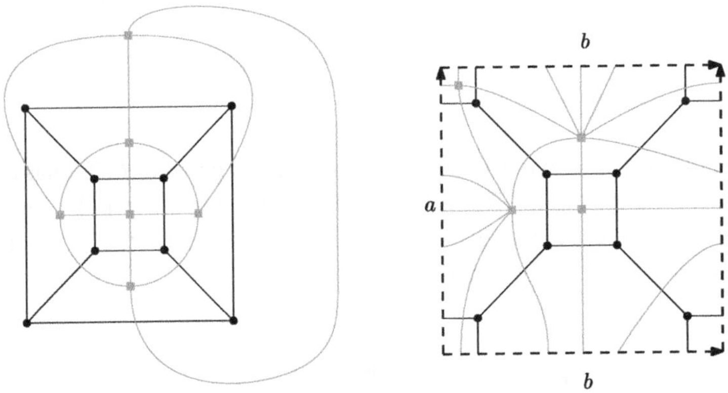

Figure 1.7: Embeddings of the cube in the plane and on the torus and the respective dual graphs. The planar dual of the cube is the octahedron.

Note that the geometric dual of a simple graph may have loops and multiple edges.

Remark 1.8.7 (Platonic solids). In the following, the Platonic solids *octahedron, dodecahedron,* and *icosahedron* will appear, which together with the three-dimensional cube Q_3 and the *tetrahedron* (as graph isomorphic to K_4) make up the five *platonic graphs*. From the values of d and d^*, it is clear that the tetrahedron is self-dual, Q_3 and the octahedron as well as the dodecahedron and the icosahedron are dual. These are the only plane graphs, which are, as we say, *completely regular*; this means that they are d-regular (all vertices have degree d) and their geometric duals are d^*-regular (which is equivalent to saying that the regions of the plane representation are all bounded by d^* edges). This can be proved with the Euler formula; see Theorem 1.8.4.

For convenience, we will first give the combinatorial description of these five platonic graphs. Here again, $|F|$ denotes the number of faces, d is the degree, d^* is the number of edges around one region, which is equal to the degree of the geometric dual graph, always in a plane representation.

d	d^*	$\lvert V\rvert$	$\lvert E\rvert$	$\lvert F\rvert$	
3	3	4	6	4	Tetrahedron
3	4	8	12	6	Cube
4	3	6	12	8	Octahedron
3	5	20	30	12	Dodecahedron
5	3	12	30	20	Icosahedron

Remark 1.8.8 (Completely regular graphs on surfaces). Completely regular graphs can also be studied for surfaces other than the plane or sphere, e. g., for the torus and also for nonorientable surfaces of higher genus like the Möbius strip, the projective plane, the Klein bottle and so on. The interesting thing is that this topological question can be formulated algebraically, and this is a possible clue to the characterization of completely regular graphs on these surfaces. The starting point in all cases would be the Euler–Poincaré formula; this shows which graphs could be completely regular on the surface under consideration, but it does not give embeddings. The problem is completely solved for the torus. More information can be found [White 2001], and may be, in [Liu 1995]. Many interesting results can be found at www.omeyer.gmx-home.de/on_completely_regular.pdf.

1.9 Comments: homomorphisms produce models

Ordinary homomorphisms are widely used. Half-strong homomorphisms were called "full" in P. Hell [37], and in G. Sabidussi [80]; and they were called "partially adjacent" by S. Antohe and E. Olaru [1].

Surjective locally strong homomorphisms appeared in the book by A. Pultr and V. Trnkova [Pultr/Trnkova 1980]. As far as we know, the term "quasi-strong" has not been used yet. Strong homomorphisms were first introduced by K. Culik in [15], under the name homomorphism. Metric homomorphisms can be found in the aforementioned paper by P. Hell. Egamorphisms are also called weak homomorphisms, e. g., in [Imrich/Klavžar 2000].

We would like to point out a more general phenomenon. Homomorphisms generate an image of a given object. This is the basis of the main principle of model building: we can view homomorphisms as the modeling tool and the homomorphic image as the model. When we use isomorphisms, all the information is retained. Since a model is usually thought of as a simplification, an isomorphic image is not really the kind of model one usually needs. So, in modeling, we want to suppress certain information about the original object, because in order to analyze the system it is helpful to first simplify the structure. To investigate different questions, we may wish to suppress different parts of the structure. Specializing this idea to graphs, strong homomorphisms reduce the number of points but maintain the structure in the sense that they reflect edges. Quasi-strong, locally strong, and half-strong homomorphisms reflect edges to a lesser and lesser extent in each step down to ordinary homomorphisms, which do not reflect edges at all.

Now let us also look back on the homomorphism theorem. One important aspect is that it produces an epi-mono factorization for every homomorphism. This is exploited in the following way. We start with one endomorphism f of G, which by the induced congruence ϱ_f defines the epi-part of the epi-mono factorization, the natural surjection $G \to G/\varrho_f$. If we now consider all possible embeddings of this factor graph into

G, we obtain all possible endomorphisms with the induced congruence ϱ_f. This principle can be used to find all endomorphisms of an object G. This is done, e. g., when we prove that the set LEnd P_n for a path of length n is a monoid if and only if $n = 3$ or $n + 1$ is prime; see Section 9.3.

Recall that the homomorphism theorem gives especially nice approaches to group and ring homomorphisms. In these two cases (categories), induced congruences are uniquely described by subobjects, namely normal subgroups in groups, also called normal divisors, and ideals in rings. These objects are much easier to handle than congruence relations; thus the investigation of homomorphisms in these categories is – to some extent – easier. For example, every endomorphism of a group A is determined by the factor group A/N, where N is a normal subgroup of A, and all possible embeddings of A/N into A. Nothing similar can be done for semigroups or in any of the graph categories (which will be introduced later).

2 Graphs and matrices

Matrices are very useful for describing and analyzing graphs. In this chapter, we shall present most of the common matrices for graphs and apply them to investigate various aspects of graph structures, such as isomorphic graphs, number of paths, or connectedness, and even endomorphisms and eigenvalues. All of this analysis is based on the so-called adjacency matrix.

We also define another important matrix, the so-called incidence matrix, which we will use later when discussing cycle and cocycle spaces.

2.1 Adjacency matrix

The definition of the adjacency matrix is the same for directed and undirected graphs, which may have loops and multiple edges.

Definition 2.1.1. Let $G = (V, E, p)$ where $V = \{x_1, \ldots, x_n\}$ is a graph. The $n \times n$ matrix $A(G) = (a_{ij})_{i,j=1,\ldots,n}$ defined by

$$a_{ij} := \left|\{e \in E \mid p(e) = (x_i, x_j)\}\right|$$

is called the ***adjacency matrix*** of G.

Example 2.1.2 (Adjacency matrices). We show the "divisor graph" of 6 and some multiple graph, along with their adjacency matrices.

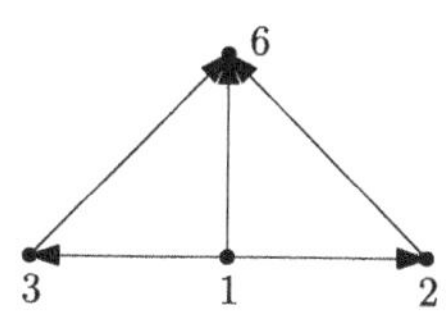

$$A(G) = \begin{pmatrix} 0 & 1 & 1 & 1 \\ 0 & 0 & 0 & 1 \\ 0 & 0 & 0 & 1 \\ 0 & 0 & 0 & 0 \end{pmatrix} \quad \begin{matrix} 1 \\ 2 \\ 3 \\ 6 \end{matrix}$$

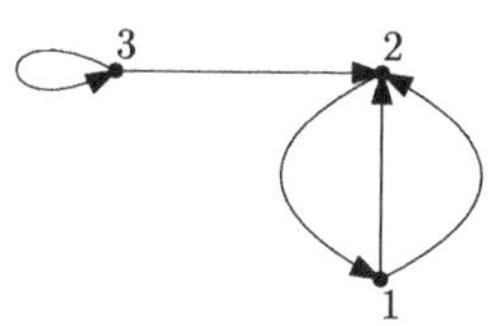

$$A(G) = \begin{pmatrix} 0 & 2 & 0 \\ 1 & 0 & 0 \\ 0 & 1 & 1 \end{pmatrix}$$

Remark 2.1.3. There exists a bijective correspondence between the set of all graphs with finitely many edges and n vertices and the set of all $n \times n$ matrices over $\mathbb{N}_0$.

It is clear that if the matrix $A(G)$ is symmetric, then the graph G is symmetric (i. e., undirected) and vice versa.

If G is simple, i. e., if it does not have multiple edges, then we can define $A(G)$ by

$$a_{ij} := \begin{cases} 1 & \text{if } (x_i, x_j) \in E, \\ 0 & \text{otherwise.} \end{cases}$$

https://doi.org/10.1515/9783110617368-002

Proposition 2.1.4. *For all* $x_i \in V$ *with* $A(G) = (a_{ij})_{i,j \in |V|=n}$, *we have*

$$\operatorname{indeg}(x_i) = \sum_{j=1}^{n} a_{ji}, \quad \textit{column sum of column } i;$$

$$\operatorname{outdeg}(x_i) = \sum_{j=1}^{n} a_{ij}, \quad \textit{row sum of row } i.$$

In the symmetric case, one has

$$\deg(x_i) = \sum_{j=1}^{n} a_{ij} = \sum_{j=1}^{n} a_{ji}.$$

Example 2.1.5 (Adjacency matrix and vertex degrees). This example illustrates that the row sums of $A(G)$ are the outdegrees of the vertices and the column sums are the indegrees.

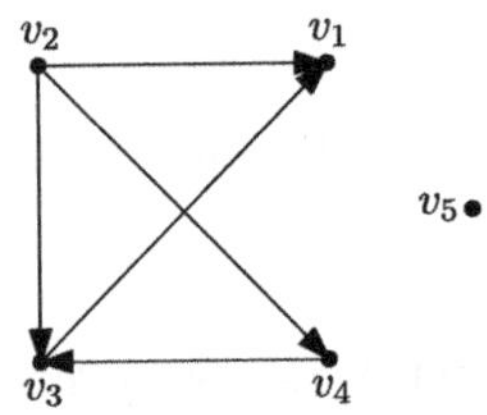

	v_1	v_2	v_3	v_4	v_5	row sum
v_1	0	0	0	0	0	0
v_2	1	0	1	1	0	3
v_3	1	0	0	0	0	1
v_4	0	0	1	0	0	1
v_5	0	0	0	0	0	0
column sum	2	0	2	1	0	

Isomorphic graphs and the adjacency matrix

The next theorem gives a simple formal description of isomorphic graphs. It does not contribute in an essential way to a solution of the so-called *isomorphism problem*, which describes the problem of testing two graphs for being isomorphic. This turns out to be a difficult problem if one wants to construct, e. g., all (nonisomorphic) graphs of a given order.

Theorem 2.1.6. *Let $G = (V, E)$ and $G' = (V', E')$ be two simple graphs with $n = |V|$. The homomorphism*

$$f : G = (V, E) \to G'$$

is an isomorphism if and only if there exists a matrix P such that

$$A(G') = P\,A(G)\,P^{-1},$$

where P is an $n \times n$ row permutation matrix which comes from the identity matrix I_n upon performing row permutations corresponding to f.

Proof. For "⇒," suppose $G \cong G'$, i. e., that G' comes from G by permutation of the vertices. Then, in $A(G)$, rows and columns are permuted correspondingly. Thus $A(G') = P\,A(G)\,P^{-1}$, where P is the corresponding row permutation matrix. Left multiplication by P then permutes the rows and right multiplication by P^{-1} permutes the columns.

For "⇐," suppose $A(G') = P\,A(G)\,P^{-1}$ where P is a permutation matrix. Then there exists a mapping $f : V \to V'$ with

$$(x_i, x_j) \in E, \text{ i. e., } a_{ij} = 1 \Leftrightarrow a_{f(i), f(j)} = 1, \text{ i. e., } (x_{f(i)}, x_{f(j)}) \in E'. \qquad \square$$

Example 2.1.7 (Isomorphisms and adjacency matrices). It is apparent that the graphs G and G' are isomorphic. The matrix P describes the permutation of vertex numbers which leads from $A(G)$ to $A(G')$, i. e., $A(G') = P\,A(G)\,P^{-1}$.

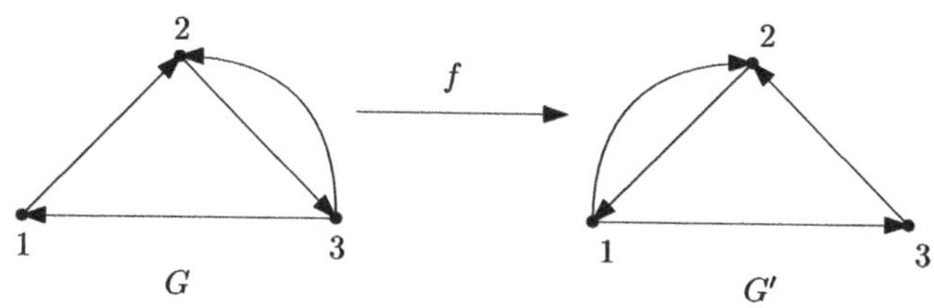

$$P = \begin{pmatrix} 0 & 1 & 0 \\ 0 & 0 & 1 \\ 1 & 0 & 0 \end{pmatrix}, \quad P^{-1} = {}^tP,$$

$$A(G) = \begin{pmatrix} 0 & 1 & 0 \\ 0 & 0 & 1 \\ 1 & 1 & 0 \end{pmatrix}, \quad A(G') = \begin{pmatrix} 0 & 1 & 0 \\ 1 & 0 & 1 \\ 1 & 0 & 0 \end{pmatrix}.$$

Components and the adjacency matrix

Simple matrix techniques enable restructuring of the adjacency matrix of a graph according to its geometric structure.

Theorem 2.1.8. *The graph G has s (weak) components $G_1, \dots, G_s$ if and only if there exists a permutation matrix P with*

$$P\,A(G)\,P^{-1} = \begin{pmatrix} A(G_1) & & & 0 \\ & A(G_2) & & \\ & & \ddots & \\ 0 & & & A(G_s) \end{pmatrix}$$

(block diagonal form)*.*

Proof. Weak connectedness defines an equivalence relation on V, so we get a decomposition of V into $V_1, \dots, V_s$. These vertex sets induce subgraphs $G_1, \dots, G_s$. Renumber G so that we first get all vertices in G_1, then all vertices in G_2, and so on. Note that there are no edges between different components. □

Theorem 2.1.9. *The directed graph G has the strong components $G_1, \dots, G_s$ if and only if there exists a permutation matrix P with*

$$P\,A(G)\,P^{-1} = \begin{pmatrix} A(G_{i_1}) & & & * \\ & A(G_{i_2}) & & \\ & & \ddots & \\ 0 & & & A(G_{i_s}) \end{pmatrix}$$

(Frobenius form, block triangular form)*.*

Proof. If we have the strong components, select G_{i_1} so that no arrows end in G_{i_1}. Then select G_{i_2} so that except for arrows starting from G_{i_1}, no arrows end in G_{i_2}. Note that there may be no arrows ending in G_{i_2}. Next, select G_{i_3} so that except for arrows starting from G_{i_1} or from G_{i_2}, no arrows end in G_{i_3}. Continue in this fashion. Observe that the numbering inside the diagonal blocks is arbitrary. The vertices of G have to be renumbered correspondingly. □

Example 2.1.10 (Frobenius form).

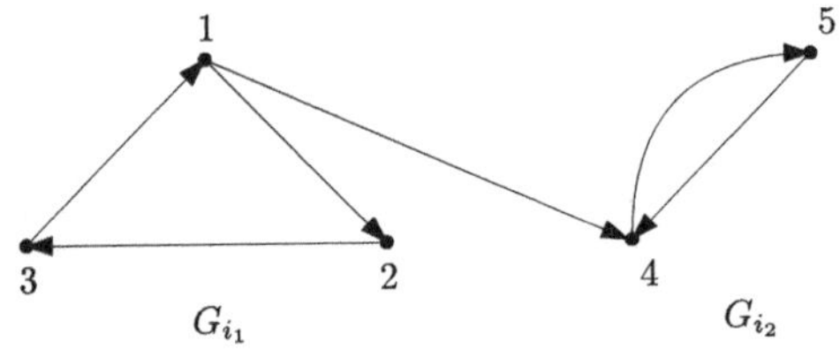

$$\begin{pmatrix} 0 & 1 & 0 & 1 & 0 \\ 0 & 0 & 1 & 0 & 0 \\ 1 & 0 & 0 & 0 & 0 \\ 0 & 0 & 0 & 0 & 1 \\ 0 & 0 & 0 & 1 & 0 \end{pmatrix}$$

Adjacency list

The adjacency list is a tool that is often used when graphs have to be represented in a computer, especially if the adjacency matrix has many zeros.

Definition 2.1.11. The ***adjacency list*** $A(x)$ of the vertex $x \in G$ in the directed case consists of all successors of x, i. e., the elements of $\text{out}(x)$ in arbitrary order. In the undirected case, it consists of all neighbors of x in arbitrary order.

The ***adjacency list*** of the graph G is $A(x_1); A(x_2); \ldots$ for $x_i \in G$.

Example 2.1.12. The adjacency list of the graph from Example 2.1.10 is

$$A(1) = 2,4; \quad A(2) = 3; \quad A(3) = 1; \quad A(4) = 5; \quad A(5) = 4.$$

If the graph G has multiple edges, then the outsets in its adjacency list may contain certain elements several times; in this case, we get so-called multisets.

2.2 Incidence matrix

The incidence matrix relates vertices with edges, so multiple edges are possible but loops have to be excluded completely. It will turn out to be useful later when we consider cycle and cocycle spaces. Its close relation to linear algebra becomes clear in Theorem 2.2.3. We give its definition now, although most of this section relates to the adjacency matrix.

Definition 2.2.1. Take $G = (V, E, p)$, with $V = \{x_1, \ldots, x_n\}$ and $E = \{e_1, \ldots, e_m\}$. The $n \times m$ matrix $B(G)$ over $\{-1, 0, 1\}$ where

$$b_{ij} := \begin{cases} 1 & \text{if } x_i = o(e_j) \\ -1 & \text{if } x_i = t(e_j) \\ 0 & \text{otherwise} \end{cases}$$

or, in the undirected case,

$$b_{ij} := \begin{cases} 1 & \text{if } x_i \in e_j \\ 0 & \text{otherwise} \end{cases}$$

is called the ***(vertex–edge) incidence matrix*** of G.

Example 2.2.2 (Incidence matrix). Here, we present the incidence matrix of the divisor graph of 6; see Example 2.1.2. The matrix is the inner part of the table.

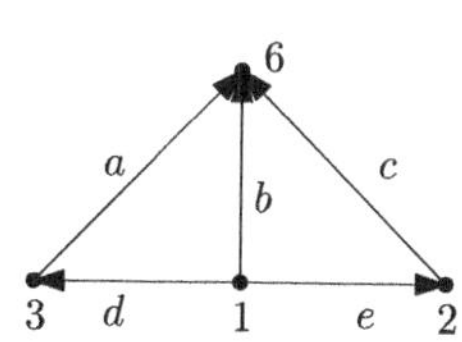

	a	b	c	d	e
1	0	1	0	1	1
2	1	0	0	−1	0
3	0	0	1	0	−1
6	−1	−1	−1	0	0

Theorem 2.2.3. *Let G be a graph with n vertices and s (weak) components, and without loops. Then $B(G)$ has rank $n - s$ over $\mathbb{Z}_2$ in the undirected case, over $\mathbb{R}$ otherwise. If $s = 1$ any $n - 1$ rows of $B(G)$ are linearly independent.*

Proof. We number the vertices according to Theorem 2.1.8 (block diagonal form), and get $B(G)$ also in block diagonal form. Its rank is the sum of the ranks of the blocks. So we consider $s = 1$. Addition of the row vectors gives the zero vector; therefore, the rows are linearly dependent, i. e., we have $\operatorname{rank}(B(G)) \leq n - 1$.

If we delete one row, i. e., one vertex, then the sum of the remaining row vectors is obviously not zero. This holds also for any subset of rows. □

2.3 Distances in graphs

We now consider reachability and distances in graphs. For each graph, these can again be represented by matrices.

Definition 2.3.1. Take $G = (V, E)$ with $V = \{x_1, \ldots, x_n\}$. The $n \times n$ matrix $R(G)$ with

$$r_{ij} := \begin{cases} 1 & \text{if there exists a nontrivial } x_i, x_j \text{ path} \\ 0 & \text{otherwise} \end{cases}$$

is called the ***reachability matrix*** of G.

The reachability matrix also shows the strong components of a graph.

Note that there may be a problem with the diagonal. In the definition, we have $r_{ii} = 1$ if and only if x_i lies on a cycle. It is also possible to set all diagonal elements to 0 or 1. This choice can be made when the graph models a problem that allows us to decide whether a vertex can be reached from itself if it lies on a cycle.

Definition 2.3.2. Take $G = (V, E)$ and use the notation from Definition 1.1.5. The matrix $D(G)$ with

$$d_{ij} := \begin{cases} \infty & \text{if } F(x_i, x_j) = \emptyset \text{ and } i \neq j \\ 0 & \text{if } i = j \\ d(x_i, x_j) & \text{otherwise} \end{cases}$$

is called the ***distance matrix*** of G. The (i, j)th element of the distance matrix is the distance from vertex x_i to vertex x_j, and is infinity if no path exists.

The adjacency matrix and paths

A simple but surprising observation is that the powers of the adjacency matrix count the number of paths from one vertex to another. We start with an example.

Example 2.3.3 (Powers of the adjacency matrix).

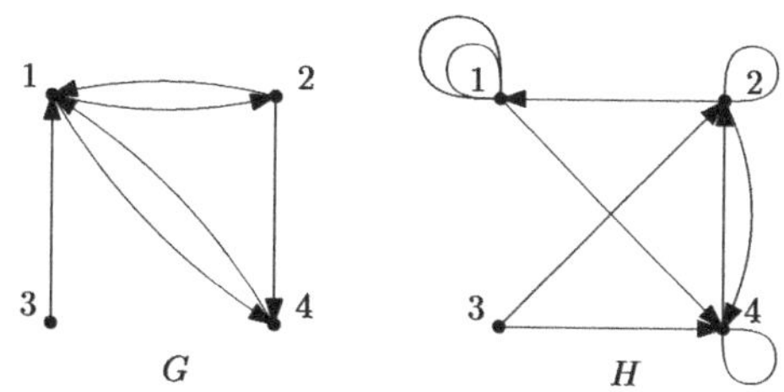

$$A(G)^2 = \begin{pmatrix} 0 & 1 & 0 & 1 \\ 1 & 0 & 0 & 1 \\ 1 & 0 & 0 & 0 \\ 1 & 0 & 0 & 0 \end{pmatrix}^2 = \begin{pmatrix} 2 & 0 & 0 & 1 \\ 1 & 1 & 0 & 1 \\ 0 & 1 & 0 & 1 \\ 0 & 1 & 0 & 1 \end{pmatrix} = A(H)$$

Theorem 2.3.4. *Take $G = (V, E, p)$ and let $a_{ij}^{(r)}$ be an entry of $(A(G))^r$. Then $a_{ij}^{(r)}$ is the number of x_i, x_j paths of length r in G.*

Proof. The result follows from the formula for the second power,

$$a_{ij}^{(2)} = \sum_{k=1}^{n} a_{ik}a_{kj},$$

together with induction. This is the formula for the entries in the product of matrices. □

Remark 2.3.5. Note that forming the second power of an adjacency matrix can be generalized to taking the product of two adjacency matrices of the same size. The result can be interpreted as a graph containing as its edges the corresponding paths of length two. A similar method works for products of more than two matrices. In all cases, the resulting graph depends on the numbering.

If, conversely, we start from a given graph G and construct the graph G^2 of paths of length two, and then perform the corresponding steps with $A(G)$, we automatically get the matrix product $A(G)^2$ without having to know its definition from linear algebra.

The adjacency matrix, the distance matrix, and circuits

The following remark and two theorems are obvious.

Remark 2.3.6. If $|V| = n$, then the length of a simple path in G is at most n. If the length equals n, then the path is a circuit.

Theorem 2.3.7. *Let G be a graph with n vertices. The elements of the distance matrix D(G) can be obtained from the powers of A(G) as follows:*
(a) $d_{ii} = 0$ *for all i;*
(b) d_{ij} *is the smallest* $r \in \mathbb{N}$ *with* $a_{ij}^{(r)} > 0$ *and* $r < n$, *if such an r exists;*
(c) $d_{ij} = \infty$ *otherwise.*

For the elements of the reachability matrix R(G), we have:
(a) $r_{ii} = 0$ *for all i;*
(b) $r_{ij} = 1$ *if and only if there exists* $r < n$ *with* $a_{ij}^{(r)} > 0$;
(c) $r_{ij} = 0$ *otherwise.*

Theorem 2.3.8. *The graph G contains no circuits if and only if* $a_{ii}^{(r)} = 0$ *in* $(A(G))^r$ *for* $r \leq n$ *and for all i.*

2.4 Endomorphisms and commuting graphs

We briefly discuss two aspects of the adjacency matrix which have not gained much attention so far.

Definition 2.4.1. Let f be a transformation of the finite set $\{1, \ldots, n\}$, i. e., a mapping of the set into itself. Define the ***transformation matrix*** $T(f) = (t_{ij})_{i,j \in \{1,\ldots,n\}}$ of f by setting its ith row t_i to be $\sum_{f(j)=i} e_j$, where e_j is the jth row of the identity matrix I_n and $\overline{0}$ is the row of zeros with n elements.

This means that the ith row of T consists of the sum of rows e_j such that j is mapped onto i by f.

For the following, start by verifying some small examples.

Exerceorem 2.4.2.
(1) The transformation f is an endomorphism of the graph G with vertex set $\{x_1, \ldots, x_n\}$ and adjacency matrix $A(G)$ if and only if the (i,j)th entry of $T(f)A(G)^t T(f)$ being nonzero implies that the (i,j)th entry of $A(G)$ is non-zero.
(2) The transformation f is an endomorphism of the graph G with vertex set $\{x_1, \ldots, x_n\}$ and incidence matrix $B(G)$ if and only if the jth column of $T(f)B(G)$ having nonzero entries implies that there exists a column of $B(G)$ which has the same nonzero entries in the same places.

Definition 2.4.3. We say that G and H (with the same number of vertices) are ***commuting graphs*** if there exist labelings of the graphs such that their adjacency matrices commute, i. e., $A(G)A(H) = A(H)A(G)$.

Theorem 2.4.4. *The graph G commutes with* K_n *if and only if G is a regular graph; it commutes with* $K_{n,n}$ *if G is a regular subgraph of* $K_{n,n}$.

Proof. See A. Heinze [36]. In addition, in this paper a construction of new commuting graphs starting with two pairs of commuting graphs is given. □

Question. Can you find a counterexample for the open "only if" part of the theorem? Construct some positive examples and some negative ones.

2.5 The characteristic polynomial and eigenvalues

The possibility of representing graphs by their adjacency matrices naturally leads to the idea of applying the theory of eigenvalues to graphs. As the eigenvalues of a matrix are invariant with respect to permutation of columns and rows, we can expect that they are suitable for describing properties of graphs which are invariant under renaming of the vertices, i. e., invariant under automorphisms. Note that "eigenvalue" is a partial translation of the German word "Eigenwert."

In this section, we investigate how the eigenvalues of the adjacency matrix reflect the geometric and combinatorial properties of a graph. The definitions are valid for both directed and undirected graphs, but our results are focused mainly on undirected graphs and, correspondingly, symmetric matrices. Here, the theory is relatively simple and many interesting results have been obtained. For directed graphs and nonsymmetric matrices, things become much more complicated. The interested reader can consult monographs on this topic, such as [Cvetković et al. 1979].

We will return to this topic in Chapter 5 and in Chapter 8.

Now let F be a field and G an undirected graph with $|V(G)| = n$.

The following definition of the characteristic polynomial is for both directed and undirected graphs. Note that the coefficients can be determined by the entries of the matrix $A(G)$ by using the determinant. This principle from linear algebra is adapted for graphs in Theorem 2.5.8 and thereafter.

Definition 2.5.1. Let $A(G)$ be the adjacency matrix of G. The polynomial of degree n in the indeterminate t over the field F given by

$$\operatorname{chapo}(G) = \operatorname{chapo}(G;t) := \det(tI_n - A(G)) = t^n + a_{n-1}t^{n-1} + \cdots + a_1 t + a_0,$$

is called the ***characteristic polynomial*** of the graph G. Here, det denotes the determinant and I_n denotes the n-row identity matrix. The zeros $\lambda \in F$ of $\operatorname{chapo}(G)$ are called the ***eigenvalues*** of the graph G. We denote by $m(\lambda)$ the ***multiplicity*** of λ.

Remark 2.5.2. An element $\lambda \in F$ is an eigenvalue of the graph G if and only if there exists a vector $v \in F^n$, $v \neq 0$, with $A(G)v = \lambda v$, i. e., v is an eigenvector of $A(G)$. We say v is an ***eigenvector*** of the graph G for λ.

The characteristic polynomial $\operatorname{chapo}(G)$ is independent of the numbering of the vertices of G. The characteristic polynomial of a matrix is invariant even under arbitrary basis transformations.

We now define the spectrum of a graph to be the sequence of its eigenvalues together with their multiplicities. It is quite surprising that for graphs that represent chemical CH-molecules there exists a correspondence between the spectrum of the graph and the chemical spectrum of the molecule; see, e. g., [Cvetković et al. 1979].

Definition 2.5.3. Let λ_i, $i = 1,\dots,n$, be the zeros of chapo(G) in natural order. We set $\lambda(G) := \lambda_1 < \dots < \lambda_p =: \Lambda(G)$. The ***spectrum*** of a graph G is the set of eigenvalues of $A(G)$ together with their multiplicities:

$$\operatorname{Spec}(G) = \begin{pmatrix} \lambda & \cdots & \lambda_i & \cdots & \Lambda \\ m(\lambda) & \cdots & m(\lambda_i) & \cdots & m(\Lambda) \end{pmatrix}.$$

The largest eigenvalue Λ is called the ***spectral radius*** of G.

The next theorem follows immediately from Theorem 2.1.8 and the properties of the characteristic polynomial.

Theorem 2.5.4. *If G has the components $G_1,\dots,G_r$, then*

$$\operatorname{chapo}(G) = \operatorname{chapo}(G_1)\cdots\operatorname{chapo}(G_r).$$

The set of all eigenvectors of an eigenvalue λ of a graph G together with the zero-vector is called the *eigenspace* of λ, denoted by $\operatorname{Eig}(G,\lambda_i)$.

The following two theorems are not true for directed graphs, i. e., for nonsymmetric matrices. For the proofs, we need several results from linear algebra.

Theorem 2.5.5. *Over $F = \mathbb{R}$, all zeros of the characteristic polynomial* chapo(G) *of the undirected graph G are irrational or integer. Moreover, $A(G)$ is diagonalizable, i. e.,* $\dim(\operatorname{Eig}(G,\lambda_i)) = m(\lambda_i)$, *and has an orthonormal basis of eigenvectors of $A(G)$.*

Proof. As G is undirected, $A(G)$ is symmetric (cf. Remark 2.1.3). Symmetric matrices have only real eigenvalues (since they are *self-adjoint* with respect to the standard scalar product over $\mathbb{R}$; i. e., $\langle v, Av\rangle = \langle Av, v\rangle$ for all $v, w \in \mathbb{R}^n$.) So $A(G)$ is a real symmetric matrix, and this is equivalent to the existence of an orthonormal basis of eigenvectors of $A(G)$.

Finally, we prove that $\lambda_i \in \mathbb{Q}$ implies $\lambda_i \in \mathbb{Z}$. Suppose that chapo$(G; \frac{r}{s}) = 0$ for $r, s \in \mathbb{Z}$ with greatest common divisor $(r,s) = 1$. Then chapo$(G; \frac{r}{s}) = a_0 + a_1(\frac{r}{s}) + \dots + a_n(\frac{r}{s})^n = 0$ with $a_n = 1$, which implies that $a_0 s^n + a_1 r s^{n-1} + \dots + a_n r^n = 0$. Since r and s have greatest common divisor 1, we get $s | a_n$, and so $a_n = 1$ implies $s = 1$. Thus $\frac{r}{s} = r \in \mathbb{Z}$. □

Theorem 2.5.6. *Take an undirected, simple graph G without loops and with eigenvalues λ_i. Then*

$$\sum_{i=1}^{n} \lambda_i = 0, \quad \sum_{i=1}^{n} \lambda_i^2 = 2|E(G)| \quad \textit{and} \quad \sum_{i=1}^{n} \lambda_i^3 = 6 \cdot \textit{number of triangles}.$$

Proof. The trace of a matrix is the sum of its diagonal elements. Therefore, we have $\operatorname{trace}(A(G)) = 0$, since G has no loops. As $A(G)$ is diagonalizable, and since it is symmetric, we get $\operatorname{trace}(\operatorname{Diag}(A)) = \sum_{i=1}^{n} \lambda_i$, where $\operatorname{Diag}(A)$ is a diagonal form of $A(G)$ which has the eigenvalues as its diagonal elements. We use the fact that the trace is invariant under similarity transformations; this is true for the coefficients of chapo(G) and so, in particular, for the coefficient of t^{n-1} in chapo(G), which by Vieta's theorem is $\sum_{i=1}^{n} \lambda_i$. Thus $\sum_{i=1}^{n} \lambda_i = 0$.

Using Theorem 2.3.4 on the powers of the adjacency matrix, we obtain that $\operatorname{trace}(A(G)^2)$ = sum of the vertex degrees, which is always equal to $2|E(G)|$. Diagonalizability of $A(G)$ then implies that $\sum_{i=1}^{n} \lambda_i^2 = \operatorname{trace}(A(G)^2)$. □

Exercise 2.5.7. Prove the statement about the number of triangles in Theorem 2.5.6.

In line with the preceding theorem, we can interpret the coefficients of the characteristic polynomial in terms of the number of cycles of the graph. In principle, this can be done for all coefficients, but here we present the result only for four coefficients and prove it for three of them; cf. [Biggs 1996], Proposition 2.3 on page 8. For the complete result see, e. g., [Behzad et al. 1979] Theorem 10.22 and the proof in H. Sachs [81].

We come back to this in Section 8.2. The theorem itself comes back as Corollary 8.2.4.

Theorem 2.5.8. *The coefficients of the characteristic polynomial of a simple, undirected graph G without loops have the following properties:*
- $a_{n-1} = 0$;
- $-a_{n-2} = |E(G)|$, *the number of edges;*
- $-a_{n-3}$ *is twice the number of triangles in G;*
- a_{n-4} *is the number of pairs of disjoint edges minus twice the number of quadrangles.*

Proof. Since the diagonal elements of $A(G)$ are all zero, we get $a_{n-1} = 0$; see the previous theorem.

We use the fact from the theory of matrices that the coefficients of the characteristic polynomial of A can be expressed in terms of the principal minors of A; in what follows, we show this for the first coefficients. A principal minor is the determinant of a submatrix obtained by taking a subset of the rows and the same subset of columns.

A principal minor with two rows and columns with a nonzero entry must be of the form $\begin{vmatrix} 0 & 1 \\ 1 & 0 \end{vmatrix}$. There is one such minor for each pair of adjacent vertices of G, and each has value -1. Thus $a_{n-2} = -1|E(G)|$.

There are essentially three possible nontrivial principal minors with three rows and columns, namely

$$\begin{vmatrix} 0 & 1 & 0 \\ 1 & 0 & 0 \\ 0 & 0 & 0 \end{vmatrix}, \quad \begin{vmatrix} 0 & 1 & 1 \\ 1 & 0 & 0 \\ 1 & 0 & 0 \end{vmatrix}, \quad \begin{vmatrix} 0 & 1 & 1 \\ 1 & 0 & 1 \\ 1 & 1 & 0 \end{vmatrix}.$$

Only the last one is nonzero, with value 2. This minor corresponds to three mutually adjacent vertices of G. This means that a_{n-3} is twice the number of triangles in G. □

Example 2.5.9 (Characteristic polynomials and eigenvalues).

Graph	Adjacency matrix	Characteristic polynomial	Eigenvalues
K_2	$\begin{pmatrix} 0 & 1 \\ 1 & 0 \end{pmatrix}$	$\operatorname{chapo}(K_2) = t^2 - 1$	$-1, 1$
P_2	$\begin{pmatrix} 0 & 0 & 1 \\ 0 & 0 & 1 \\ 1 & 1 & 0 \end{pmatrix}$	$\operatorname{chapo}(P_2) = t^3 - 2t$	$-\sqrt{2}, 0, \sqrt{2}$
K_4	$\begin{pmatrix} 0 & 1 & 1 & 1 \\ 1 & 0 & 1 & 1 \\ 1 & 1 & 0 & 1 \\ 1 & 1 & 1 & 0 \end{pmatrix}$	$\operatorname{chapo}(K_4) = t^4 - 6t^2 - 8t - 3$	$-1, -1, -1, 3$
$C_4 = K_{2,2}$	$\begin{pmatrix} 0 & 1 & 0 & 1 \\ 1 & 0 & 1 & 0 \\ 0 & 1 & 0 & 1 \\ 1 & 0 & 1 & 0 \end{pmatrix}$	$\operatorname{chapo}(C_4) = t^4 - 4t^2$	$-2, 0, 0, 2$
$K_{2,3}$	$\begin{pmatrix} 0 & 0 & 1 & 1 & 1 \\ 0 & 0 & 1 & 1 & 1 \\ 1 & 1 & 0 & 0 & 0 \\ 1 & 1 & 0 & 0 & 0 \\ 1 & 1 & 0 & 0 & 0 \end{pmatrix}$	$\operatorname{chapo}(K_{2,3}) = t^5 - 6t^3$	$-\sqrt{6}, 0, 0, 0, \sqrt{6}$
$K_{4,4}$		$\operatorname{chapo}(K_{4,4}) = t^8 - 16t^6$	$-4, 0, 0, 0, 0, 0, 0, 4$

Proposition 2.5.10. *We have*

$$\operatorname{Spec}(K_n) = \begin{pmatrix} -1 & n-1 \\ n-1 & 1 \end{pmatrix}.$$

Proof. Here and later, we will also use the following notation for determinants:

$$\begin{vmatrix} t & -1 & \cdots & -1 & -1 \\ -1 & t & \ddots & -1 & -1 \\ \vdots & \ddots & \ddots & \ddots & \vdots \\ -1 & -1 & \ddots & t & -1 \\ -1 & -1 & \cdots & -1 & t \end{vmatrix} \overset{\text{(subtract row 1 from the others)}}{=}$$

$$\begin{vmatrix} t & -1 & \cdots & \cdots & -1 \\ -1-t & t+1 & 0 & \cdots & 0 \\ \vdots & 0 & \ddots & \ddots & \vdots \\ \vdots & \vdots & \ddots & \ddots & 0 \\ -1-t & 0 & \cdots & 0 & t+1 \end{vmatrix} \overset{\text{(add columns } 2,\dots,n \text{ to column 1)}}{=}$$

$$\begin{vmatrix} -(n-1)+t & -1 & \cdots & \cdots & -1 \\ 0 & t+1 & 0 & \cdots & 0 \\ \vdots & \ddots & \ddots & \ddots & \vdots \\ \vdots & & \ddots & \ddots & 0 \\ 0 & \cdots & \cdots & 0 & t+1 \end{vmatrix} = (-(n-1)+t)(t+1)^{n-1},$$

and this gives the statement. □

Theorem 2.5.11. *We have*

$$\operatorname{Spec}(K_{p,q}) = \begin{pmatrix} -\sqrt{pq} & 0 & \sqrt{pq} \\ 1 & p+q-2 & 1 \end{pmatrix}.$$

Proof. Several proofs of this result can be found in the chapter *On the eigenvalues of a graph* by A. J. Schwenk and R. J. Wilson, in [Beineke/Wilson 1978]. We demonstrate the following version.

The matrix of the bipartite graph $K_{p,q}$ has the form

$$\begin{pmatrix} 0 & J \\ {}^tJ & 0 \end{pmatrix} = A(K_{p,q}),$$

where J is a $p \times q$ matrix formed from ones. This matrix has only two linearly independent rows, i. e., the eigenvalue 0 has multiplicity $m(0) = p + q - 2$. Now Theorem 2.5.6 implies that $\Lambda = -\lambda$ and, using the fact that $|E| = p + q$ for $K_{p,q}$, Theorem 2.5.6 gives $\Lambda = -\lambda = \sqrt{pq}$. Then the characteristic polynomial is

$$\operatorname{chapo}(K_{p,q}) = (t^2 - pq)t^{p+q-2}.$$ □

Exercise 2.5.12. Prove that the converses of both results are also true, i. e., complete graphs and complete bipartite graphs are characterized by their spectra, within the family of complete graphs and complete bipartite graphs, respectively.

Exercise 2.5.13. Verify Theorems 2.5.6 and 2.5.8 for the graphs in Examples 2.5.9 and in Exerceorem 2.6.5 and for the cospectral graphs in Example 2.7.2.

2.6 Circulant graphs

The so-called circulant graphs generalize, e. g., cycles and complete graphs. Because of the circulant structure of their adjacency matrices, the computation of the characteristic polynomial is simpler than usual. Note, however, that the eigenvalues will not, in general, be real.

Definition 2.6.1. An $n \times n$ matrix S is called a ***circulant matrix*** if its entries satisfy

$$s_{ij} = s_{1j-i+1},$$

where the indices are reduced modulo n and thus belong to the set $\{1, \dots, n\}$.

In other words, row i of S can be obtained from row 1 of S via a circular shift of $i-1$ steps. Thus every circulant matrix is determined by its first row.

Remark 2.6.2. Let W denote the circulant matrix with first row $(0, 1, 0, \dots, 0)$, and let S be the general circulant matrix with first row $(s_1, \dots, s_n)$. Calculations give

$$S = \sum_{j=1}^{n} s_j W^{j-1} = s_1 W^0 + s_2 W^1 + \dots + s_n W^{n-1}.$$

As chapo$(W) = t^n - 1$, we get the eigenvalues $1, \omega, \omega^2, \dots, \omega^{n-1}$, where $\omega = \exp \frac{2\pi i}{n}$, the nth roots of unity. They are pairwise distinct, so we get that W is diagonalizable.

The eigenvalues of S are then determined by

$$\lambda_r = \sum_{j=1}^{n} s_j \omega^{(j-1)r}, \quad r = 0, 1, \dots, n-1.$$

In particular, for the circulant matrix

$$A = \begin{pmatrix} 0 & a_2 & \dots & \dots & a_n \\ a_n & 0 & a_2 & \dots & a_{n-1} \\ \vdots & \ddots & \ddots & \ddots & \vdots \\ \vdots & & \ddots & \ddots & a_2 \\ a_2 & \dots & \dots & a_n & 0 \end{pmatrix}$$

we get the eigenvalues

$$\lambda_r = \sum_{j=1}^{n} a_j \omega^{(j-1)r}, \quad r = 0, \dots, n-1.$$

Thus $\lambda_0 = \sum_{j=1}^{n} a_j = \sum_{j=2}^{n} a_j$ and $\lambda_r = \sum_{j=1}^{n-1} a_{j+1} \omega^{jr}$ for $r \neq 0$; see [Biggs 1996], page 16, and, e. g., page 594 of [Brieskorn 1985].

Definition 2.6.3. A ***circulant graph*** is an undirected graph whose vertices can be arranged so that $A(G)$ is a circulant matrix.

The adjacency matrix of a circulant graph is symmetric with zeros on the diagonal, and we have $a_i = a_{n-i+2}$ for $2 \leq i \leq n$ according to Definition 2.6.1.

Theorem 2.6.4 ([Cvetković et al. 1979] Section 2.6, p. 72 ff.). *The following properties hold:*

(a)

$$\operatorname{Spec}(K_n) = \begin{pmatrix} -1 & n-1 \\ n-1 & 1 \end{pmatrix}.$$

(b)

$$\operatorname{Spec}(C_n) = \begin{cases} \begin{pmatrix} -2 & 2\cos\frac{(n-2)\pi}{n} & \cdots & 2\cos\frac{2\pi}{n} & 2 \\ 1 & 2 & \cdots & 2 & 1 \end{pmatrix} & \textit{for } n \textit{ even}, \\ \begin{pmatrix} 2\cos\frac{(n-1)\pi}{n} & \cdots & 2\cos\frac{2\pi}{n} & 2 \\ 2 & \cdots & 2 & 1 \end{pmatrix} & \textit{for } n \textit{ odd}. \end{cases}$$

(c)

$$\operatorname{Spec}(K_{2_1,\dots,2_s}) = \begin{pmatrix} -2 & 0 & 2s-2 \\ s-1 & s & 1 \end{pmatrix}.$$

(d) *P_{n-1} has the simple eigenvalues*

$$\lambda_j = 2\cos\frac{\pi j}{n+1}, \quad j = 1, \dots, n.$$

Proof. (a) Compare with Proposition 2.5.10. As K_n is circulant, we get

$$\Lambda =: \lambda_0 = n-1, \quad \lambda_{r\neq 0} = \sum_{j=1}^{n-1} \omega^{jr} = -1,$$

since $1 + \omega^r + \dots + \omega^{(n-1)r} = 0$.

(b) The circuit C_n is a circulant graph and the first row of $A(C_n)$ is $(0, 1, 0, \dots, 0, 1)$. Therefore,

$$\begin{aligned} \lambda_r = \omega^r + \omega^{(n-1)r} &= e^{\frac{2\pi i}{n} r} + e^{\frac{2\pi i(n-1)}{n} r} \\ &= e^{\frac{2\pi i r}{n}} + \underbrace{e^{2\pi r}}_{=1} e^{-\frac{2\pi i r}{n}} = 2\cos\frac{2\pi r}{n}. \end{aligned}$$

(c) Again $K_{2,\dots,2}$ is a circulant graph. The first row of the adjacency matrix has length $2s$ and contains 0 at positions 1 and $s+1$ and 1 elsewhere; cf. [Biggs 1996] page 17.

(d) We already know the characteristic polynomials of paths. To determine the eigenvalues, one can use the following determinant, the so-called continuant (see, e. g., p. 595 of [Brieskorn 1985] just mentioned in Remark 2.6.2):

$$\begin{vmatrix} a_1 & 1 & 0 & \dots & 0 \\ 1 & \ddots & \ddots & \ddots & \vdots \\ 0 & \ddots & \ddots & \ddots & 0 \\ \vdots & \ddots & \ddots & \ddots & 1 \\ 0 & \dots & 0 & 1 & a_n \end{vmatrix}. \qquad \square$$

For the following, see *On the eigenvalues of a graph* by A. J. Schwenk and R. J. Wilson, Section 6, in the book [Beineke/Wilson 1978], as well as Table 4 in the Appendix of [Cvetković et al. 1979].

Exerceorem 2.6.5. We have:
(a) $\operatorname{chapo}(W_p) = (t-1+\sqrt{p})(t-1-\sqrt{p})\prod_{i=1}^{p-2}(t-2\cos\frac{2\pi i}{p-1})$, where W_p is the wheel with $p-1$ spokes; that is, using again the notation for the join as mentioned in Section 1.8, formally introduced in Chapter 4, $W_p = C_{p-1} + K_1$.
In particular, for the tetrahedron $W_3 = K_4 = C_3 + K_1$ we have $\operatorname{chapo}(K_4) = (t+1)(t-3)(t+1)^2$.
(b) $\operatorname{chapo}(Q_n) = \prod_{i=0}^{n}(t+n-2i)^{\binom{n}{i}}$, where Q_n is the n-dimensional cube, see Definition 4.3.4. In particular, $\operatorname{chapo}(Q_3) = (t-3)(t-1)^3(t+1)^3(t+3)$.
(c) chapo (octahedron) $= (t-4)t^3(t+2)^2$.
(d) chapo (dodecahedron) $= (t-3)(t^2-5)^3(t-1)^5t^4(t+2)^4$.
(e) chapo (icosahedron) $= (t-5)(t^2-5)^3(t+1)^5$.

2.7 Eigenvalues and the combinatorial structure

As the spectrum of a graph is independent of the numbering of its vertices, there was once the hope that the spectrum could describe the structure of a graph up to isomorphism; however, this soon turned out to be wrong.

Cospectral graphs

The smallest pair of cospectral graphs (i. e., nonisomorphic graphs with the same spectrum) was found with the graphs $K_{1,4}$ and $C_4 \bigcup K_1$. Since the second graph is not connected, the next step was to seek connected cospectral graphs; this was achieved with two graphs with six vertices. Nevertheless, there exist classes of graphs which are characterized by their spectra, e. g., complete graphs or completely bipartite graphs, as we saw in the previous section.

Definition 2.7.1. Nonisomorphic graphs with the same spectrum are said to be ***cospectral.***

Example 2.7.2 (Cospectral graphs).
(a) We have

$$\operatorname{Spec}(K_{1,4}) = \operatorname{Spec}(C_4 \bigcup K_1) = \begin{pmatrix} -2 & 0 & 2 \\ 1 & 3 & 1 \end{pmatrix}$$

with characteristic polynomial $t^3(t^2-4)$.

(b) The graphs G_1 and G_2 are the smallest connected cospectral graphs; they have the characteristic polynomial

$$t^6 - 7t^4 - 4t^3 + 7t^2 + 4t - 1 = (t-1)(t+1)^2(t^3 - t^2 - 5t + 1)$$

(c) There exist two cospectral trees with eight vertices and characteristic polynomial $t^8 - 7t^6 + 9t^4 = t^4(t^4 - 7t^2 + 9)$:

See A. Mowshowitz [64].

Remark 2.7.3.
(a) For every k there exist cospectral k-tuples of regular, connected graphs.
(b) Almost all (cf. Remark 7.2.14) trees with a given number of vertices are cospectral; i. e.,

$$\lim_{p\to\infty} \frac{s_p}{t_p} = 0,$$

where s_p is the number of trees with p vertices which are not cospectral to any other tree with p vertices, and t_p is the number of trees with p vertices. See A. J. Schwenk and R. J. Wilson [83].
(c) Compare also Remark 2.7.6.

Eigenvalues, diameter, and regularity

The following theorem reveals an interesting connection between eigenvalues and the combinatorial structure of the graph. It is also interesting because of its proof, which uses some linear algebra in a quite tricky way. We may say that computations are done in the so-called *adjacency algebra*.

Theorem 2.7.4. *If the undirected, connected graph G has exactly p different eigenvalues, then* $\operatorname{diam}(G) < p$.

Proof. Because of Theorem 2.5.5, there exists a basis of eigenvectors of $A = A(G)$. The minimal polynomial $\operatorname{mipo}(G;t)$ of A is the normalized polynomial of minimal degree such that $\operatorname{mipo}(G;A) = 0$. Now the minimal polynomial $\operatorname{mipo}(G;t)$ of A has only simple zeros and is of degree p. This implies that a nontrivial linear combination of $I = A^0, A, A^2, \ldots, A^p$ is 0, namely $\operatorname{mipo}(G;A) = 0$.

Now suppose that $\mathrm{diam}(G) = q$ and let $x = x_0, \ldots, x_q = y$ be a simple x, y path with q edges in G; i. e., for any $i \leq q$ there exists a path of length i from x_0 to x_i but no shorter path. Then A^i has at the $(0, i)$ position an entry greater than zero, and all $I = A^0, A, A^2, \ldots, A^{i-1}$ have a zero entry there; so A^i is linearly independent of $I, A, \ldots, A^{i-1}$. Thus $I, A, \ldots, A^q$ are linearly independent. This implies $\mathrm{diam}(G) = q < p$. □

Theorem 2.7.5. *If G is a d-regular connected graph, then d is a simple eigenvalue of G with eigenvector $u = {}^t(1, \ldots, 1)$ such that $|\lambda| \leq d$ for all other eigenvalues λ of G.*

Proof. It is clear that $Au = du$ for $u := {}^t(1, \ldots, 1)$. Therefore, d is an eigenvalue corresponding to the eigenvector u.

Let $x = {}^t(x_1, \ldots, x_n)$ be any nonzero vector with $Ax = dx$, and suppose that x_j is an entry of x with the largest absolute value. Now $(Ax)_j = dx_j$ implies $\sum' x_i = dx_j$, where $\sum'$ denotes summation over those d vertices v_i which are adjacent to v_j. Then maximality of x_j implies that $x_i = x_j$ for all these vertices. Choosing another one of the x_i and using connectedness of G, we can show that all entries of x are equal. Thus x is a multiple of u. Therefore, the eigenspace of d has dimension 1, and thus d is simple.

Suppose now that $Ay = \lambda y$ with $y \neq 0$, and let y_i denote an entry of y with largest absolute value. By the previous argument, we have $\sum' y_i = \lambda y_j$, and so $|\lambda||y_j| = |\sum' y_i| \leq \sum' |y_i| \leq d|y_j|$. Thus $|\lambda| \leq d$. □

Automorphisms and eigenvalues

Remark 2.7.6. For all finite groups $A_1, \ldots, A_n$, there exist families of cospectral graphs $G_1, \ldots, G_n$ with $A_i \cong \mathrm{Aut}(G_i)$ for $i = 1, \ldots, n$, compare [Cvetković et al. 1979], Theorem 5.13 on page 153 and Corollary on page 160. In L. Babai [7], this statement is generalized to endomorphism monoids.

Theorem 2.7.7. *Let G be undirected with an eigenvalue λ of multiplicity one, and let v be an eigenvector corresponding to λ. If P is the matrix of an automorphism of G, then*

$$Pv = \pm v.$$

In the directed case, we have $Pv = \mu v$ where $\mu \in \mathbb{C}$ with $|\mu| = 1$.

Proof. If v is an eigenvector corresponding to λ, then Pv is also an eigenvector corresponding to λ, as $APv = PAv = P\lambda v = \lambda Pv$ for the permutation matrix P which describes the automorphism. Now multiplicity one implies that $\dim \mathrm{Eig}(G, \lambda) = 1$ and, therefore, we get that $Pv = \mu v$ for $\mu \in \mathbb{C}$. As P describes an automorphism, we have $P^r = I$ for some $r \in \mathbb{N}$. Consequently, $|\mu| = 1$, and thus $\mu = \pm 1$ if G is undirected. □

Theorem 2.7.8. *Let G be undirected. If G has an automorphism $p \neq \mathrm{id}$, such that $p^2 \neq \mathrm{id}$, then G has at least one eigenvalue with multiplicity greater than one. In other words, if all* *eigenvalues of G are simple, then* Aut G *consists entirely of involutions, i. e.,* $p^2 = \mathrm{id}_{\mathrm{Aut}(G)}$ *for all* $p \in \mathrm{Aut}(G)$.

Proof. If all eigenvalues have multiplicity one, then $P^2v = v$ for all eigenvectors of G by Theorem 2.7.7, because $Pv = \pm v$ where P denotes the matrix of p. Since all eigenvectors span $\mathbb{R}^n$ with $|V| = n$, we get that $P^2v = v$ for all $v \in \mathbb{R}^n$. Therefore, P^2 is the identity matrix and $p^2 = \mathrm{id}_{\mathrm{Aut}(G)}$. □

Exercise 2.7.9. Control the results of Theorem 2.7.5 for the graphs in Theorem 2.6.4 and Theorem 2.6.5 and for nonregular graphs.

2.8 Comments

For further research, we recommend looking at Remark 2.3.5, concerning the product of graphs, and Section 2.4, on the representation of endomorphisms by transformation matrices.

Since square matrices have determinants and permanents, these concepts can be applied to graphs. So the value of the determinant can be related to the combinatorial structure of the graph. Note that the permanent of (the adjacency matrix of) a digraph counts the number of cycle covers, i. e., of 2-factors of the digraph; references to this can be found on the internet, e. g., en.wikipedia.org/wiki/Permanent.

In Section 5.3, we will study the spectra of line graphs. Several other questions concerning eigenvalues and the automorphism group are discussed in Chapter 8.

One subject that we do not touch on at all is the so-called Laplacian eigenvalues of graphs. See, e. g., B. Mohar [62]; also see [Bapat 2011] and T. Bıyıkoğlu, J. Leydold and P. F. Stadler [10]. We take an edge-weighted graph G and let $A(G)$ be the $n \times n$ weighted adjacency matrix. Take the $n \times n$ diagonal matrix $D(G)$ where the vertex degrees are the diagonal elements. Then $L(G) := D(G) - A(G)$ is called the *Laplacian matrix* of G.

There are other polynomials for graphs, e. g., the so-called chromatic polynomial $\mathrm{chropo}(G, k)$. This has a purely combinatorial meaning. Evaluating it for an integer k gives the number of k-colorings of G. Of course, its eigenvalues can also be investigated; see [Tutte 1998].

3 Categories and functors

Here, we adopt the categorical point of view on graphs and their various types of morphisms. This is helpful in distinguishing the resulting different structures and gives some new insights in various constructions with graphs, taken up in the next chapter.

This chapter provides a short introduction to category theory. Categories play an important, albeit mostly hidden, role in many branches of mathematics; it is also useful in many parts of informatics. In what follows, we will consider categories of graphs and, therefore, introduce those concepts which will be used for graph categories; we will also give examples of various categories which can be constructed using graphs. The advantage of the graph-based approach to categories and functors is that the often very abstract concepts can be made quite concrete and understandable in this context. Most of this chapter follows [Kilp et al. 2000]; more information on categories and functors can be found, e. g., in [Herrlich/Strecker 1973].

3.1 Categories

The concept of a category serves to describe objects (which may but do not have to be sets) together with their morphisms (which may but do not have to be mappings). Moreover, this concept enables us to describe, for example, the class of all sets, which is not a set. This, a fortiori, is the case for the class of all graphs.

Definition 3.1.1. A ***category*** $\boldsymbol{C}$ consists of the following data:

1. A class $\operatorname{Ob}\boldsymbol{C}$, the ***C-objects***; if A is a $\boldsymbol{C}$-object, then we write $A \in \operatorname{Ob}\boldsymbol{C}$ or simply $A \in \boldsymbol{C}$.
2. A set $\boldsymbol{C}(A,B)$ or $\operatorname{Mor}_{\boldsymbol{C}}(A,B)$ for every pair (A,B) of $\boldsymbol{C}$-objects, such that

$$\boldsymbol{C}(A,B) \bigcap \boldsymbol{C}(C,D) = \emptyset$$

   for all $A,B,C,D \in \boldsymbol{C}$ with $(A,B) \neq (C,D)$. The elements of $\boldsymbol{C}(A,B)$ are called ***C-morphisms*** from A to B. For $f \in \boldsymbol{C}(A,B)$, we call A the ***domain (source)*** and B the ***codomain (tail, sink)*** of f and write $f : A \to B$ or $A \xrightarrow{f} B$.
3. A composition of morphisms, i. e., a partial relation as follows: for any three objects $A,B,C \in \boldsymbol{C}$ there exists a mapping, the so-called ***law of composition***

$$\circ : \begin{cases} \boldsymbol{C}(A,B) \times \boldsymbol{C}(B,C) \to \boldsymbol{C}(A,C) \\ \quad (f,g) \qquad\qquad \mapsto g \circ f \end{cases}$$

   (the symbol $\circ$ is often omitted), such that the following properties hold:
   (ass) the associativity law $h \circ (g \circ f) = (h \circ g) \circ f$ for the composition of morphisms, whenever all necessary compositions are defined;

https://doi.org/10.1515/9783110617368-003

(id) there exist ***identity morphisms***, which behave like neutral elements with respect to the composition of morphisms, i. e., for every object $A \in \boldsymbol{C}$ there exists a morphism $\mathrm{id}_A \in \boldsymbol{C}(A,A)$ such that

$$f \circ \mathrm{id}_A = f \quad \text{and} \quad \mathrm{id}_A \circ g = g$$

for all $B, C \in \boldsymbol{C}, f \in \boldsymbol{C}(A,B)$ and $g \in \boldsymbol{C}(C,A)$.

The union of all morphism sets of a category $\boldsymbol{C}$ will in general be a class and not a set. This is called the ***class of morphisms*** of $\boldsymbol{C}$, denoted by $\mathrm{Morph}(\boldsymbol{C})$.

Categories with sets and mappings, I

If the objects of a category are sets and the morphisms are mappings, then Definition 3.1.1 turns into the following.

A category consists of the following data:

1. A class of sets.
2. A set $\mathrm{Map}(A,B)$ of mappings from A to B for every pair of sets A, B. According to the definition of mappings we automatically get

$$\mathrm{Map}(A,B) \bigcap \mathrm{Map}(A',B') = \emptyset \quad \text{for } (A,B) \neq (A',B').$$

   (If two mappings have different domains or codomains, they are already different.)
3. For any two mappings $f \in \mathrm{Map}(A,B)$ and $g \in \mathrm{Map}(B,C)$, where A, B, C are sets, a composition of mappings $g \circ f \in \mathrm{Map}(A,C)$ for which the following hold automatically:
   (ass) associativity;
   (id) the existence of identity mappings, i. e., for every set A and $a \in A$ a mapping $\mathrm{id}_A \in \mathrm{Map}(A,A)$ with $\mathrm{id}_A(a) = a$ that satisfies the conditions required above.

Constructs, and small and large categories

Definition 3.1.2. A category $\boldsymbol{C}$ is called a ***construct*** or a ***concrete category*** if its objects are (structured) sets, its morphisms are (structure-preserving) mappings between the respective sets, and the composition law is the composition of these mappings. A category $\boldsymbol{C}$ is said to be ***small*** if $\mathrm{Ob}\,\boldsymbol{C}$ is a set; otherwise it is said to be ***large***.

Theorem 3.1.3. *If $\boldsymbol{C}$ is a category, then $\boldsymbol{C}^{\mathrm{op}}$ is also a category, where*

$$\mathrm{Ob}\,\boldsymbol{C}^{\mathrm{op}} := \mathrm{Ob}\,\boldsymbol{C};$$
$$\boldsymbol{C}^{\mathrm{op}}(A,B) := \boldsymbol{C}(B,A); \quad \textit{and}$$

$$g \bullet f := f \circ g \quad \text{for } f \in \mathbf{C}^{\mathrm{op}}(A,B) = \mathbf{C}(B,A),$$
$$g \in \mathbf{C}^{\mathrm{op}}(B,C) = \mathbf{C}(C,B).$$

The category $\mathbf{C}^{\mathrm{op}}$ is called the ***opposite (dual)*** *category to $\mathbf{C}$. It comes from $\mathbf{C}$ by "inverting all arrows."*

Question. Why is $\boldsymbol{Set}^{\mathrm{op}}$ not a concrete category?

Special objects and morphisms

Definition 3.1.4. An object T of a category $\boldsymbol{C}$ is said to be ***terminal*** if $\boldsymbol{C}(A,T)$ contains exactly one element for every $A \in \boldsymbol{C}$. We say that an object I of a category $\boldsymbol{C}$ is ***initial*** if $\boldsymbol{C}(I,A)$ contains exactly one element for every $A \in \boldsymbol{C}$.

Remark 3.1.5. We say that initial and terminal objects are ***categorically dual*** as T is terminal in $\boldsymbol{C}$ if and only if it is initial in $\boldsymbol{C}^{\mathrm{op}}$.

In any category, we can define isomorphisms without using concepts like injective or surjective and without using that the objects have "elements," which will not be the case if the objects are not sets. Moreover, we will introduce notions that imitate injectivity and surjectivity without using elements. In some concrete categories, however, these turn out to be a little weaker than injectivity and surjectivity.

Definition 3.1.6. A morphism $f \in \boldsymbol{C}(A,B)$ with $A,B \in \boldsymbol{C}$ is called an ***isomorphism*** if there exists a morphism $g \in \boldsymbol{C}(B,A)$ with the properties that $f \circ g = \mathrm{id}_B$ and $g \circ f = \mathrm{id}_A$.

A morphism $f \in \boldsymbol{C}(A,B)$ with $A,B \in \boldsymbol{C}$ is called a ***monomorphism*** if it is left cancellable, i. e., for all morphisms $g,h \in \boldsymbol{C}(C,A)$ with $f \circ g = f \circ h$ we get $g = h$.

A morphism $f \in \boldsymbol{C}(A,B)$ with $A,B \in \boldsymbol{C}$ is called an ***epimorphism*** if it is right cancellable, i. e., for all morphisms $g,h \in \boldsymbol{C}(B,C)$ with $g \circ f = h \circ f$ we get $g = h$.

Proposition 3.1.7. *Terminal objects of a category are always isomorphic to each other, and so are initial objects.*

Proof. Take two terminal objects T_1 and T_2 of $\boldsymbol{C}$. Then by definition there exist morphisms $f : T_1 \to T_2$ and $g : T_2 \to T_1$. Therefore, id_{T_2} and $f \circ g$ are morphisms in $\boldsymbol{C}(T_2, T_2)$, and $|\boldsymbol{C}(T_2, T_2)| = 1$ implies $\mathrm{id}_{T_2} = f \circ g$. Analogously, we prove that $\mathrm{id}_{T_1} = g \circ f$. Consequently, f and g are isomorphisms.

The statement for initial objects can be derived from the result for terminal objects by going to the opposite category. □

Categories with sets and mappings, II

Exercise 3.1.8. Prove that in the category $\boldsymbol{Set}$ terminal objects are the one-element sets (which are all isomorphic), and the empty set is the initial object. In the category

Set and in various graph categories, monomorphisms are injective, epimorphisms are surjective and vice versa. Moreover, mappings that are both surjective and injective (which are then said to be bijective) are isomorphisms in ***Set*** but not in the category of graphs with graph homomorphisms, cf. Remark 1.4.4.

Categories with graphs

The following category $\boldsymbol{Path}_G$ plays a role in object-oriented programming in informatics.

Example 3.1.9 (Small nonconcrete categories).

(a) Every directed graph G defines a small category $\boldsymbol{Path}_G$, with object set V consisting of all vertices of G. If x and y are two vertices, then $\boldsymbol{Path}_G(x,y)$, the set of all morphisms from x to y, consists of all x,y paths. The composition of morphisms is the concatenation of paths.
If $a : x \to y$ and $b : y \to z$ are two nontrivial paths, then $b \circ a = ab$ is the path which is generated by traversing first a and then b. If we have $a = (e_1,\dots,e_n)$ and $b = (e_{n+1},\dots,e_m)$, then

$$b \circ a = ab = (e_1,\dots,e_n,e_{n+1},\dots,e_m).$$

This implies that

$$(e_1,\dots,e_n) = e_1 \cdots e_n = e_n \circ \cdots \circ e_1.$$

The trivial paths are the identities, i. e., for $a : x \to y$, we get

$$a \circ \mathrm{id}_x = \mathrm{id}_x \circ a = a,$$
$$\mathrm{id}_y \circ a = a \circ \mathrm{id}_y = a.$$

Thus, all requirements for a category are fulfilled by $\boldsymbol{Path}_G$.

(b) See the examples in Remarks 3.2.6 and 3.2.11.

Example 3.1.10 (A small construct). The set $\boldsymbol{Gra}_4$ of all graphs with four vertices and edge-preserving mappings of these graphs as morphisms is a small concrete category.

Example 3.1.11 (Noncategories).

(a) Ordered sets with antitone mappings ($x \le y \Rightarrow f(x) \ge f(y)$) and the composition of mappings do not form a category, since the composition of two antitone mappings is not antitone.

(b) Graphs with half-, locally, or quasi-strong graph homomorphisms do not form a category, since the composition of two such morphisms is not necessarily of the same kind.

Example 3.1.12 (Large constructs). For the following categories, the composition law is always the composition of mappings.

Gra	graphs	graph homomorphisms
SGra	graphs	strong graph homomorphisms
CGra	graphs	graph comorphisms
EGra	graphs	graph egamorphisms
SEGra	graphs	strong graph egamorphisms

Note that the categories ***EGra*** and ***SEGra*** turn into ***Gra*** and ***SGra*** if all graphs have a loop at every vertex.

Other categories

Example 3.1.13 (Large constructs). The composition law is always the composition of mappings.

Set	sets	mappings
Sgr	semigroups	semigroup homomorphisms
Mon	monoids	monoid homomorphisms
Grp	groups	group homomorphisms
Ab	Abelian groups	group homomorphisms
Rng	rings	ring homomorphisms
Field	fields	field homomorphisms
S-***Act***	left S-acts, $S \in$ ***Sgr***	left act homomorphisms
Act-S	right S-acts, $S \in$ ***Sgr***	right act homomorphisms
R-***Mod***	left R-modules, $R \in$ ***Rng***	left module homomorphisms
Mod-R	Right R-modules, $R \in$ ***Rng***	right module homomorphisms
F-***Vec***	F-vector spaces, $F \in$ ***Field***	linear mappings
Top	topological spaces	continuous mappings
Ord	ordered sets	isotone (order-preserving) maps
Top$^\circ$	topological spaces	open mappings

Example 3.1.14 (In ***Sgr***, epimorphisms may not be surjective). By $i : (\mathbb{N}, +) \hookrightarrow (\mathbb{Z}, +)$ denote the natural embedding, which of course is not surjective. But it is an epimorphism, as every homomorphism starting in $(\mathbb{Z}, +)$ is uniquely determined by its value on 1.

Example 3.1.15 (Large categories, not concrete over ***Set***).

(a) The category ***Rel*** has as objects all sets, and for sets $A, B \in$ ***Rel*** the morphism set ***Rel***$(A, B) := \mathscr{P}(A \times B)$ is the set of all binary relations between A and B; the composition is the composition of relations.

(b) If $\boldsymbol{C}$ is a concrete category with at least two objects, then the dual category $\boldsymbol{C}^{\mathrm{op}}$ is not concrete in general.

Example 3.1.16 (Small ("strange") categories, not concrete over ***Set***).

(a) If $(M, \cdot, 1)$ is a monoid, let $\operatorname{Ob} \boldsymbol{M} := \{1\}$ and $\boldsymbol{M}(1,1) := M$, i. e., the category $\boldsymbol{M}$ has exactly one object, morphisms are the monoid elements, and the composition in $\boldsymbol{M}$ is monoid multiplication.
(b) Objects of the category $\mathbb{Z}$-***Mat*** are all natural numbers. Morphisms from $m \in \mathbb{Z}$ to $n \in \mathbb{Z}$ are all $m \times n$ matrices over $\mathbb{Z}$. Composition of morphisms is matrix multiplication.
(c) Take $\operatorname{Ob} \boldsymbol{P} := \mathscr{P}(X)$, the power set of X. Set $\mathbf{P}(A,B) := \{(A,B)\}$, i. e., it is a one-element set if $A \subseteq B$, and empty otherwise. Composition of morphisms is defined via $(A,B) \circ (B,C) := (A,C)$.
(d) For every ordered set $(P, \leq)$, take the objects of the category $\overline{\boldsymbol{P}}$ to be the elements of the set P; the morphism sets are all one-element or empty sets, i. e., $\overline{\mathbf{P}}(x,y) := \{(x,y)\}$ if $x \leq y$, and empty otherwise. The composition law is $(x,y) \circ (y,z) := (x,z)$. The previous example is the special case where $(P, \leq) = (\mathscr{P}(X), \subseteq)$.

3.2 Products & Co.

In addition to terminal and initial objects, we define some other objects, which together with certain morphisms form the so-called coproducts, products, and tensor products. The definitions are given axiomatically, i. e., in a very abstract form. Consequently, they are not constructive, since we only formulate which properties they must have, if they exist.

Coproducts

The idea behind the concept of a coproduct is to describe the characteristic properties of unions of sets categorically, i. e., without using sets and elements.

Definition 3.2.1. Let $(C_i)_{i \in I}$ be a nonempty family of objects in $\mathbf{C}$. The pair $((u_i)_{i \in I}, C)$ with $C \in \mathbf{C}$ and $u_i \in \mathbf{C}(C_i, C)$ is called the ***coproduct*** of the $(C_i)_{i \in I}$, if $((u_i)_{i \in I}, C)$ solves the following *universal problem*.

For all $((k_i)_{i \in I}, T)$ with $T \in \mathbf{C}$ and $k_i \in \mathbf{C}(C_i, T)$, there exists exactly one $k \in \mathbf{C}(C, T)$ such that the following diagram is commutative for all $i \in I$:

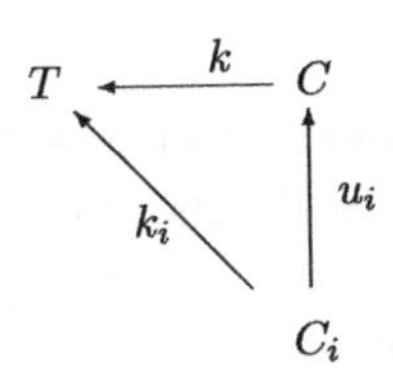

As usual, we write $C = \coprod_{i\in I} C_i$. The morphism u_i is called the ith ***injection***. We also write $[(k_i)_{i\in I}] = k$ and say that k is ***coproduct induced*** by $(k_i)_{i\in I}$.

Exercise 3.2.2. Direct sums of vector spaces (with the natural injections) turn out to be coproducts, which are not unions of the vector spaces, however. Recall that for a field F and a set I, the elements of the direct sum (i. e., the coproduct) of the F-vector spaces V_i, $i \in I$, consist of the $|I|$-tuples $(v_{i\in I})$ which have at most finitely many components unequal zero. Prove that these vector spaces together with the natural injections satisfy the properties of the coproduct.

More examples of coproducts in various graph categories are given in the next chapter. The following concept looks very abstract. It will turn out, also in the next chapter, that the amalgam of two graphs with a common subgraph is the result of gluing together the two graphs along the common subgraph.

Definition 3.2.3. Let H, G_1, and G_2 be objects, and let $m_1 : H \to G_1$ and $m_2 : H \to G_2$ be monomorphisms in the category $\mathbf{C}$. We call this constellation an ***amalgam situation***. The pair $((u_1, u_2), G_1 \coprod_{(H,(m_1,m_2))} G_2)$ is called an ***amalgam (amalgamated coproduct)*** of G_1 and G_2 with respect to $(H, (m_1\ m_2))$ if:
(a) $u_1 : G_1 \to G_1 \coprod_{(H,(m_1,m_2))} G_2$ and $u_2 : G_2 \to G_1 \coprod_{(H,(m_1,m_2))} G_2$ are morphisms such that $u_1 m_1 = u_2 m_2$, i. e., the square in the diagram below is commutative; and
(b) $((u_1, u_2), G_1 \coprod_{(H,(m_1,m_2))} G_2)$ solves the following *universal problem* in $\mathbf{C}$.
For every pair $((f_1, f_2), Q)$, where $f_1 : G_1 \to G$ and $f_2 : G_2 \to G$ with $f_1 m_1 = f_2 m_2$, i. e., making the external rectangle commutative, there exists exactly one morphism $f : G_1 \coprod_{(H,(m_1,m_2))} G_2 \to G$ such that both triangles in the diagram are commutative.

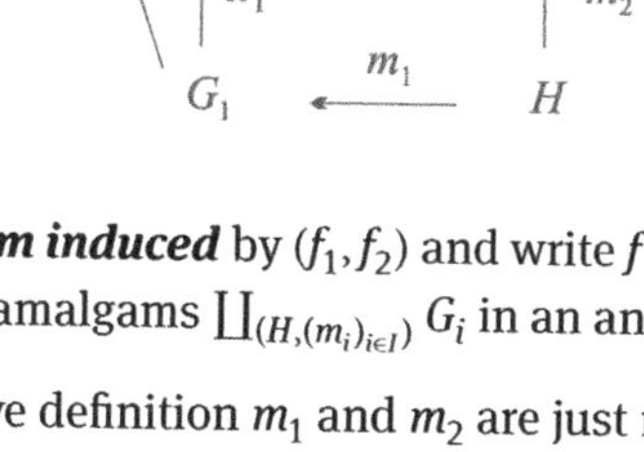

We say that f is ***amalgam induced*** by (f_1, f_2) and write $f = [(f_1, f_2)^H]$.
We can define multiple amalgams $\coprod_{(H,(m_i)_{i\in I})} G_i$ in an analogous way.

Remark 3.2.4. If in the above definition m_1 and m_2 are just morphisms in $\mathbf{C}$, we get a so-called ***pushout***. If, in addition, $G_1 = G_2$, then the pushout is called the ***coequalizer*** of (m_1, m_2).

Exercise 3.2.5. Coproducts as well as amalgams and pushouts are unique up to isomorphism in any category in which they exist.

The idea of the proof is to assume the existence of two coproducts where each plays the role of T with respect to the other; the role of the k_i is then taken by the corresponding injections. The uniqueness of k in all these situations provides the isomorphism, similar to the situation in Proposition 3.1.7.

Remark 3.2.6 (The coproduct as initial object in a new category). We take two objects G_1 and G_2 in $\boldsymbol{C}$ and consider a new category $\boldsymbol{C}^{(G_1,G_2)}$ whose objects are triples (f_1, f_2, G), where f_1 and f_2 are morphisms in $\boldsymbol{C}$ which end in G and start, respectively, in G_1 and G_2. For two such triples (f_1, f_2, G) and (h_1, h_2, H), a morphism in this category is a morphism f in $\boldsymbol{C}$ such that $fh_1 = f_1$, and similarly with index 2. Now the universal property of the coproduct implies that the coproduct $G_1 \coprod G_2$ is the *initial object* in this new category.

In a suitably chosen category $\boldsymbol{C}^{(G_1,G_2,H)}$, the pushout becomes the initial object.

Products

The following two definitions are categorically dual to the definitions of the coproduct and the amalgam. Formally, this means that the new ones can be obtained from the old ones by reversing all arrows and exchanging mono and epi. The motivating idea comes from direct products of vector spaces and Cartesian products of sets, with the same goal as for the definition of coproducts.

Again, more examples of products in various graph categories will be presented in the next chapter.

Definition 3.2.7. Let $(P_i)_{i \in I}$ be a nonempty family of objects in $\boldsymbol{C}$. The pair $(P, (p_i)_{i \in I})$ with $P \in \boldsymbol{C}$ and $p_i \in \boldsymbol{C}(P, P_i)$ is called the ***product*** of $(P_i)_{i \in I}$ if it solves the following *universal problem* in $\boldsymbol{C}$.

For all $(Q, (q_i)_{i \in I})$ with $Q \in \boldsymbol{C}$ and $q_i \in \boldsymbol{C}(Q, P_i)$, there exists exactly one $q \in \boldsymbol{C}(Q, P)$ such that the following diagram is commutative for all $i \in I$:

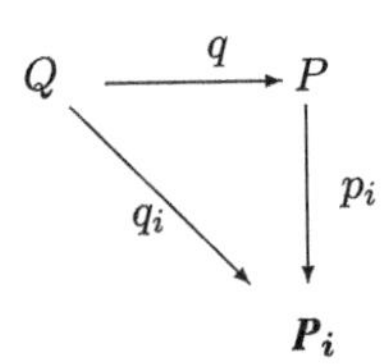

We write $P = \prod_{i \in I} P_i$. The morphism p_i is called the ith ***projection***. We also write $\langle (q_i)_{i \in I} \rangle = q$ and call q the ***product induced*** by $(q_i)_{i \in I}$.

Definition 3.2.8. Let G_1, G_2 and H be objects, and let $n_1 : G_1 \to H$ and $n_2 : G_2 \to H$ be epimorphisms in the category $\boldsymbol{C}$. We call this constellation a ***coamalgam situation***. The pair $(G_1 \prod^{((n_1,n_2),H)} G_2, (p_1, p_2))$ is called the ***coamalgam (coamalgamated product)*** of G_1 and G_2 with respect to $((n_1, n_2), H)$ if:

(a) $p_1 : G_i \prod^{((n_1,n_2),H)} G_2 \to G_1$ and $p_2 : G_i \prod^{((n_1,n_2),H)} G_2 \to G_2$ are morphisms such that $n_1 p_1 = n_2 p_2$, i. e., the square in the diagram below is commutative; and

(b) $(G_1 \prod^{((n_1,n_2),H)} G_2, (p_1, p_2))$ solves the following universal problem in $\boldsymbol{C}$.
For every pair $(G, (f_1, f_2))$, where $f_1 : G \to G_1$ and $f_2 : G \to G_2$ with $n_1 f_1 = n_2 f_2$ (i. e., making the exterior rectangle commutative), there exists exactly one morphism

$$f : G \to G_1 \prod^{((n_1,n_2),H)} G_2$$

such that both triangles in the diagram are commutative.

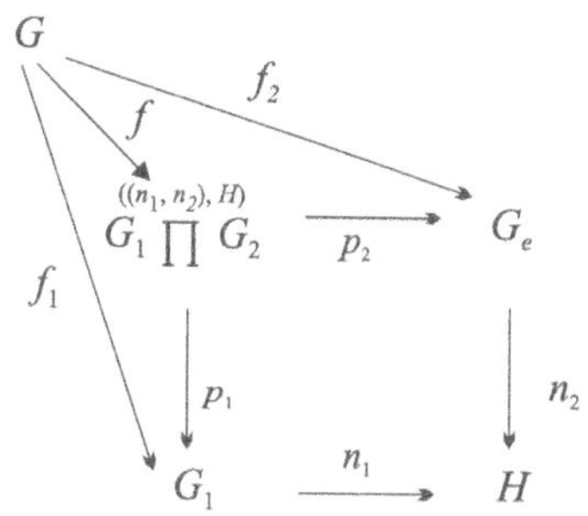

We say that f is ***coamalgam induced*** by (f_1, f_2) and write $\langle (f_1, f_2)_H \rangle = f$. Multiple coamalgams $\prod^{((n_i)_{i \in I}, H)} G_i$ can be defined in an analogous way.

Remark 3.2.9. If in the above definition n_1 and n_2 are just morphisms in $\boldsymbol{C}$, we get a so-called ***pullback***. Moreover, if in this situation $G_1 = G_2$, the pullback is called the ***equalizer*** of (n_1, n_2). Further, we observe that a subobject $W \subseteq G_1 \prod G_2$ is called a ***subdirect product*** of G_1 and G_2 if $p_i(W) = G_i$ for $i = 1, 2$. So a coamalgam is a special subdirect product.

Theorem 3.2.10. *Products, as well as coamalgams, pullbacks, and equalizers, are unique up to isomorphism in every category where they exist.*

Proof. This is an exercise which can also be done by the categorical dualization of Exercise 3.2.5. □

Remark 3.2.11 (The product as terminal object in a new category). As for coproducts and amalgams, we take another step toward abstraction. Now take two objects G_1 and G_2 in the category $\boldsymbol{C}$ and consider a new category $\boldsymbol{C}_{(G_1,G_2)}$ whose objects are triples (G, f_1, f_2), where f_1 and f_2 are morphisms in $\boldsymbol{C}$ which start in G and end, respectively, in G_1 and G_2. For two such triples (G, f_1, f_2) and (H, h_1, h_2), a morphism in the new category is a morphism f such that $h_1 \circ f = f_1$, and similarly with index 2. Now the universal property of the product implies that the product $G_1 \prod G_2$ is the *terminal object* in the new category.

In a suitably modified category $\boldsymbol{C}_{(G_1,G_2,H)}$, the coamalgam will be the terminal object.

Tensor products

We observe that tensor products can be defined only in concrete categories, since in the definition we have to use that the "tensor factors" have elements, i. e., they are sets. Again, tensor products are, in every category where they exist, unique up to isomorphism. Consequently, every tensor product of two factors is also the terminal object in a suitably defined category (compare Remarks 3.2.6 and 3.2.11).

Definition 3.2.12. Let $\boldsymbol{C}$ be a concrete category and let $A, B, C \in \boldsymbol{C}$. A mapping from the Cartesian product of the sets A and B into the set C, i. e., $f : A \times B \to C$, is called a ***bimorphism*** from $A \times B$ to C if for every $a \in A$ and every $b \in B$ we have $f(a, \cdot) \in \boldsymbol{C}(B, C)$ and $f(\cdot, b) \in \boldsymbol{C}(A, C)$.

Definition 3.2.13. Take $A, B \in \boldsymbol{C}$. The pair (τ, T), where $T \in \boldsymbol{C}$ and $\tau : A \times B \to T$, is a bimorphism. It is called the ***tensor product*** of A and B in $\boldsymbol{C}$ if (τ, T) solves the following *universal problem.*

For all $X \in \boldsymbol{C}$ and all bimorphisms $\xi : A \times B \to X$, there exists exactly one morphism $\xi^* \in \boldsymbol{C}(T, X)$ such that the following diagram is commutative:

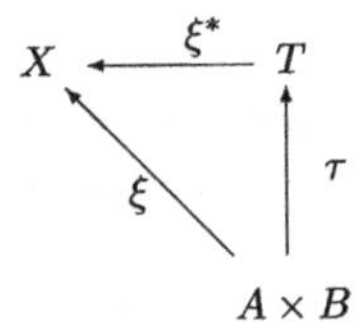

We write $T = A \otimes B$ and call ξ^* the ***tensor product induced*** by ξ.

Exercise 3.2.14. Tensor products are unique up to isomorphism in every category where they exist.

Categories with sets and mappings, III

Exerceorem 3.2.15. In the category of sets and mappings, the disjoint union $A \dot{\bigcup} B$ of two sets A and B with the natural injections u_1 and u_2 is the coproduct. The induced mapping is obtained as $k(x) = k_1(x)$ for $x \in A$ and $k(x) = k_2(x)$ for $x \in B$.

The Cartesian product $A \times B$ of two sets A and B with the natural projections p_1 and p_2 is the product. The induced mapping is $q(x) = (q_1(x), q_2(x))$.

The Cartesian product $A \times B$ of two sets A and B with the mapping $\tau = \mathrm{id}_{A \times B}$ is the tensor product; here, we have $\xi^* = \xi$.

The amalgam over a common subset $A \cap B = H$ of the sets A and B is the (non-disjoint) union of A and B. This is possible also if $n_1(A) = n_2(B) = H \subsetneq A \cap B$. Corresponding to the idea of the amalgam, we can take alternatively the disjoint union and then identify the elements of the common subset H.

The coamalgam of the sets A and B with respect to a common image set H consist of those pairs $(a,b) \in A \times B$ with $n_1(a) = n_2(b)$.

For the proofs, all properties of the respective definitions must be shown directly in the concrete situation, in particular the properties of the induced mappings.

3.3 Functors

Functors are to categories what mappings are to sets. In addition, for algebraic categories there exists a dualism between homomorphisms and antihomomorphisms, i. e., mappings which preserve the multiplication (say) and mappings which reverse the multiplication (e. g., forming^{-1}). This is modeled in the relations between categories by the concepts of covariant and contravariant functors.

Covariant and contravariant functors

We define connections between categories that preserve or reverse compositions of morphisms, which—remember—do not have to be mappings.

Definition 3.3.1. Let $\boldsymbol{C}$ and $\boldsymbol{D}$ be categories. Let $F : \boldsymbol{C} \to \boldsymbol{D}$ be an assignment of a unique object $F(A) \in \boldsymbol{D}$ to an object $A \in \boldsymbol{C}$ and a unique morphism $F(f)$ in $\boldsymbol{D}$ to a morphism $f : A \to A'$ in $\boldsymbol{C}$. We formulate the following two pairs of conditions, (1) and (2) or (1) and (2*):

(1) $F(\mathrm{id}_A) = \mathrm{id}_{F(A)}$ for $A \in \boldsymbol{C}$; we say that F ***preserves identities***.

(2) $F(f) : F(A) \to F(A')$ and $F(f_2 f_1) = F(f_2)F(f_1)$ for $f_1 \in \boldsymbol{C}(A_1, A_2)$ and $f_2 \in \boldsymbol{C}(A_2, A_3)$, where $A_1, A_2, A_3 \in \boldsymbol{C}$; we say that F ***preserves*** the composition of morphisms.

(2*) $F(f) : F(A') \to F(A)$ and $F(f_2 f_1) = F(f_1)F(f_2)$ for $f_1 \in \boldsymbol{C}(A_1, A_2)$ and $f_2 \in \boldsymbol{C}(A_2, A_3)$, where $A_1, A_2, A_3 \in \boldsymbol{C}$; we say that F ***reverses*** the composition of morphisms.

If F satisfies (1) and (2), we call F a ***covariant functor***. In this case, we have

$$F(\mathrm{Mor}_{\boldsymbol{C}}(A_1, A_2)) \subseteq \mathrm{Mor}_{\boldsymbol{D}}(F(A_1), F(A_2)).$$

If F satisfies (1) and (2*), we call F a ***contravariant functor***. In this case, we have

$$F(\mathrm{Mor}_{\boldsymbol{C}}(A_1, A_2)) \subseteq \mathrm{Mor}_{\boldsymbol{D}}(F(A_2), F(A_1)).$$

We call F a ***functor*** if a specification of the *variance* is not necessary.

Composition of functors

Like mappings, functors can be composed if they “fit together.”

Definition 3.3.2. Let $\boldsymbol{C}$, $\boldsymbol{D}$, and $\boldsymbol{E}$ be categories and let $F : \boldsymbol{C} \to \boldsymbol{D}$ and $G : \boldsymbol{D} \to \boldsymbol{E}$ be functors. The ***composition*** GF or $G \circ F$ of the functors F and G is defined by $(GF)(A) = G(F(A))$ and $(GF)(f) = G(F(f))$ for $A, A' \in \boldsymbol{C}$ and $f \in \operatorname{Mor}_{\boldsymbol{C}}(A, A')$.

Remark 3.3.3. With this definition, $GF : \boldsymbol{C} \to \boldsymbol{E}$ is a functor. Here, GF is covariant if F and G are both covariant or both contravariant. Otherwise, GF is contravariant.

Special functors—examples

Definition 3.3.4. A category $\boldsymbol{C}$ is called a ***subcategory*** of the category $\boldsymbol{D}$ if every object from $\boldsymbol{C}$ is an object of $\boldsymbol{D}$ and if $\boldsymbol{C}(A, A') \subseteq \boldsymbol{D}(A, A')$. This means that there exists a functor $I_{\boldsymbol{D}}^{\boldsymbol{C}} : \boldsymbol{C} \to \boldsymbol{D}$ defined by $I_{\boldsymbol{D}}^{\boldsymbol{C}}(A) = A$ for $A \in \boldsymbol{C}$ and $I_{\boldsymbol{D}}^{\boldsymbol{C}}(f) = f$ for $f \in \boldsymbol{C}(A, A')$. This functor is called an ***inclusion functor***. Let $F : \boldsymbol{D} \to \boldsymbol{E}$ be any functor; then we call $FI_{\boldsymbol{D}}^{\boldsymbol{C}} : \boldsymbol{C} \to \boldsymbol{E}$ the ***restriction*** of F to the subcategory $\boldsymbol{C}$ of $\boldsymbol{D}$. For $\boldsymbol{C} = \boldsymbol{D}$, we call $I_{\boldsymbol{C}}^{\boldsymbol{C}}$ the ***identity functor*** on $\boldsymbol{C}$, written as $\operatorname{Id}_{\boldsymbol{C}}$.

Since the inclusion functor is covariant, the restriction of F preserves the variance of F; cf. Remark 3.3.3.

Definition 3.3.5.

(a) Let $\boldsymbol{C}$ be a concrete category. For $A \in \boldsymbol{C}$, we denote by $\lfloor A \rfloor \in \boldsymbol{Set}$ the so-called *underlying set* of the object A. For $f \in \operatorname{Mor}_{\boldsymbol{C}}(A_1, A_2)$, where $A_1, A_2 \in \boldsymbol{C}$, we denote by $\lfloor f \rfloor : \lfloor A_1 \rfloor \to \lfloor A_2 \rfloor$ the mapping in $\boldsymbol{Set}$ "under" f. In this way, $\lfloor\ \ \rfloor : \boldsymbol{C} \to \boldsymbol{Set}$ becomes a covariant functor, the ***forgetful functor*** of $\boldsymbol{C}$ into $\boldsymbol{Set}$.

(b) The transfer from a category $\boldsymbol{C}$ to the opposite (dual) category $\boldsymbol{C}^{\mathrm{op}}$ is a contravariant functor, the ***op*** or ***dualization functor***. We write $-^{\mathrm{op}} : \boldsymbol{C} \to \boldsymbol{C}^{\mathrm{op}}$.

Mor functors

We now consider three Mor functors for a category $\boldsymbol{C}$.

Definition 3.3.6. Let $A, A', B, B' \in \boldsymbol{C}$ be objects. Defining

$$\begin{aligned} & \operatorname{Mor}_{\boldsymbol{C}}(\ \ , B) : \boldsymbol{C} \to \boldsymbol{Set} \\ \text{with} \quad & \operatorname{Mor}_{\boldsymbol{C}}(A, B) := \boldsymbol{C}(A, B) \in \boldsymbol{Set} \\ \text{and} \quad & \operatorname{Mor}_{\boldsymbol{C}}(f, B) : \operatorname{Mor}_{\boldsymbol{C}}(A', B) \to \operatorname{Mor}_{\boldsymbol{C}}(A, B) \quad \text{for } f : A \to A' \\ \text{where} \quad & \operatorname{Mor}_{\boldsymbol{C}}(f, B) = \beta \circ f \quad \text{for } \beta \in \operatorname{Mor}_{\boldsymbol{C}}(A', B) \end{aligned}$$

gives the ***contravariant* Mor *functor***.

Analogously, we define the ***covariant* Mor *functor***

$$\operatorname{Mor}_{\boldsymbol{C}}(A,\ \ ) : \boldsymbol{C} \to \boldsymbol{Set},$$

where now $\mathrm{Mor}_{\boldsymbol{C}}(A,g) : \mathrm{Mor}_{\boldsymbol{C}}(A,B) \to \mathrm{Mor}_{\boldsymbol{C}}(A,B')$ is given by $\mathrm{Mor}_{\boldsymbol{C}}(A,g)(\alpha) = g \circ \alpha$ for $g : B \to B'$ and $\alpha \in \mathrm{Mor}_{\boldsymbol{C}}(B,B')$. Combining the two, we get

$$\mathrm{Mor}_{\boldsymbol{C}}(\ ,\ ) : \boldsymbol{C}^{\mathrm{op}} \times \boldsymbol{C} \to \boldsymbol{Set},$$

the **Mor *functor*** in two variables.

The following diagram shows the situation for the contravariant Mor functor:

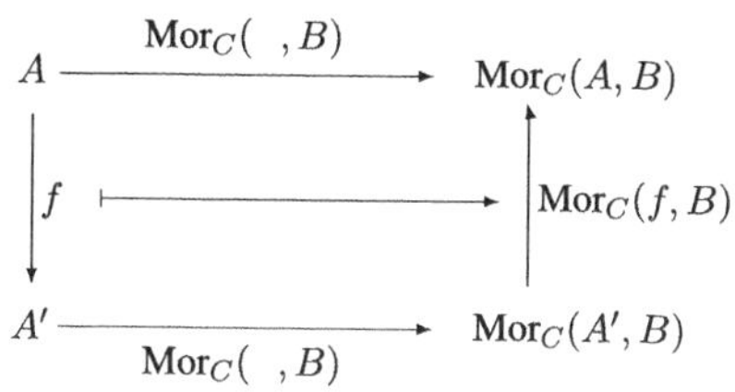

Other examples of functors can be obtained from the coproducts, products, and tensor product, if we fix “one component.” We will make this concrete for graphs in the next chapter.

Properties of functors

The following properties of functors model injective, surjective, and bijective mappings. For functors, these properties can be considered separately for objects and for morphisms.

Definition 3.3.7. Let $\boldsymbol{C}$ and $\boldsymbol{D}$ be categories. A covariant functor $F : \boldsymbol{C} \to \boldsymbol{D}$ is said to be:

- ***faithful*** if the mapping

$$\mathrm{Mor}_{\boldsymbol{C}}(A,A') \to \mathrm{Mor}_{\boldsymbol{D}}(F(A),F(A'))$$

  is injective for all $A, A' \in \boldsymbol{C}$;
- ***full*** if the mapping

$$\mathrm{Mor}_{\boldsymbol{C}}(A,A') \to \mathrm{Mor}_{\boldsymbol{D}}(F(A),F(A'))$$

  is surjective for all $A, A' \in \boldsymbol{C}$;
- a ***full embedding*** if F is full and faithful;
- ***dense*** (or *representative*) if for every $B \in \boldsymbol{D}$ there exists an $A \in \boldsymbol{C}$ such that $F(A)$ is isomorphic to B;
- an ***injector*** if F is a faithful functor which is injective up to isomorphisms with respect to objects, i. e., $F(A) \cong F(A')$ implies $A \cong A'$;

- a ***surjector*** if F is a full functor which is surjective with respect to objects, i. e., for every $B \in \boldsymbol{D}$ there exists an $A \in \boldsymbol{C}$ such that $F(A) = B$.

Definition 3.3.8. If $\boldsymbol{C}$ is a subcategory of $\boldsymbol{D}$ so that the inclusion functor is full, then $\boldsymbol{C}$ is called a ***full subcategory*** of $\boldsymbol{D}$.

Preservation and reflection of properties by functors provides useful information when investigating categories with the help of functors.

Definition 3.3.9. We say that a functor $F : \boldsymbol{C} \to \boldsymbol{D}$ ***preserves*** a property $\mathcal{P}$ of a morphism f in $\boldsymbol{C}$ if $F(f)$ in $\boldsymbol{D}$ also has the property $\mathcal{P}$. We say that F ***reflects*** a property $\mathcal{P}$ if f has $\mathcal{P}$ in $\boldsymbol{C}$ whenever $F(f)$ has $\mathcal{P}$ in $\boldsymbol{D}$. Analogous definitions can be made with respect to properties of objects.

It is clear that every functor preserves commutative diagrams.

On the level of mappings, we know this same principle: graph homomorphisms preserve edges, while graph comorphisms reflect edges.

If we look for more analogies between mappings and functors, the existence of the identity functor on every category suggests that for a functor $F : \boldsymbol{C} \to \boldsymbol{C}'$ there might exist a "left inverse" functor $G : \boldsymbol{C}' \to \boldsymbol{C}$ such that $G \circ F$ is the identity functor on $\boldsymbol{C}$. This would mean that the two functors $\mathrm{Id}_{\boldsymbol{C}}$ and $G \circ F$ behave similarly on objects and on morphisms. This leads to the concept of a natural transformation.

Definition 3.3.10. A ***natural transformation*** $\Theta : \mathrm{Id}_{\boldsymbol{C}} \to G \circ F$ relates the two functors so that the following rectangle is commutative for all objects $A, B \in \boldsymbol{C}$ and all morphisms $f : A \to B$ (here Θ_A is a morphism in $\boldsymbol{C}$ for every object $A \in \boldsymbol{C}$):

$$\begin{array}{ccc} A & \xrightarrow{\Theta_A} & G(F(A)) \\ \Big\downarrow f & \longmapsto & \Big\downarrow G(F(f)) \\ B & \xrightarrow[\Theta_B]{} & G(F(B)) \end{array}$$

This is the so-called condition of being natural, which can be written as

$$G(F(f))(\Theta_A(a)) = \Theta_B(f(a)) \quad \text{for all } a \in A.$$

A natural transformation Θ is called a ***natural equivalence*** if Θ_A is an isomorphism in $\boldsymbol{C}$ for every $A \in \boldsymbol{C}$. In the same way, we can define natural transformations and equivalences more generally for two functors $F_1, F_2 : \boldsymbol{C} \to \boldsymbol{D}$ instead of $\mathrm{Id}_{\boldsymbol{C}}$ and $G \circ F$.

3.4 Comments

Categories came up out of the wish to consider, for instance, all vector spaces over a fixed field. In this category, one takes linear mappings as morphisms. This is similar to the category of all sets along with the mappings between them.

The main problem is that the collection of all sets does not form a set. This might seem fascinating and possibly disturbing. This a fortiori is the case for the class of all graphs. Category language somehow gets around the problem without focusing too much attention on it: for everyday use, we just ignore the issue.

In this chapter, beyond basic definitions and usual examples, we have given several examples of "strange" categories which, nonetheless, are of interest even in informatics. We point to Remarks 3.2.6 and 3.2.11, which contain abstraction steps similar to those used in informatics.

In what follows, we will use the language of categories and functors in many places. In Chapter 4 for example, we first construct all usual unions and products of graphs in the elementary way. And then we interpret and classify the constructions as sums and products in different graph categories from a "higher" viewpoint - the "Categorical Imperative".

The concepts of natural transformation and natural equivalence, introduced in Section 3.3, do not seem "natural" at all. They are very abstract and seem quite artificial. But they are useful in Section 4.6 on diamond products and power products. These products, considered as functors, turn out to be "left inverses" to tensor functors and product functors, respectively, in certain graph categories.

Natural equivalence is known from linear algebra. There one proves that a vector space is naturally isomorphic to its double dual. A finite-dimensional vector space is also isomorphic to its dual, but this isomorphism is not natural.

It may be worthwhile to have a look at End functors which, e. g., start in graph categories and go to the category of monoids. Problems arise since this is actually a functor in two variables, contravariant in the first and covariant in the second; see Definition 3.3.6. This is probably the reason that, so far, there has been no real progress in this direction.

4 Binary graph operations

In set theory and many other areas—not just in mathematics—one can generate new objects from old via binary operations such as unions and Cartesian products, analogous to producing new numbers by addition or multiplication. Owing to the rich structure of graphs, there are several variants for each construction, and we will present these separately and in detail.

We will first consider four forms of unions of graphs, followed by eight forms of products. All constructions will be described directly in the definitions and can be used independently of any categorical considerations; but whenever possible we will also provide the categorical descriptions of the constructions (they solve so-called universal problems). This will make the structural differences clearer.

If we choose very special categories, the unions become *initial objects* and the products *terminal objects*; compare with Remarks 3.2.6 and 3.2.11, for example.

4.1 Unions

In this section, the vertex sets of the new graphs will be the unions of the vertex sets of the old graphs.

The union

Definition 4.1.1. Let $G_1 = (V_1, E_1)$ and $G_2 = (V_2, E_2)$ be graphs with disjoint vertex sets, i. e., $V_1 \bigcap V_2 = \emptyset$. The ***union*** (or ***coproduct***) of G_1 and G_2 is defined to be

$$G_1 \bigcup G_2 := (V_1 \bigcup V_2, E_1 \bigcup E_2).$$

The mappings $u_i := \mathrm{id}_{G_1 \bigcup G_2}|_{G_i}$, $i \in \{1, 2\}$, are called the ***natural injections.***

The following theorem shows that this construction in the category ***Gra*** satisfies the properties that we formulated for the coproduct in general categories. It also contains the statement that the union of two sets with the usual injections is the coproduct in the category ***Set***.

Recall from linear algebra that the proof for the coproduct in the category of F-vector spaces is quite different; see Exercise 3.2.2.

Theorem 4.1.2. *The pair* $((u_1, u_2), G_1 \bigcup G_2)$ *is the (categorical) coproduct in* **Gra** *and in* **EGra***; i. e.,*

(a) *The natural injections* $u_1 : G_1 \to G_1 \bigcup G_2$ *and* $u_2 : G_2 \to G_1 \bigcup G_2$ *are morphisms.*
(b) $((u_1, u_2), G_1 \bigcup G_2)$ *solves the following universal problem.*

https://doi.org/10.1515/9783110617368-004

For all graphs G and for all morphisms $f_1 : G_1 \to G$ and $f_2 : G_2 \to G$ there exists exactly one morphism f such that following diagram is commutative:

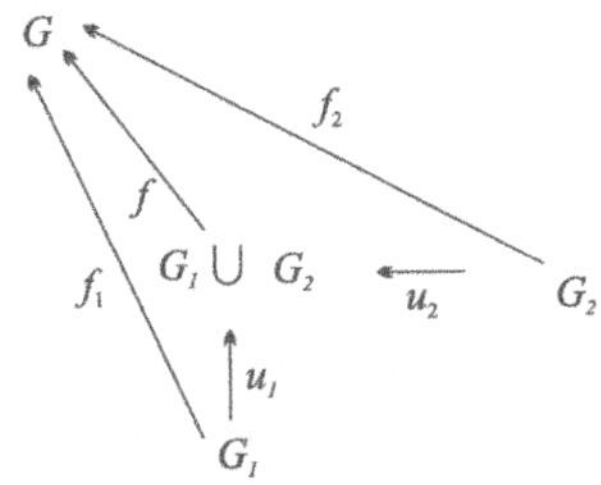

*Here, i. e., in the categories **Gra** and **EGra**, we have*

$$f : \begin{cases} G_1 \bigcup G_2 & \to G \\ x_i & \mapsto f_i(x_i) \quad \text{for } x_i \in G_i,\ i \in \{1,2\}. \end{cases}$$

*We write $G_1 \coprod G_2$ and, analogously, $\coprod_{i \in I} G_i$ for multiple unions. Moreover, we write $[(f_1,f_2)] = f$ and say that f is **coproduct induced** by (f_1,f_2).*

Proof. We formulate the proof for the category ***Gra***. The only difference in ***EGra*** arises when f_1 or f_2 is in ***EGra*** but not in ***Gra***; but in that case, clearly the coproduct-induced f, defined as in ***Gra***, is also in ***EGra***.

From the construction, it becomes clear that $((u_1,u_2), G_1 \bigcup G_2)$ is independent of G and f_1, f_2. We define $f(x_1) := f_1(x_1)$ for $x_1 \in G_1$ and $f(x_2) := f_2(x_2)$ for $x_2 \in G_2$. Since V_1 and V_2 are disjoint, f is correctly defined and the diagram is commutative.

To prove uniqueness of f, suppose that there exists a g with the same properties. Then

$$g(x_i) = (u_i \circ g)(x_i) = f_i(x_i) = (u_i \circ f)(x_i) = f(x_i) \quad \text{for all } x_i,\ i = 1,2.$$

The proof up to this point is not needed if we know that the disjoint union together with the injections is the coproduct in the category ***Set***. But we have to show that u_1 and u_2 are graph homomorphisms, which is clear from their definition, and that f is a graph homomorphism. So take $x_1, x_1' \in V_1$; then

$$\begin{aligned}(x_1,x_1') \in E_1 \bigcup E_2 &\Rightarrow (x_1,x_1') \in E_1 \\ &\Rightarrow (f_1(x_1), f_1(x_1')) \in E(G) \\ &\Rightarrow (f(x_1), f(x_1')) \in E(G),\end{aligned}$$

since by hypothesis f_1 is a graph homomorphism; similarly for edges from E_2. □

Example 4.1.3 (Coproducts in ***SGra***?). The injections u_i are always strong, but f is not strong in general, even if the f_i are strong. Thus $((u_i)_{i \in I}, \bigcup_{i \in I} G_i)$ is not the coproduct in the category ***SGra***, consisting of graphs with strong graph homomorphisms.

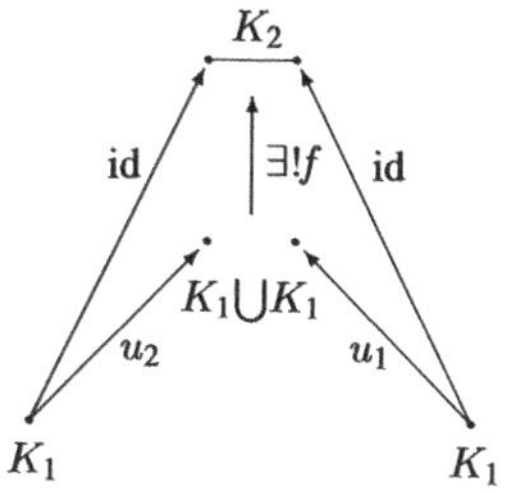

It is clear that f is not strong in this situation.

The join

The following definition of the join is given for undirected graphs. For directed graphs, several variations are possible.

Definition 4.1.4. Let $G_1 = (V_1, E_1)$ and $G_2 = (V_2, E_2)$, where $V_1 \bigcap V_2 = \emptyset$. The ***join*** of G_1 and G_2, denoted by $G_1 + G_2$, is defined to be the union $G_1 \bigcup G_2$ plus all edges between vertices from G_1 and vertices from G_2. Formally, this means

$$G_1 + G_2 := (V_1 \bigcup V_2\,,\ E_1 \bigcup E_2 \bigcup \{(x_1, x_2) \mid x_i \in V_i,\ i = 1, 2\}).$$

Example 4.1.5 (Join).

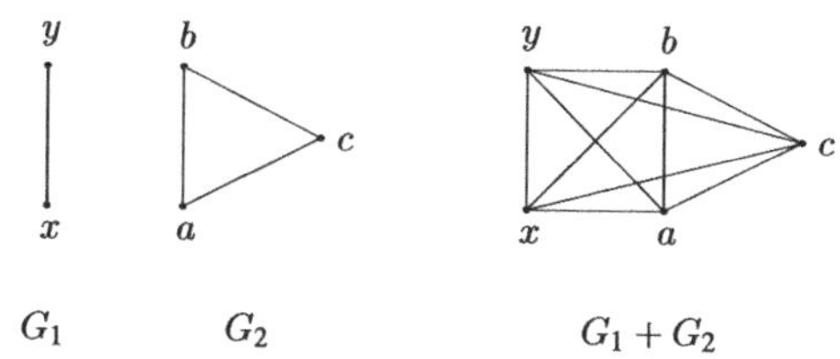

Corollary 4.1.6. *We have $G_1 \bigcup G_2 \subseteq G_1 + G_2$, i. e., the union is a (nonstrong) subgraph of the join.*

Exerceorem 4.1.7. In the category ***CGra***, we have $G_1 \coprod G_2 \cong ((u_1, u_2), G_1 + G_2)$, i. e., in this category the join together with the injections is the coproduct.

The edge sum

The following definition of the edge sum is valid for both undirected and directed graphs. The definition of the edge sum requires that the two graphs have the same vertex set. The edge sum is obtained by laying one graph on top of the other.

Definition 4.1.8. Let $G_1 = (V, E_1)$ and $G_2 = (V, E_2)$ be graphs. The ***edge sum*** is defined to be

$$G_1 \oplus G_2 := (V, E_1 \bigcup E_2).$$

Example 4.1.9 (Edge sum).

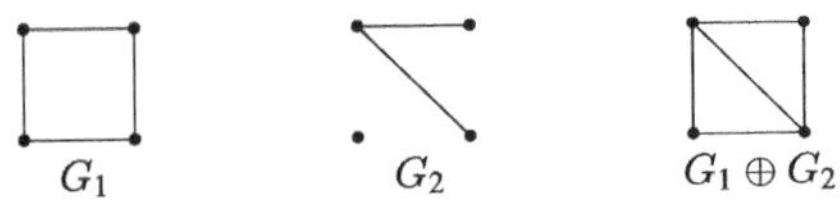

For graphs with different vertex sets, we modify the construction as follows.

Definition 4.1.10. Take the graphs $G_1 = (V_1, E_1)$ and $G_2 = (V_2, E_2)$ and set $V_1 \bigcap V_2 = V$. The ***generalized edge sum*** is defined to be

$$G_1 \overline{\oplus} G_2 := (G_1 \bigcup \overline{K}_{|V_2 \setminus V|}) \oplus (G_2 \bigcup \overline{K}_{|V_1 \setminus V|}),$$

where $\overline{K}_n$ is the totally disconnected graph with n vertices.

We interpret the construction as follows: add to G_1 the vertices of G_2 which do not belong to G_1, and add to G_2 the vertices of G_1 which do not belong to G_2. Call the results G_1' and G_2'; then form their edge sum. This gives the generalized edge sum. The problem with this construction is that we have to say which vertices of the graphs are considered equal. The following example shows that there may be several possibilities. It is clear that there is no difference between directed and undirected graphs in this case.

Example 4.1.11 (Generalized edge sum).

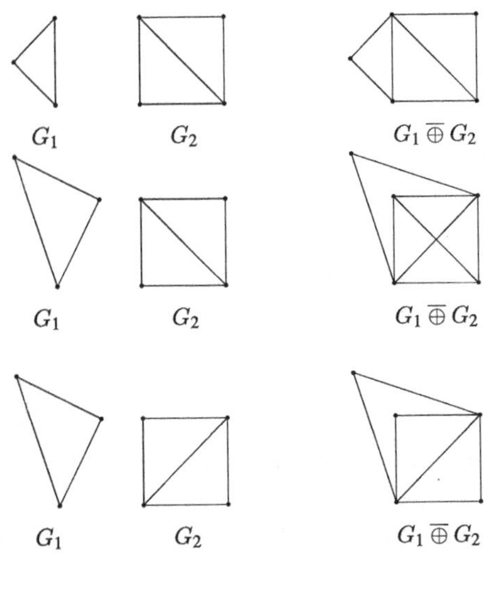

These difficulties are circumvented by making the following definition.

Definition 4.1.12. Let $H = (V, E)$, $G_1 = (V_1, E_1)$ and $G_2 = (V_2, E_2)$ be graphs, and let $m_1 : H \to G_1$ and $m_2 : H \to G_2$ be injective strong graph homomorphisms. The ***amalgam (amalgamated coproduct, pushout)*** of G_1 and G_2 with respect to $(H, (m_1\, m_2))$ is defined by

$$V(G_1 \coprod_{(H,(m_1,m_2))} G_2) := (V_1 \setminus m_1(H)) \bigcup V \bigcup (V_2 \setminus m_2(H))$$

and

$$\begin{aligned} &E(G_1 \coprod_{(H,(m_1,m_2))} G_2) \\ &\quad := \{(x_i, y_i) \in E_i \mid x_i, y_i \in V_i \setminus m_i(H),\ i = 1,2\} \\ &\quad \bigcup\{(x, z) \mid z \in V,\ x_i \in V_i \setminus m_i(H),\ (x_i, m_i(z)) \in E_i,\ i = 1,2\} \\ &\quad \bigcup\{(z, z') \mid z, z' \in V,\ (m_i(z), m_i(z')) \in E_i,\ i = 1,2\}. \end{aligned}$$

Again, we define multiple amalgams $\coprod_{(H,(m_i)_{i\in I})} G_i$ analogously.

In practice, we consider H as a common subgraph of G_1 and G_2 and form the union in such a way that we paste together the two graphs along H.

Remark 4.1.13. Formally, we get the same result if we define

$$G_1 \coprod_{(H,(m_1,m_2))} G_2 := (G_1 \coprod G_2)/\mu$$

where, for $x, y \in G_1 \coprod G_2$, we set

$$x\,\mu\,y \quad \text{if } \exists z \in H \text{ with } m_1(z) = x,\ m_2(z) = y \text{ or } x = y.$$

This implies that $(x_\mu, y_\mu) \in E(G_1 \coprod_{(H,(m_1,m_2))} G_2)$ if there exists $i \in \{1,2\}$, $x' \in x_\mu \bigcap G_i$, $y' \in y_\mu \bigcap G_i$ with $(x', y') \in E_i$.

Example 4.1.14 (Amalgam).

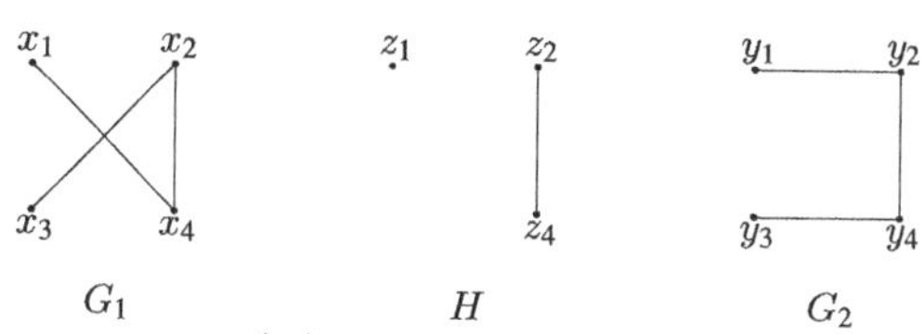

$$m_1(z_1) = x_3, \qquad m_2(z_1) = y_1$$
$$m_1(z_2) = x_2, \qquad m_2(z_2) = y_2$$
$$m_1(z_4) = x_4, \qquad m_2(z_4) = y_4$$

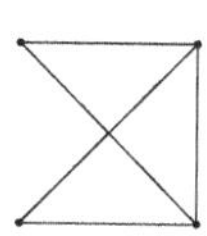

$G_1 \coprod_{(H,(m_1,m_2))} G_2$

Theorem 4.1.15. *The amalgam $G_1 \coprod_{(H,(m_1,m_2))} G_2$ has the properties of the categorically defined amalgam in **Gra** and in **EGra**; i. e.:*

(a) *the (codomain-modified) natural injections $u_1 : G_1 \to G_1 \coprod_{(H,(m_1,m_2))} G_2$ and $u_2 : G_2 \to G_1 \coprod_{(H,(m_1,m_2))} G_2$ are graph homomorphisms with $u_1 m_1 = u_2 m_2$, i. e., the square is commutative; and*

(b) *$((u_1, u_2), G_1 \coprod_{(H,(m_1,m_2))} G_2)$ solves the following* universal problem *in **Gra** and in **EGra**.*
For all graphs G and all morphisms $f_1 : G_1 \to G$ and $f_2 : G_2 \to G$ with $f_2 m_2 = f_1 m_1$, i. e., which make the exterior quadrangle commutative, there exists exactly one morphism $f : G_1 \coprod_{(H,(m_1,m_2))} G_2 \to G$ such that the triangles are commutative.

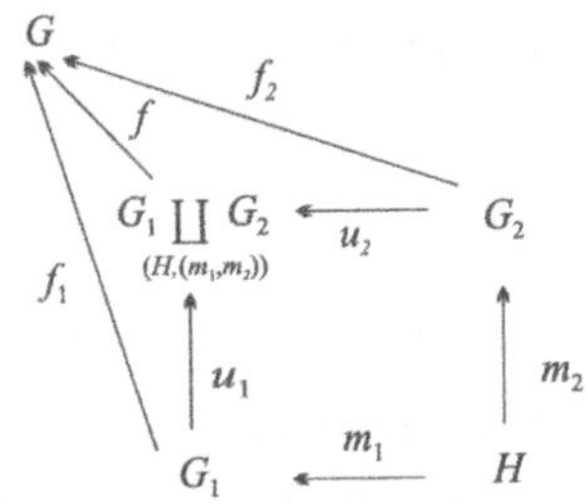

*Here, i. e., in the category **Gra**, one has $f(x_i) = f_i(x_i)$ for $i = 1, 2$.*

Proof. For $i = 1, 2$, define

$$u_i(x_i) := \begin{cases} x_i & \text{if } x_i \in V_i \backslash m_i(H) \\ z & \text{if } m_i(z) = x_i \text{ for } z \in H. \end{cases}$$

It is clear that these are graph homomorphisms and that for $z \in H$ we have $u_1(m_1(z)) = z = u_2(m_2(z))$, as for sets.

As for the coproduct, we define

$$f(x_i) := f_i(x_i) \quad \text{for } i = 1, 2.$$

Now f is well-defined as for sets, since by hypothesis we have

$$f(x) = f_1(m_1(z)) = f_2(m_2(z)) = f(y) \quad \text{if } \begin{cases} m_1(z) = x \in G_1 \\ m_2(z) = y \in G_2. \end{cases}$$

As for the coproduct, we get that f is a graph homomorphism. For the mappings f_1, f_2, f, u_1, u_2, we show commutativity by calculation as for sets; we also show uniqueness of f. In ***EGra*** we get the same results. □

Exercise 4.1.16. For $H = \emptyset$, the amalgam becomes the coproduct, i. e.,

$$G_1 \coprod_{\emptyset} G_2 \cong G_1 \coprod G_2.$$

Corollary 4.1.17. *The generalized edge sum $G_1\bar{\oplus}G_2$ is an amalgam $G_1\coprod_{(H,(m_1,m_2))} G_2$ with $H = (V_1 \bigcap V_2, \emptyset)$ and the injections $m_i : V_1 \bigcap V_2 \to V_i$, $i = 1, 2$, where $m_i : V_1 \bigcap V_2 \to V_i$ is defined by $m_i = \mathrm{id}_{V_i}|_{V_1 \bigcap V_2}$.*

Proof. By construction of the amalgam, we get

$$V(G_1 \coprod\nolimits_{(H,(m_1,m_2))} G_2) = (V_1 \setminus V_2) \bigcup (V_1 \bigcap V_2) \bigcup (V_2 \setminus V_1),$$
$$E(G_1 \coprod\nolimits_{(H,(m_1,m_2))} G_2) = \{(x, y) \mid (x, y) \in E(G_1) \bigcup E(G_2)\}.$$

□

Remark 4.1.18. For $G_1 = (V, E_1)$ and $G_2 = (V, E_2)$, the edge sum $G_1 \oplus G_2$ is the amalgam $G_1 \coprod_{(H,(m_1,m_2))} G_2$ with $H = (V, \emptyset)$ and the identity injections m_1 and m_2.

Exercise 4.1.19. Construct the amalgam of two graphs in the categories ***CGra*** and ***EGra***. You can take the graphs from Example 4.1.14.

4.2 Products

In this section, we consider binary graph operations for which the vertex set of the result is the Cartesian product of the vertex sets of the "factors." We proceed in the same way as for the union of the vertex sets, i. e., we give the definitions of the new graphs and describe the constructions by their categorical properties.

The cross product

The cross product is defined in the same way for directed and undirected graphs. Note that in the literature the names enclosed in parentheses are also used. We choose to use the term "cross product" because it is suggested by the structure of this product in the first example.

Definition 4.2.1. The ***cross product (categorical product, conjunction)*** of the graphs $G_i = (V_i, E_i)$, $i = 1, 2$, is defined to be

$$G_1 \times G_2 := \left(V_1 \times V_2,\ \{((x, y), (x', y')) \mid (x, x') \in E_1 \text{ and } (y, y') \in E_2\}\right).$$

Multiple cross products can be defined analogously. In the pictures, we will mostly label vertices simply as xx' instead of (x, x').

Example 4.2.2 (Cross product).

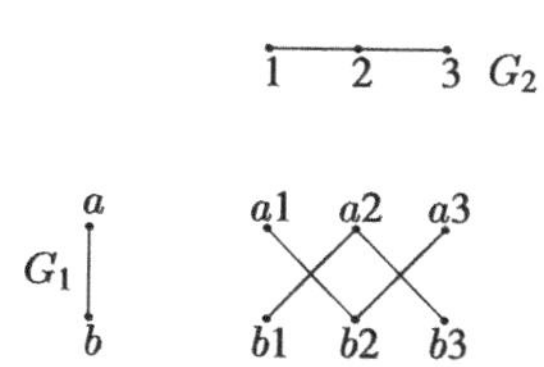

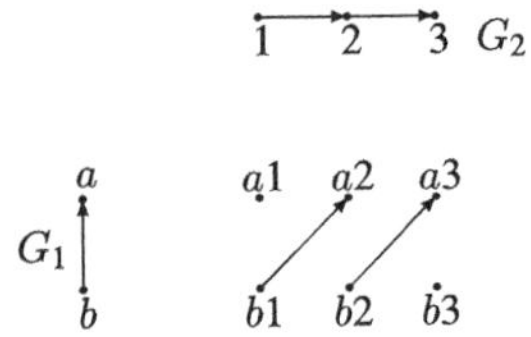

Theorem 4.2.3. *The cross product together with the natural projections $p_1 : G_1 \times G_2 \to G_1$ and $p_2 : G_1 \times G_2 \to G_2$ form the categorical product in the category **Gra**; i. e.:*

(a) *p_1 and p_2 are morphisms;*

(b) *$(G_1 \times G_2, (p_1, p_2))$ solves the following* universal problem *in the category **Gra**.*
For all graphs G and all morphisms $f_1 : G \to G_1$ and $f_2 : G \to G_2$, there exists exactly one morphism $f : G \to G_1 \times G_2$ such that the following diagram is commutative:

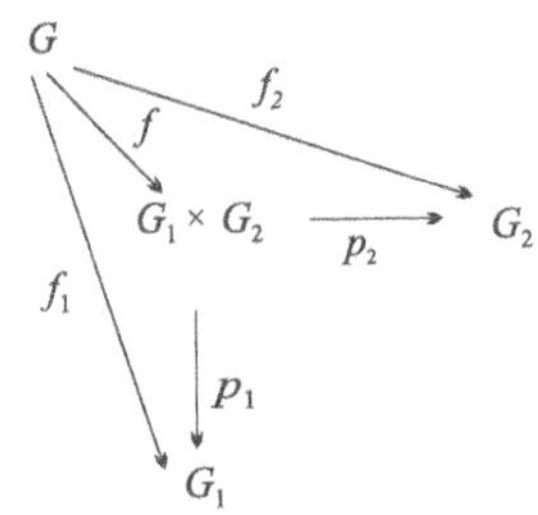

We write $G_1 \times G_2$ and, analogously, $\prod_{i \in I} G_i$ for multiple products. Moreover, we write $f =: \langle (f_1, f_2) \rangle$ and say that f is ***product induced*** by (f_1, f_2).

Here, i. e., in the category ***Gra***, we have $f(x) = (f_1(x), f_2(x))$ for all $x \in G$.

Proof. This is left as an exercise: turn around the arrows and replace injections by projections in the corresponding proof for the coproduct. □

Remark 4.2.4. The cross product $G_1 \times G_2$ corresponds to the so-called ***Kronecker product*** of the adjacency matrices,

$$A(G_1 \times G_2) = A(G_1) \times A(G_2)$$

where, for $i, j \in \{1, \dots m\}$ and $k, \ell \in \{1, \dots, n\}$, we define

$$A(G_1) \times A(G_2) = (a_{ij}) \times (b_{k\ell}) = \begin{pmatrix} a_{11}(b_{k\ell}) & \cdots & a_{1m}(b_{k\ell}) \\ \vdots & \ddots & \vdots \\ a_{m1}(b_{k\ell}) & \cdots & a_{mm}(b_{k\ell}) \end{pmatrix}$$

with

$$a_{ij}(b_{k\ell}) = \begin{pmatrix} a_{ij}b_{11} & \cdots & a_{ij}b_{1n} \\ \vdots & \ddots & \vdots \\ a_{ij}b_{n1} & \cdots & a_{ij}b_{nn} \end{pmatrix}.$$

In this way, we obtain a $mn \times mn$ matrix where mn is the number of vertices of $G_1 \times G_2$.

The coamalgamated product

The next definition, categorically dual to Definition 3.2.3, we give formally, which means that:
- all "arrows" for the morphisms are reversed; and
- injective and surjective are exchanged.

Moreover, we see again that the categorical description of the cross product is categorically dual to the categorical description of the union.

Definition 4.2.5. Let $G_1 = (V_1, E_1)$, $G_2 = (V_2, E_2)$, and $H = (V, E)$ be graphs, and let $n_1 : G_1 \to H$ and $n_2 : G_2 \to H$ be surjective strong graph homomorphisms. The strong subgraph of $G_1 \times G_2$ with the vertex set $\{(x_1, x_2) \in G_1 \times G_2 \mid n_1(x_1) = n_2(x_2)\}$ is called the ***coamalgam (coamalgamated product, pullback)*** of G_1 and G_2 with respect to $((n_1, n_2), H)$.

We write $G_1 \prod^{((n_1,n_2),H)} G_2$ and, analogously, $\prod^{((n_i)_{i\in I},H)} G_i$ for multiple coamalgams. Note that for the vertices of the coamalgam we have that $\{(x_1, x_2) \in G_1 \times G_2 \mid n_1(x_1) = n_2(x_2)\} = \bigcup_{z\in H} n_1^{-1}(z) \times n_2^{-1}(z)$.

Theorem 4.2.6. *The coamalgam $G_1 \prod^{((n_1,n_2),H)} G_2$ has the following properties:*
(a) *the (domain-modified) natural projections $p_1 : G_1 \prod^{((n_1,n_2),H)} G_2 \to G_1$ and $u_2 : G_1 \prod^{((n_1,n_2),H)} G_2 \to G_2$ are graph homomorphisms and we have $n_1 p_1 = n_2 p_2$, i. e., the square is commutative;*
(b) *$(G_1 \prod^{((n_1,n_2),H)} G_2, (p_1, p_2))$ solves the following* universal problem *in* ***Gra***.
For all graphs G and all morphisms $f_1 : G \to G_1$ and $f_2 : G \to G_2$ such that $n_1 f_1 = n_2 f_2$, i. e., which make the exterior quadrangle commutative, there exists exactly one morphism f such that the triangles are commutative.

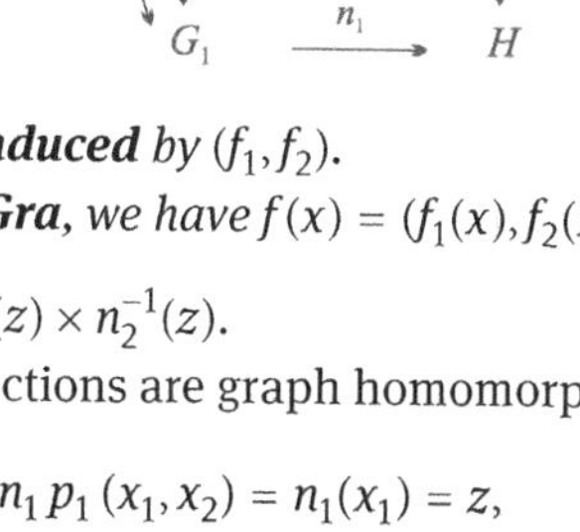

*We say that f is **coamalgam induced** by (f_1, f_2).*
*Here, i. e., in the category **Gra**, we have $f(x) = (f_1(x), f_2(x))$ for all $x \in G$.*

Proof. Take $(x_1, x_2) \in \bigcup_{z\in H} n_1^{-1}(z) \times n_2^{-1}(z)$.
(a) It is clear that the projections are graph homomorphisms. Moreover,

$$n_1 p_1 (x_1, x_2) = n_1(x_1) = z,$$
$$n_2 p_2 (x_1, x_2) = n_2(x_2) = z.$$

(b) Define

$$f(y) = (f_1(y), f_2(y)) \in G_1 \prod^{((n_1,n_2),H)} G_2.$$

Then

$$n_1 p_1 f(y) = n_1 f_1(y) = n_2 f_2(y) = n_2 p_2 f(y),$$

and thus $f(y) \in G_1 \prod^{((n_1,n_2),H)} G_2$. In this way both triangles become commutative, and f is unique as for sets since, again, so far we have only mappings on the vertex sets.

Furthermore, f is a graph homomorphism:

$$(y, y') \in E(G) \Rightarrow (f_1(y), f_1(y')) \in E(G_1) \quad \text{and} \quad (f_2(y), f_2(y')) \in E(G_2).$$

Consequently,

$$((f_1(y), f_2(y)), (f_1(y'), f_2(y'))) \in E(G_1 \prod^{((n_1,n_2),H)} G_2)$$

since everything lies in $G_1 \prod^{((n_1,n_2),H)} G_2$, which by definition is a strong subgraph of $G_1 \prod G_2$. □

Remark 4.2.7. The definitions of the mappings f (including correctness and uniqueness) as well as their commutativity properties have been proved as for sets and mappings. Since graphs and graph homomorphisms are sets (the vertex sets) and mappings, the coproducts, amalgams, products, and coamalgams must have the required properties. Consequently, also the injections, projections, and induced morphisms are the same mappings. The only additional steps in the proofs are to show that injections, projections and induced morphisms belong to the category in question.

Corollary 4.2.8. *For $H = K_1^{(1)}$, the coamalgam $G_1 \prod^{((n_1,n_2),H)} G_2$ turns into the cross product, i. e., we have $G_1 \prod^{K_1^{(1)}} G_2 = G_1 \times G_2$.*

Proof. For $H = K_1^{(1)}$, we always have $n_1 f_1 = n_2 f_2$ for all f_1, f_2. Thus the formulation of the above theorem is the categorical description of the product. □

Example 4.2.9 (Coamalgam).

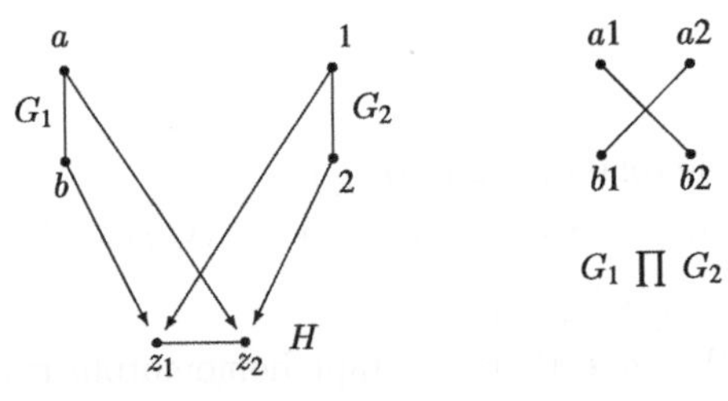

The coamalgam is the strong subgraph of $G_1 \prod G_2$ with the vertices $n_1^{-1}(z_1) \times n_2^{-1}(z_1) = \{b1\}$ and $n_1^{-1}(z_2) \times n_2^{-1}(z_2) = \{a2\}$; that is, it consists of the edge (a_2, b_1).

Exercise 4.2.10. The cross product is not the product in the category ***CGra***. By Remark 4.2.7, the projections or the induced morphism will not be in ***CGra***.

The disjunction of graphs

Definition 4.2.11. The ***disjunction*** of the graphs G and H is defined to be

$$G \vee H := (V(G) \times V(H), \{\{(x,y), (x',y')\} \mid \{x,x'\} \in E(G) \text{ or } \{y,y'\} \in E(H)\}).$$

Exercise 4.2.12. In ***CGra***, the disjunction $(G \vee H, (p_1, p_2))$ is the categorical product of G and H. We have to show that the induced morphism and the injections belong to ***CGra***.

Example 4.2.13 (Disjunction).

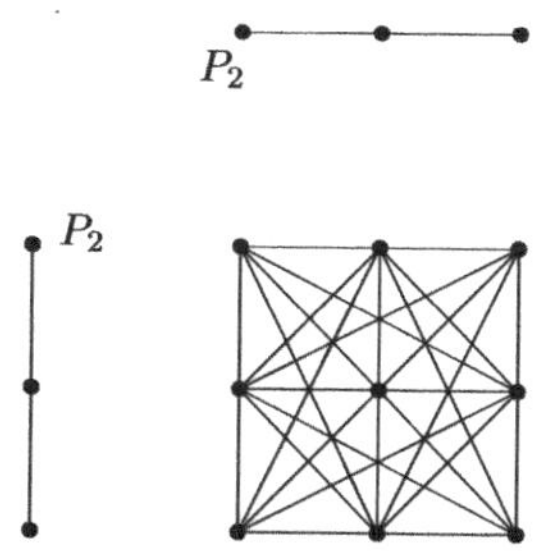

Exactly the edges between any two nonadjacent vertices on the outer boundary of the square in the picture do not exist.

Exercise 4.2.14. Find the construction of the coamalgam of two graphs in the category ***CGra***. Start with an example.

4.3 Tensor products and the product in *EGra*

After the product and the coamalgam, which have similar categorical characterizations, we now consider constructions that we can describe as tensor products. Moreover, we give the product in ***EGra***.

The box product

Here, we again have the same definitions for directed and undirected graphs. Alternative names for the box product are given in parentheses. We decided to use the name "box product" because that is what is suggested by the structure of the graph in the first example. The graphs are the same ones as in Example 4.2.2.

Definition 4.3.1. The ***box product (product, Cartesian product, Cartesian sum)*** of the graphs $G_1 = (V_1, E_1)$ and $G_2 = (V_2, E_2)$ is defined to be

$$G_1 \square G_2 := (V_1 \times V_2, \{((x,y),(x,y')) \mid x \in G_1,\ (y,y') \in E_2\} \\ \bigcup \{((x,y),(x',y)) \mid y \in G_2,\ (x,x') \in E_1\}).$$

Remark 4.3.2. The box product $G_1 \square G_2$ has the adjacency matrix $(A(G_1) \times I_2) + (I_1 \times A(G_2))$, where I_i denotes the identity matrix with the size of G_i, for $i = 1, 2$, and $\times$ denotes the Kronecker product (see Remark 4.2.4) and $+$ the sum of the matrices (cf. [Cvetković et al. 1979], Section 2.5 on p. 67). This construct is called the ***Kronecker sum*** of the two matrices.

Example 4.3.3 (Box product).

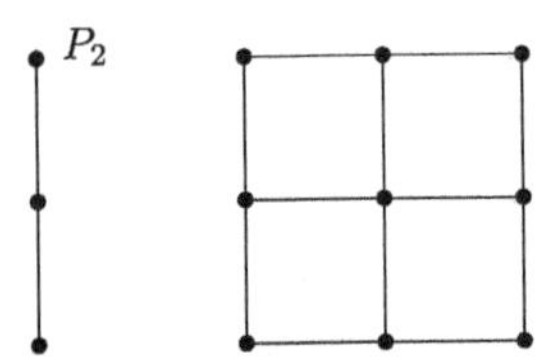

Recall that mappings that start in two-fold Cartesian products and which componentwise are morphisms in the respective category, like τ and ξ, are called *bimorphisms*; cf. Definition 3.2.12. The most famous box products are "cubes." In Chapter 1, we introduced the 3-dimensional cube. Now we generalize this to n-dimensional cubes using the box product.

Definition 4.3.4. The graph given by $Q_1 = K_2$ and $Q_n = Q_{n-1} \square K_2$ for $n > 1$ is called the ***n-cube***.

With this definition, it is easy to draw the four-dimensional cube in two-dimensional space. It has eight three-dimensional cubes as "faces." Should we throw it into four-dimensional space, it would fall on one of the three-dimensional faces. With some practice, one can imagine the five-dimensional cube, and so on.

Theorem 4.3.5. *The box product $G_1 \square G_2$ together with the identity mapping $\tau : V_1 \times V_2 \to G_1 \square G_2$ is the tensor product in the categories* **Gra** *and* **EGra***; i. e.:*

(a) *for every $x \in V_1$ the mapping $\tau(x, \ ) : G_2 \to G_1 \square G_2$ is a morphism, and for every $y \in V_2$ the mapping $\tau(\ , y) : G_1 \to G_1 \square G_2$ is a morphism, i. e., τ is a bimorphism;*

(b) *$(\tau,\ G_1 \square G_2)$ solves the following* universal problem *in* **Gra** *and in* **EGra**.
For every graph X and every bimorphism $\xi : V_1 \times V_2 \to X$, there exists exactly one morphism $\xi^* : G_1 \square G_2 \to X$ such that the following diagram is commutative:

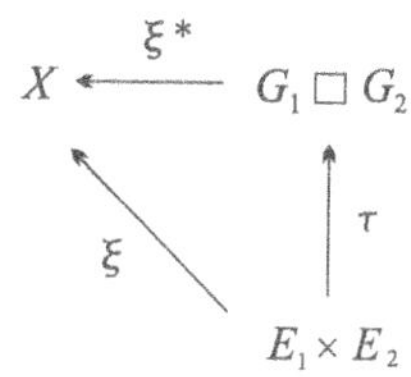

We say that ξ^* is tensor product induced by ξ.

Here, i. e., in the categories **Gra** *and* **EGra**, *one has $\xi^* = \xi \circ \tau^{-1}$.*

Proof. It is clear that $\xi^* = \xi\tau^{-1}$ makes the diagram commutative and is uniquely determined as for sets.

We have to show that ξ^* is a graph homomorphism. Take $((x_1, x_2), (x_1', x_2')) \in E(G_1 \,\Box\, G_2)$, i. e.,

$$[(x_1, x_1') \in E(G_1) \wedge x_2 = x_2'] \vee [(x_2, x_2') \in E(G_2) \wedge x_1 = x_1'].$$

Consider

$$\begin{aligned}(\xi^*(x_1, x_2),\ \xi^*(x_1', x_2')) &= (\xi\tau^{-1}(x_1, x_2),\ \xi\tau^{-1}(x_1', x_2')) \\ &= (\xi(x_1, x_2),\ \xi(x_1', x_2')).\end{aligned}$$

Now, $x_2 = x_2'$ and $(x_1, x_1') \in E(G_1) \xrightarrow{\xi \text{ bimorph}} (\xi(x_1, x_2), \xi(x_1', x_2)) \in E(X)$, and $x_1 = x_1'$ and $(x_2, x_2') \in E(G_2) \xrightarrow{\xi \text{ bimorph}} (\xi(x_1, x_2), \xi(x_1, x_2')) \in E(X)$. □

Example 4.3.6 (Box product in ***CGra***). The box product is not the tensor product in the category ***CGra***.

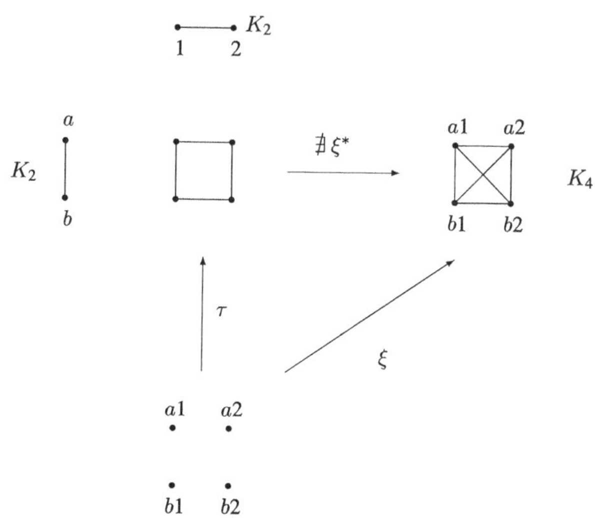

We see that τ is a bicomorphism, since the embeddings

$$\tau(a,\ ) : K_2 \to K_2 \square K_2 \quad \text{and} \quad \tau(b,\ ) : K_2 \to K_2 \square K_2$$
$$\tau(\ ,1) : K_2 \to K_2 \square K_2 \quad \text{and} \quad \tau(\ ,2) : K_2 \to K_2 \square K_2$$

are graph comorphisms.

We choose $X = K_4$ and define ξ by the embeddings

$$\xi(a,\ ) : K_2 \to K_4 \quad \text{and} \quad \xi(b,\ ) : K_2 \to K_4,$$
$$\xi(\ ,1) : K_2 \to K_4 \quad \text{and} \quad \xi(\ ,2) : K_2 \to K_4,$$

according to the labeling of the vertices, which are graph comorphisms. Then ξ is a bicomorphism.

But the induced mapping ξ^* is not a graph comorphism, as $(\xi^*(a1), \xi^*(b2))$ is an edge without a preimage.

Exercise 4.3.7. The box product is not the product in the category ***CGra***. Here, the projections from the box product are graph homomorphisms but not graph comorphisms. To see this, consider the above example for the box product. Here, we have

$$(p_2(a1), p_2(a2)) = (1,2) \in E(G_2) \quad \text{but} \quad (p_1(a1), p_1(a2)) = (a) \notin E(G_1).$$

The boxcross product

Now we consider the edge sum of the cross product and the box product. This so-called boxcross product also has a categorical meaning: it is the product in the category ***EGra***.

Definition 4.3.8. The ***boxcross product (strong product, normal product)*** is defined to be

$$G_1 \boxtimes G_2 := (G_1 \times G_2) \oplus (G_1 \square G_2).$$

Exercise 4.3.9. The boxcross product $(G_1 \boxtimes G_2, (p_1, p_2))$ together with the natural projections constitute the product in the category ***EGra***. Again, we have to show that the induced mapping and the projections are the category ***EGra***.

Example 4.3.10 (Boxcross product).

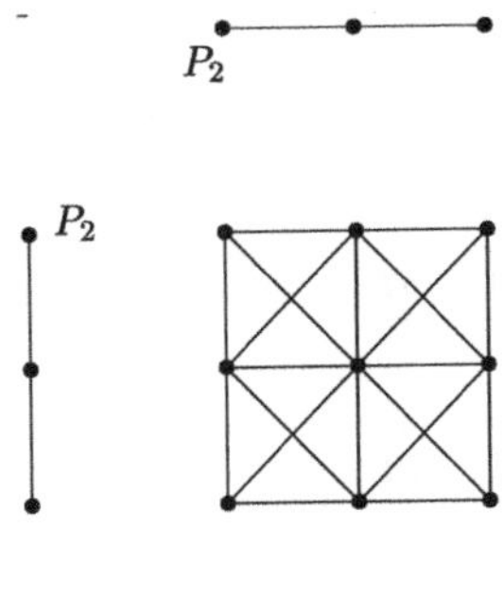

It is easy to see that the projections are not comorphisms.

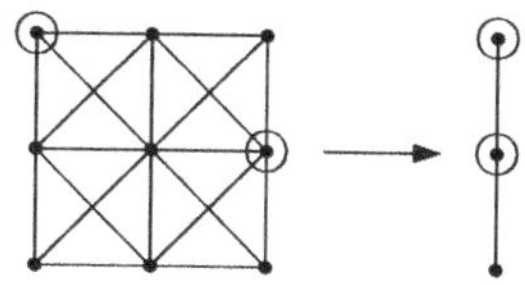

In the preimage under p_2, the edge between the encircled vertices does not exist!

Exercise 4.3.11. Find the construction of the coamalgam of two graphs in the category ***EGra***. Start with an example.

The complete product

The following definition is the same for directed and for undirected graphs.

Definition 4.3.12. The ***complete product (join product)*** is defined by

$$G \boxast H := (G \square H) \oplus (K_{|G|} \times K_{|H|}).$$

Example 4.3.13 (Complete product).

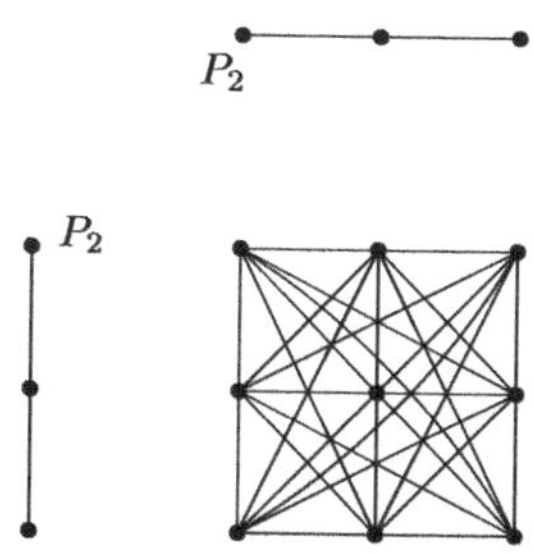

Exercise 4.3.14. The complete product together with the identity mapping $\tau : V(G) \times V(H) \to G \boxast H$ is the tensor product but not the product in the category ***CGra***.

Synopsis of the results

Corollary 4.3.15. *We summarize in a table which of the compositions between graphs play which categorical role in the respective categories.*

	Gra	***EGra***	***CGra***
Coproduct	*Union*	*Union*	*Join*
Product	*Cross product*	*Boxcross product*	*Disjunction*
Tensor product	*Box product*	*Box product*	*Complete product*

Corollary 4.3.16. *In* ***SGra*** *and* ***SEGra****, coproducts, products and tensor products do not exist.*

Proof. This follows from the fact that the category ***SGra*** is the intersection of the categories ***Gra*** and ***CGra***. Now, all three constructions are different in these two categories, but they would have to coincide on the intersection. A similar argument can be used for ***SEGra***. □

Product constructions as functors in one variable

All product constructions define covariant functors in the respective categories. We make this concrete for the box product.

Example 4.3.17 (Tensor functors). For the box product and a fixed $G \in \boldsymbol{Gra}$, we get the functor

$$\begin{array}{rccc} G \,\square\, - : & \boldsymbol{Gra} & \longrightarrow & \boldsymbol{Gra} \\ & H_1 & \longmapsto & G \,\square\, H_1 \\ & \Big\downarrow f & \longmapsto & G \,\square\, f := \Big\downarrow \quad \begin{array}{c} (x,y) \\ \Big\downarrow \\ (x,f(y)) \end{array} \\ & H_2 & \longmapsto & G \,\square\, H_2 \,. \end{array}$$

It is an easy exercise to see that the properties of a functor hold.

The respective functors could also be considered in the first variable.

4.4 Lexicographic products and the corona

The lexicographic products are also graphs built on the Cartesian product of the vertex sets of two (or more) graphs. They do not have a categorical description. This is also true of the corona and its generalizations.

Lexicographic products

For directed and undirected graphs, we have the same definitions. After Example 4.4.4, we will give a practical method for constructing lexicographic and generalized lexicographic products.

Definition 4.4.1. The ***lexicographic product (composition)*** of G_1 and G_2 is defined to be

$$\begin{aligned} G_1[G_2] := (V_1 \times V_2, \{&((x,y),(x',y')) \mid (x,x') \in E_1\} \\ &\cup \{\{(x,y),(x,y')\} \mid x \in V_1,\ (y,y') \in E_2\}). \end{aligned}$$

Remark 4.4.2. The lexicographic product $G_1[G_2]$ has the adjacency matrix $(A(G_1)\times J_2)+(I_1\times A(G_2))$, where J_2 denotes the matrix of ones of the same size as G_2, I_1 is the identity matrix of the same size as G_1, and $\times$ denotes the Kronecker product (cf. Remark 4.2.4) and + the sum of the matrices (cf. [Cvetković et al. 1979], Section 2.5 on p. 71).

Definition 4.4.3. Let $G = (V,E)$ and let $(H_x)_{x\in G}$ be graphs with $H_x = (V_x, E_x)$. The ***generalized lexicographic product (G-join)*** of G with $(H_x)_{x\in G}$ is defined to be

$$
\begin{aligned}
G[(H_x)_{x\in G}] :=(&\{(x,y_x) \mid x \in V,\ y_x \in V_x\},\\
&\{((x,y_x),\ (x',y'_x)) \mid (x,x') \in E\}\\
&\bigcup \{((x,y_x),\ (x,y'_x)) \mid x \in V,\ (y_x,y'_x) \in E_x\}).
\end{aligned}
$$

Example 4.4.4 (Lexicographic products and a generalized lexicographic product).

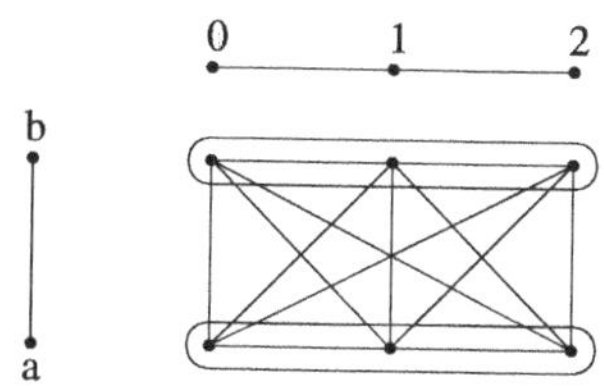

$K_2[P_2]$

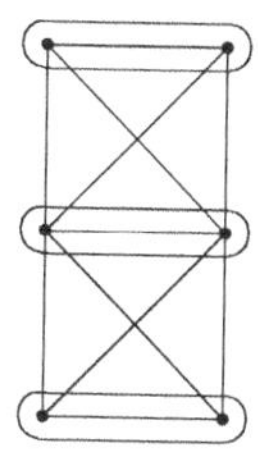

$P_2[K_2]$

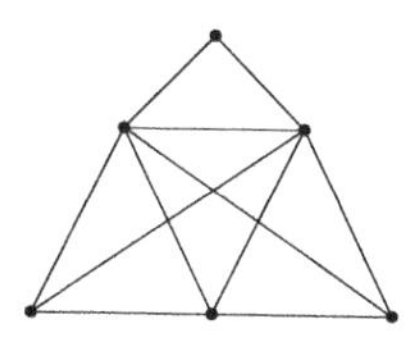

$P_2[(K_1, K_2, P_2)]$

Construction 4.4.5. We can operationalize the definition of $G_1[G_2]$ as follows. Take the first graph G_1, pump up its vertices to bubbles and insert the second graph G_2 in each bubble. And if (x,y) is an edge in G_1, we connect all vertices of G_2 in the bubble of x to all vertices of G_2 in the bubble of y.

We proceed analogously for the generalized lexicographic product $G[(H_x)_{x\in G}]$. Now we insert the graph H_x in the pumped-up vertex $x \in G$. And if (x,y) is an edge in G, we connect all vertices of H_x to all vertices in H_y.

Exercise 4.4.6. We have $G[H] \oplus [G]H = G \vee H$ and

$$K_{n,m} = K_2[(\overline{K}_n, \overline{K}_m)], \qquad K_{n_1,\dots,n_r} = K_r[(\overline{K}_{n_1}, \dots, \overline{K}_{n_r}]).$$

The corona

We mention the corona only briefly, since it is a construction by accident. It originated from a statement about automorphism groups which turned out to be false for lexicographic products. This was the equation in Exercise 4.4.10 with the lexicographic product instead of the corona.

As for the join, different variants are possible for directed graphs.

The corona $G_1 \lhd G_2$ was defined by Frucht and Harary as the following graph. Take one copy of G_1 and n_1 copies of G_2, where n_1 denotes the number of vertices of G_1. Now connect the ith vertex of G_1 by edges with all the vertices of the ith copy of G_2.

Definition 4.4.7. Let $G_1 = (V_1, E_1)$ and $G_2 = (V_2, E_2)$ be graphs. The ***corona of G_1 and G_2*** is defined to be

$$G_1 \lhd G_2 := (V_1 \bigcup (V_1 \times V_2), E_1 \bigcup \{(x, (x,y)) \mid x \in V_1,\ y \in V_2\}$$
$$\bigcup \{((x,y), (x,y')) \mid x \in V_1,\ (y,y') \in E_2\}).$$

Remark 4.4.8. The corona is generalized in the same way as the lexicographic product; for each vertex of G_1 one takes different graphs instead of one G_2, in analogy to the generalized lexicographic product. The notation is, for instance, $K_2 \lhd [P_2, K_2 \bigcup K_1]$, as shown in Example 4.4.9.

Example 4.4.9 (Coronas).

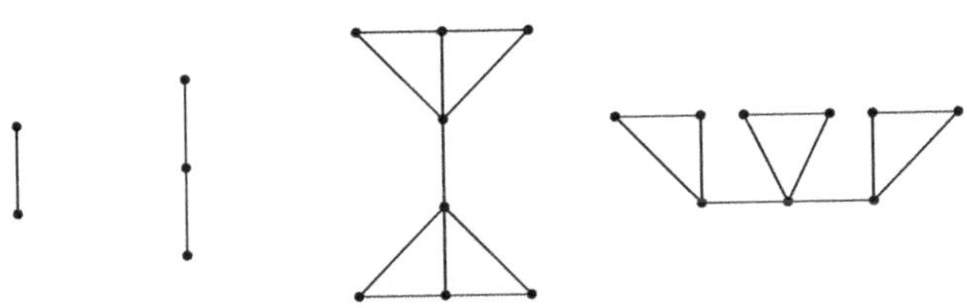

G_1 $\quad$ G_2 $\quad$ $G_1 \lhd G_2$ $\quad$ $G_2 \lhd G_1$

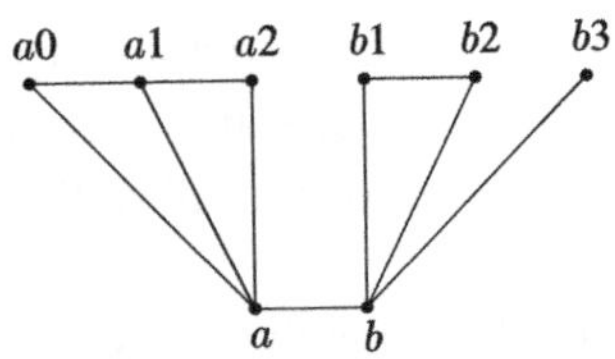

$K_2 \lhd [P_2, K_2 \bigcup K_1]$

Exerceorem 4.4.10. Prove that $\mathrm{Aut}(G \lhd H) = (\mathrm{Aut}(G) \wr \mathrm{Aut}(H) | G)$, using the notation for the wreath product from Chapter 9. What can you say if Aut is replaced by LEnd or QEnd or SEnd?

4.5 Algebraic properties

In this section, we do some algebra on a "higher level," i. e., we compose not elements but entire graphs and look at some algebraic properties of these compositions, such as commutativity.

Remark 4.5.1. The following relations are valid for subgraphs:

$$\begin{array}{cccc} & G_1[G_2] \supseteq & & \supseteq G_1 \square G_2 \\ G_1 \boxast G_2 \supseteq G_1 \vee G_2 \supseteq & & G_1 \boxtimes G_2 & \\ & [G_1]G_2 \supseteq & & \supseteq G_1 \times G_2 \end{array}$$

Remark 4.5.2. All operations except for the lexicographic product and the corona are commutative. The lexicographic product is commutative using the natural bijection $(x,y) \mapsto (y,x)$ if and only if both factors are complete graphs, and also in the trivial cases where $G = H$ or one factor is K_1. All operations are associative. So we always get "semigroups of graphs," in the case of the edge sum on graphs with a fixed vertex set.

For $\square, \boxtimes, \boxast, \vee$ and the lexicographic product, K_1 is the identity element; for $\bigcup$ and $+$, the empty set is the identity element. So in these cases, we even get "monoids of graphs." For the other operations, such as $\times$ and $\oplus$, identity elements do not exist.

Zero elements never exist for operations based on the union of the underlying sets; the empty set is the zero element for operations based on the Cartesian product of the underlying sets.

Using the results of the following theorem, we get "semirings of graphs" with $\bigcup$ as addition and all products except the lexicographic product. With the join $+$ as addition and $\boxast$ or $\vee$ as multiplication, we also get "semirings of graphs."

Theorem 4.5.3 (Distributivities). *Let G, H_1, and H_2 be graphs with $|G| = n$. Then (assuming $V(H_1) = V(H_2)$ for $\oplus$) the following hold:*

(1.1) $G \times (H_1 \bigcup H_2) = (G \times H_1) \bigcup (G \times H_2)$.

(1.2) $G \times (H_1 + H_2) = (G \times H_1) + (G \times H_2)$ *if and only if* $G = K_n^{(n)}$.

(1.3) $G \times (H_1 \oplus H_2) = (G \times H_1) \oplus (G \times H_2)$.

(2.1) $G \square (H_1 \bigcup H_2) = (G \square H_1) \bigcup (G \square H_2)$.

(2.2) $G \square (H_1 + H_2) = (G \square H_1) + (G \square H_2)$ *if and only if* $G = K_1$.

(2.3) $G \square (H_1 \oplus H_2) = (G \square H_1) \oplus (G \square H_2)$.

(3.1) $G \boxtimes (H_1 \bigcup H_2) = (G \boxtimes H_1) \bigcup (G \boxtimes H_2)$.

(3.2) $G \boxtimes (H_1 + H_2) = (G \boxtimes H_1) + (G \boxtimes H_2)$ *if and only if* $G = K_n$.

(3.3) $G \boxtimes (H_1 \oplus H_2) = (G \boxtimes H_1) \oplus (G \boxtimes H_2)$.

(4.1) $G \boxast (H_1 \bigcup H_2) = (G \boxast H_1) \bigcup (G \boxast H_2)$.

(4.2) $G \boxast (H_1 + H_2) = (G \boxast H_1) + (G \boxast H_2)$.

(4.3) $G \boxast (H_1 \oplus H_2) = (G \boxast H_1) \oplus (G \boxast H_2)$.

(5.1) $G \vee (H_1 \bigcup H_2) = (G \vee H_1) \bigcup (G \vee H_2)$.

(5.2) $G \vee (H_1 + H_2) = (G \vee H_1) + (G \vee H_2)$.

(5.3) $G \vee (H_1 \oplus H_2) = (G \vee H_1) \oplus (G \vee H_2)$.

(6.1a) $G[H_1 \bigcup H_2] = (G[H_1]) \bigcup (G[H_2])$ *if and only if* $G = \overline{K}_n$.

(6.2a) $G[H_1 + H_2] = (G[H_1]) + (G[H_2])$ *if and only if* $G = K_n^{(n)}$.

(6.3a) $G[H_1 \oplus H_2] = (G[H_1]) \oplus (G[H_2])$.

(6.1b) $(H_1 \bigcup H_2)[G] = (H_1[G]) \bigcup (H_2[G])$.

(6.2b) $(H_1 + H_2)[G] = (H_1[G]) + (H_2[G])$.

(6.3b) $(H_1 \oplus H_2)[G] = (H_1[G]) \oplus (H_2[G])$.

4.6 Mor constructions

This section is for specialists who like tricky constructions. To such specialists who like category theory as well, the left adjointness of these constructions to different products will be a source of fascinating and technically challenging problems.

All of the following six constructions can also be made for directed graphs. The resulting graphs will differ in the numbers of vertices and edges.

See, for comparison, Mati Kilp and Ulrich Knauer [47]. Parts (a) of Construction 4.6.1 and Theorem 4.6.4 can also be found in Definition 5.18 and as a remark before Proposition 5.19 in G. Hahn and C. Tardiff [29].

Diamond products

For the following three constructions, we will use the same symbol and the same notation. The differences will become clear from the category where the construction takes place. The definitions are the same for directed and undirected graphs.

Construction 4.6.1.

(a) The ***diamond product*** $G \diamond H$ of two graphs G and H in ***Gra*** is defined by

$$V(G \diamond H) := \textbf{\textit{Gra}}(G, H), \text{ the set of graph homomorphisms from } G \text{ to } H,$$

$$E(G \diamond H) := \{(\alpha, \beta) \mid (\alpha(x), \beta(x)) \in E(H) \text{ for all } x \in G\}.$$

(b) The ***diamond product*** $G \Diamond H$ of two graphs G and H in ***EGra*** is defined by

$$V(G \Diamond H) := \textbf{\textit{EGra}}(G,H), \text{ the set of graph egamorphisms from } G \text{ to } H,$$
$$E(G \Diamond H) := \{(\alpha,\beta) \mid (\alpha(x),\beta(x)) \in E(H) \text{ for all } x \in G\}.$$

(c) The ***diamond product*** $G \Diamond H$ of two graphs G and H in ***CGra*** is defined by

$$V(G \Diamond H) := \textbf{\textit{CGra}}(G,H), \text{the set of graph comorphisms from } G \text{ to } H,$$
$$E(G \Diamond H) := \{(\alpha,\beta) \mid \exists x \in G \text{ such that } (\alpha(x),\beta(x)) \in E(H)\}.$$

Note that these operations are highly noncommutative.

Note, moreover, that the definitions of adjacencies in (a) and (b) have the same structure, which is understandable as the two categories have the same tensor product (see Theorem 4.3.5) and the constructions are left adjoint to tensor products (see Theorem 4.6.4).

Example 4.6.2 (Diamond products).
For $G =$ 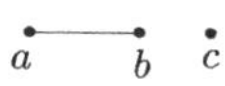

and $H =$ 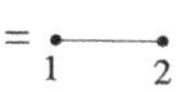

we get $G \Diamond H$ as follows:

in ***Gra***: in ***EGra***: in ***CGra***:

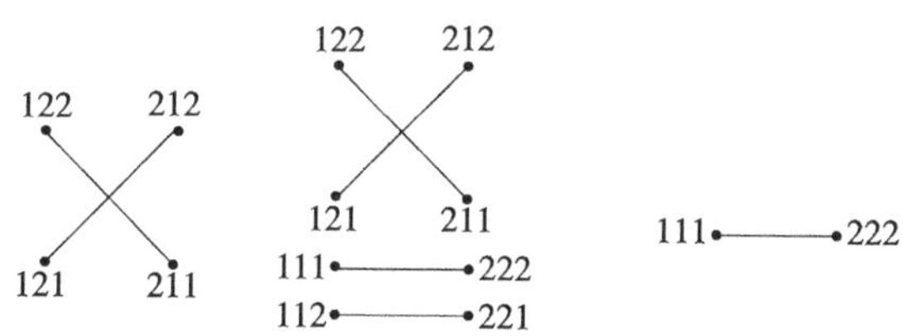

The vertex ijk denotes the morphism that maps a to i, b to j and c to k for $i,j,k \in \{1,2\}$.

Remark 4.6.3. The diamond products define covariant functors in the respective categories. So for ***Gra***, we get

$$\begin{array}{cccc} G \Diamond - : & \textbf{\textit{Gra}} & \longrightarrow & \textbf{\textit{Gra}} \\ & H_1 & \longmapsto & G \Diamond H_1 \\ & \big\downarrow f & \longmapsto & G \Diamond f := \Big\downarrow \quad \overset{\alpha}{\underset{f\alpha}{\big\downarrow}} \\ & H_2 & \longmapsto & G \Diamond H_2 \,. \end{array}$$

Considering the respective functors in the first variable, we get contravariant functors; cf. Definition 3.3.6.

Left inverses for tensor functors

In the situation described in the next theorem, one usually says that the diamond functors are left adjoint to the tensor functors; cf. Example 4.3.17. Recall Definition 3.3.10.

Theorem 4.6.4. *The diamond functors are "left inverse" to the tensor functors in one variable in* ***Gra***, ***EGra*** *and* ***CGra***.

Proof. (a) We show that there exists a *natural transformation*

$$\Theta : \mathrm{Id}_{\boldsymbol{Gra}}(-) \to (G \diamond -)(G\square-) = G \diamond (G\square-),$$

where $\mathrm{Id}_{\boldsymbol{Gra}}(-)$ is the identity functor on ***Gra***; in other words, Θ relates the two functors with respect to objects and morphisms. The following rectangle in ***Gra***, which contains the definition of $\Theta_A(a)$ for $A \in \boldsymbol{Gra}$ and $a \in A$, is commutative for all morphisms $f : A \to B$ in ***Gra***:

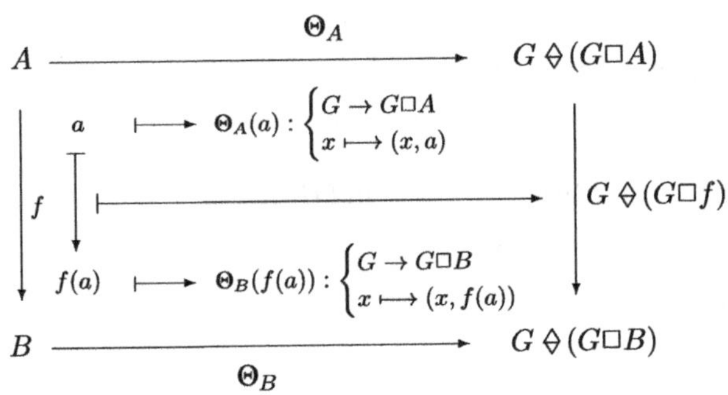

1. We compute for all $a \in A$ and all $x \in G$ that

$$\begin{aligned}(G \diamond (G \square f))(\Theta_A(a))(x) &= (G \diamond (\mathrm{id}_G \square f))(\Theta_A(a))(x)\\ &= ((\mathrm{id}_G \square f)\Theta_A(a))(x)\\ &= (\mathrm{id}_G \square f)(\Theta_A(a)(x)) = (\mathrm{id}_G \square f)(x,a)\\ &= (x,f(a)) = (\Theta_B(f(a)))(x).\end{aligned}$$

   This proves commutativity.
2. We prove that for all $a \in A$ we get $\Theta_A(a) \in V(G \diamond (G \square H))$. Since

$$(\Theta_A(a)(x), \Theta_A(a)(x')) = ((x,a),(x',a)) \in E(G \square A),$$

   for $(x,x') \in E(G)$ we have $\Theta_A(a) \in V(G \diamond (G \square A)) = \boldsymbol{Gra}(G, G \square A)$.
3. We prove that Θ_A is a morphism in ***Gra***. If $(a,a') \in E(A)$, then for all $x \in G$ we get

$$(\Theta_A(a)(x), \Theta_A(a')(x)) = ((x,a),(x,a')) \in E(G \square A)$$

by the definition of $G \,\Box\, A$. Consequently,

$$(\Theta_A(a), \Theta_A(a')) \in E(G \between (G \,\Box\, A)).$$

Thus $\Theta_A \in \boldsymbol{Gra}(A, G \between (G \,\Box\, A))$.

Putting the above together, we have that Θ is a natural transformation.

(b) Analogous to (a).

(c) We follow the scheme of the proof of (a).

1. The definition of the mapping $\Theta_A : A \to (G \between -)(G \boxast -)$ for $A \in \boldsymbol{CGra}$ and the proof of commutativity of the diagrams are the same as in (a).
2. If $(\Theta_A(a)(x), \Theta_A(a)(x')) = ((x, a), (x', a)) \in E(G \boxast A)$, then the definition of the complete product implies that $(x, x') \in E(G)$. Consequently, $\Theta_A(a) \in V(G \between (G \boxast A)) = \boldsymbol{CGra}(G, G \boxast A)$.
3. If $(\Theta_A(a), \Theta_A(a')) \in E(G \between (G \boxast A))$, i.e., there exists $x \in V(G)$ such that $(\Theta_A(a)(x), \Theta_A(a')(x)) = ((x, a), (x, a')) \in E(G \boxast A)$, then the definition of the complete product implies that $(a, a') \in E(A)$. Therefore, $\Theta_A \in \boldsymbol{CGra}(A, G \between (G \boxast A))$.

Again, we have that Θ is a natural transformation. □

Power products

For the following three constructions, we will again use the same symbol and the same notation, with the differences becoming clear from the category where the construction takes place; the definitions are also the same for directed and undirected graphs.

Construction 4.6.5.

(a) The ***power product*** $G \searrow H$ of the graphs G and H in ***Gra*** is defined by

$$\begin{aligned} V(G \searrow H) &:= \boldsymbol{Set}(G, H) = \mathrm{Map}(G, H), \text{ the set of mappings from } G \text{ to } H, \\ E(G \searrow H) &:= \{(\alpha, \beta) \mid \alpha \neq \beta,\ (\alpha(x), \beta(x')) \in E(H) \text{ for all } (x, x') \in E(G)\}. \end{aligned}$$

(b) The ***power product*** $G \searrow H$ of the graphs G and H in ***EGra*** is defined by

$$\begin{aligned} V(G \searrow H) &:= \boldsymbol{EGra}(G, H), \\ E(G \searrow H) &:= \{(\alpha, \beta) \mid (\alpha(x), \beta(x')) \in E(H) \text{ for all } (x, x') \in E(G), \\ &\qquad (\alpha(x), \beta(x)) \in E(H) \text{ for all } x \in G\}. \end{aligned}$$

(c) The ***power product*** $G \searrow H$ of the graphs G and H in ***CGra*** is defined by

$$\begin{aligned} V(G \searrow H) &:= \boldsymbol{CGra}(G, H), \\ E(G \searrow H) &:= \{(\alpha, \beta) \mid \exists\, x, x' \in G : (\alpha(x), \beta(x')) \in E(H),\ (x, x') \notin E(G)\}. \end{aligned}$$

The symbol $\searrow$ is supposed to remind us that these operations are not commutative.

Example 4.6.6 (Power products).

For $G =$ 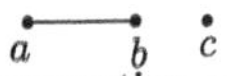 and $H =$ 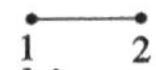 we get $G \searrow H$ as follows, where the vertex sets are the respective sets of morphisms:

in ***Gra***: in ***EGra***: in ***CGra***:

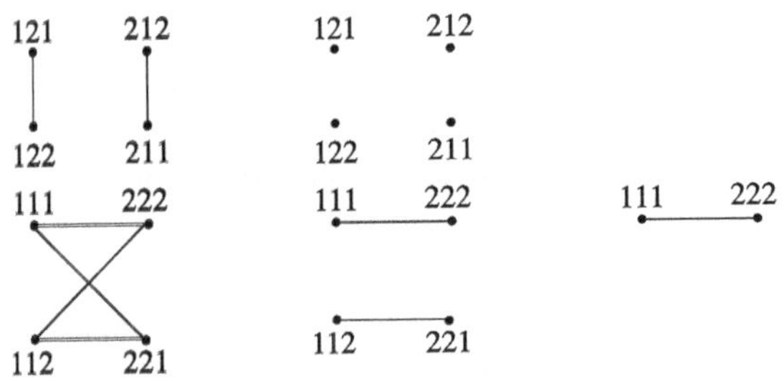

As in Example 4.6.2, the vertex ijk denotes the morphism which maps a to i, b to j, and c to k for $i, j, k \in \{1, 2\}$.

Left inverses to product functors

In the situation described in the next theorem, one usually says that the power functors are left adjoint to the product functors. Recall Definition 3.3.10 and compare with Theorem 4.6.4.

Theorem 4.6.7. *The power functors are "left inverse" to the product functors in one variable in **Gra, EGra** and **CGra**, if we consider the constructions as functors.*

Proof. The proofs for ***Gra*** and ***EGra*** follow the scheme of the proof of part (a) in Theorem 4.6.4. We prove the statement for ***CGra***.

1. The definition of the mapping $\Theta_A : A \to (G \searrow -)(G \vee -)$ for $A \in$ ***CGra*** and the proofs of commutativity of the diagrams are the same as in part (a) of Theorem 4.6.4.
2. If $(\Theta_A(a)(x), \Theta_A(a)(x')) = ((x, a), (x', a)) \in E(G \vee A)$, then the definition of the disjunction implies that $(x, x') \in E(G)$. Thus $\Theta_A(a) \in V(G \searrow (G \vee A)) =$ ***CGra***$(G, G \vee A)$.
3. If $(\Theta_A(a), \Theta_A(a')) \in E(G \searrow (G \vee A))$, i.e., there exists $x, x' \in V(G)$ such that $(\Theta_A(a)(x), \Theta_A(a')(x')) = ((x, a), (x', a')) \in E(G \vee A)$ but $(x, x') \notin E(G)$, then the definition of the disjunction implies that $(a, a') \in E(A)$. Thus $\Theta_A \in$ ***CGra***$(A, G \searrow (G \vee A))$.

Putting the above together, we have again that Θ is a natural transformation. □

Exercise 4.6.8. Determine diamond and power products of several small graphs in each of the three categories.

4.7 Comments

In this chapter, there are several exercises which the reader can use to gain familiarity with the subject.

In Sections 4.1 through 4.3, it is interesting to see how graph compositions such as sums and various products get a categorical interpretation. In particular, in each case we can see that a graph-theoretical construction satisfies universal and categorical properties. In the abstract definition of the categorical product as given in Section 3.2, we described only the abstract properties of an object with a family of morphisms, called the categorical product. We also prove that, e. g., the cross product with the projections satisfies these abstract properties in the category ***Gra*** and can therefore be called the product in this category.

The meaning of a universal construction can also be made clear in this concrete case. If we start with G_1 and G_2, then whatever graph G and whatever homomorphisms $f_1 : G \to G_1$ and $f_2 : G \to G_2$ we take, we can always find $f : G \to G_1 \times G_2$ such that the diagram is commutative; cf. Theorem 4.2.3.

Here we also get an impression of what the difference is between a categorical description—of the product, e. g., and a noncategorical definition—of the lexicographic product, for example. The latter is given only inside a given category, but not in an arbitrary abstract category. This means that we cannot take it to this or another construction in a different category by using a functor. We will resume this discussion in Chapter 11.

The Mor constructions of Section 4.6, separate from their categorical meanings, are of some interest in themselves and can be studied with respect to various algebraic or other properties, i. e., which properties of the components are inherited by the respective construction, and under what additional conditions. As far as we can see, there are many open questions.

5 Line graph and other unary graph operations

Similar to binary graph operations, new objects can also be constructed from just one graph. The formation of the complement and loop complement are unary operations with an unchanged vertex set. The same is true of the opposite graph of directed or undirected graphs; see Definition 1.1.8. Constructing the geometric dual graph for a planar graph may be considered a unary operation with a changing vertex set.

The three operations starting from Section 5.2 also give graphs with new vertex sets. They work in a natural way for undirected graphs. For directed graphs, there are several possibilities in each of the three cases; these can be formulated according to specific needs or just as a game to familiarize oneself with the concepts.

5.1 Complements, opposite graphs, and geometric duals

Definition 5.1.1. If $G = (V, E)$ is a graph without loops, we define the ***complement*** of G to be $\overline{G} = (V, \overline{E})$ where $(x, y) \in \overline{E}$ if and only if $(x, y) \notin E$, $x \neq y$. If $G = (V, E)$ is a graph, possibly with loops, we define the ***loop complement*** of G to be $\overline{G}^{\circ} = (V, \overline{E}^{\circ})$ where $(x, y) \in \overline{E}^{\circ}$ if and only if $(x, y) \notin E$.

Exercise 5.1.2. The formation of the complement and of the loop complement can be considered as covariant functors from the category ***Gra*** to the category ***CGra***.

Theorem 5.1.3. *If the graph G is d-regular with n vertices and has eigenvalues $d, d_2, \ldots, d_n$, then G and $\overline{G}$ have the same eigenvectors and $\overline{G}$ has eigenvalues $n - d - 1, -1 - d_2, \ldots, -1 - d_n$.*

Proof. See [Godsil/Royle 2001], Lemma 8.5.1 on page 172. The adjacency matrix of $\overline{G}$ is given by $A(\overline{G}) = J_n - I_n - A(G)$, where J_n is the $n \times n$ matrix consisting entirely of ones and I_n is the $n \times n$ identity matrix. Let $\{u, u_2, \ldots, u_n\}$ be an orthonormal set of eigenvectors of $A(G)$, where $u = {}^t(1, \ldots, 1)$; cf. Theorem 2.7.5. Then u is an eigenvector of $A(\overline{G})$ with the eigenvalue $n - 1 - d$, as an easy computation shows. For $2 \leq i \leq n$, the eigenvector $u_i = (u_{i_1}, \ldots, u_{i_n})$ is orthogonal to u and so $u_{i_1} + \cdots + u_{i_n} = 0$. Now we calculate

$$A(\overline{G})u_i = (J_n - I_n - A(G))u_i = 0 - 1 - d_i.$$

Therefore, u_i is an eigenvector of $A(\overline{G})$ with eigenvalue $-1 - d_i$. □

Remark 5.1.4. The opposite graph for a directed graph is defined in Definition 1.1.8; this can be seen as a contravariant functor; see Definition 3.3.5.

We note that on the category ***Path***$_G$ (cf. Example 3.1.9(a)), this functor takes a morphism which is an x, y path to a morphism which is a y, x path.

Remark 5.1.5 (Geometric dual). Recall Remark 1.8.6, i. e., the geometric dual G^* of a 2-cell embedded graph G is the graph which has the regions of the original graph G as

https://doi.org/10.1515/9783110617368-005

vertices; so it has a new vertex set, and two vertices in G^* are adjacent if and only if the two regions in G have a common edge. With other words, put a vertex into every face of the considered embedding of G and connect two such vertices if the faces have a common border.

Since different embeddings may be possible on one surface, different geometric duals will exist.

It might be interesting to consider this procedure as a functor.

As the geometric dual of a simple graph may have loops and multiple edges, we will have to use the morphism concept from Definition 1.4.1 in this case; i. e., the functor would go to the category ***EGra***. It would need some effort to define a suitable category where such a functor could start.

5.2 The line graph

We discuss this construction and its properties in some detail. In particular, we study the determinability of a graph by its line graph—can nonisomorphic graphs have isomorphic line graphs?

In Section 5.3, we will discuss eigenvalues of line graphs and how they depend on the eigenvalues of the original graph.

Definition 5.2.1. The graph $LG = (E, \{\{e,e'\} \mid e \cap e' \neq \emptyset,\ e \neq e'\})$ is called the ***line graph of G.***

Lemma 5.2.2. *We have* $|V(LG)| = |E|$ *and* $|E(LG)| = \sum_{x\in V} \binom{\deg_G(x)}{2}$.

Proof. Any two edges in G which are incident with the vertex x of G give an edge in LG; thus we have a total of $\sum_{x\in V} \binom{\deg_G(x)}{2}$ edges in LG. □

Remark 5.2.3 (Line graphs of directed graphs). A line graph of a directed graph can be constructed in several different ways. We can use the above definition unchanged, or we can join two vertices e_1 and e_2 of the line graph with an undirected edge if both edges in the original graph have a common source or a common tail. This always gives an undirected graph. We can also require that two vertices e_1 and e_2 of the line graph form an edge (e_1, e_2) if $t(e_1) = o(e_2)$ or $o(e_1) = t(e_2)$ in the original graph.

Remark 5.2.4 (The line functor). We note that L can be interpreted as a functor from the category ***Gra*** into the category ***EGra*** upon setting $Lf(e) := (f(o(e)), f(t(e)))$ where f is a morphism in ***Gra***, e on the left-hand side of the equality is a vertex in LG, and e on the right-hand side is the corresponding edge in the graph G.

Example 5.2.5 (Line graph). The line graphs of graphs on the left are shown on the right.

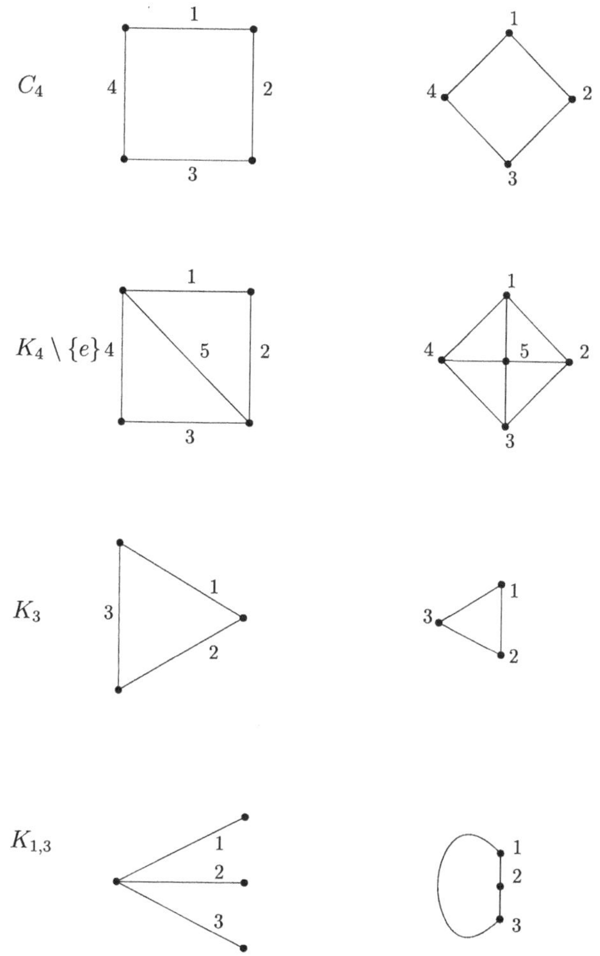

Observe that $L(K_4 \setminus \{e\}) = C_4 + K_1$.

The line graph of K_3 amalgamated with K_2 at one vertex is $K_4 \setminus \{e\}$.

LK_4 is the amalgam of $C_4 + K_1$ with itself amalgamated along C_4, which is isomorphic to $T(K_3)$ in Example 5.4.4.

We note that the complement of LK_5 is the ***Petersen graph***, also denoted by $G(5, 2)$. See Figure 5.1 for a drawing.

This is a fascinating graph which serves as example or counterexample in many different situations. There is a monograph devoted to this graph: [Holton/Sheehan 1993]. Another more involved example of the line graph construction is shown in Figure 5.2.

Lemma 5.2.6. *Take $x_0 \in G$ with $\deg_G(x_0) = 1$, and let $\{x_0, \ldots, x_\ell\}$ be a simple path such that $\deg_G(x_1) = \cdots = \deg_G(x_{\ell-1}) = 2$ and $\deg_G(x_\ell) = 1$ or $\deg_G(x_\ell) > 2$, where $\ell \geq 2$. The ℓ edges on this path form a simple path in LG of length $\ell - 1$, where the end vertices have*

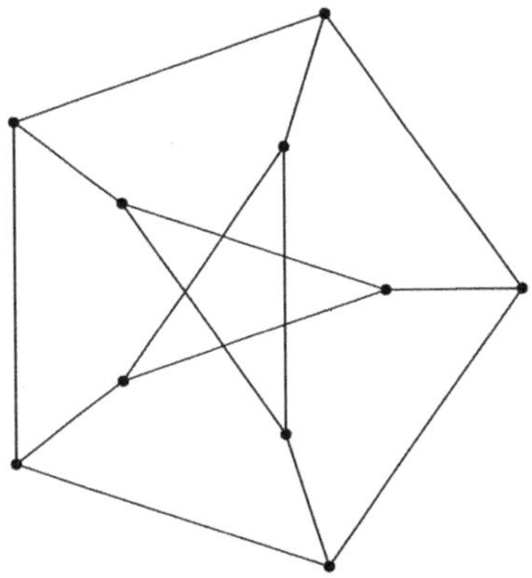

Figure 5.1: The Petersen graph.

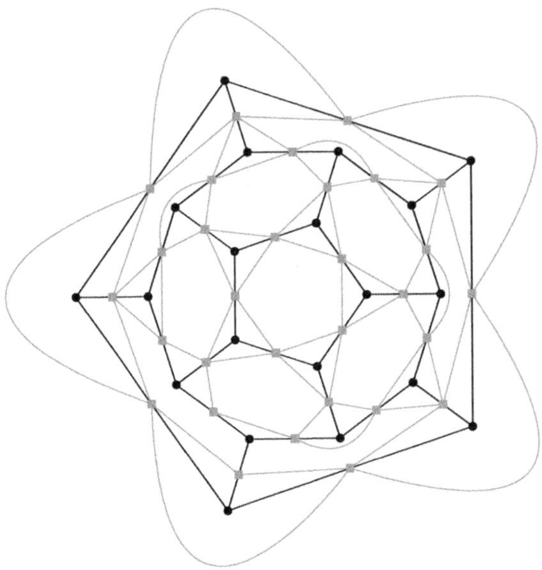

Figure 5.2: The Dodecahedron graph and its line graph—the graph of the ***Icosidodecahedron***.

the same degree properties. Conversely, each simple path of this type having length $\ell - 1$ in LG comes from such a simple path of length ℓ in G.

Proof. It is clear that the edges $e_1, \ldots, e_\ell$ of the path in G are the vertices of a path of length $\ell - 1$ in LG. If $\deg_G(x_\ell) = 1$, then $\deg_{LG}(e_\ell) = 1$. If $\deg_G(x_\ell) > 2$, then the edges $e_{\ell+1}, e_{\ell+2}, \ldots$ in G are incident with x_ℓ. Then these edges of G as vertices of LG are adjacent to e_ℓ, i. e., $\deg_{LG}(x_\ell) > 2$.

Conversely, suppose that $e_1, \ldots, e_\ell$ is a simple path of length $\ell - 1$ in LG. Then this is a simple path of length ℓ in G with the vertices $\{x_0, \ldots, x_\ell\}$. It follows that $\deg_G(x_\ell) \neq 2$ if $\deg_{LG}(e_\ell) \neq 2$, since otherwise both degrees would be 2. □

Theorem 5.2.7. *A connected graph G is isomorphic to its line graph LG if and only if it is a circuit; that is, $G \cong LG$ if and only if $G \cong C_n$ for some $n \in \mathbb{N}$.*

Proof. Suppose $G \cong LG$. Then

$$\begin{aligned} n = |V| = |E| &= |V(LG)| = |E(LG)| \\ &= \sum_{x\in V} \binom{\deg_G(x)}{2} \\ &= \frac{1}{2}\sum_{x\in V} \deg_G(x)(\deg_G(x)-1) \\ &= \frac{1}{2}\sum_{x\in V} \deg_G(x)^2 - \frac{1}{2}\sum_{x\in V}\deg_G(x) = \frac{1}{2}\sum_{x\in V}\deg_G(x)^2 - n, \end{aligned}$$

and thus $4n = \sum_{x\in V} \deg_G(x)^2$.

If $\deg_G(x) \geq 2$, then because of $4n = 2^2 n$ we get that $\deg_G(x) = 2$.

If $\deg_G(x) = 1$, then there exists a simple path of length ℓ in G, as G is connected. Since $G \cong LG$, there exists a simple path of length ℓ in LG, which corresponds to a simple path of length $\ell + 1$ in G by Lemma 5.2.6, and so on. Thus, in G there would have to exist arbitrarily long simple paths.

The converse is obvious. □

Determinability of *G* by *LG*

Here, we pose a typical question: Can G be described uniquely by LG? The answer is yes, with two exceptions. This also answers the question of under what conditions the functor L is an injector.

For the next theorem, compare the two graphs K_3 and $K_{1,3}$ in Example 5.2.5 and their line graphs.

Theorem 5.2.8. *Let G and G' be connected and simple. We have $LG \cong LG'$ if and only if $\{G, G'\} = \{K_3, K_{1,3}\}$ or $G \cong G'$. In the latter case, every isomorphism $\varphi_1 : LG \to LG'$ is induced by exactly one isomorphism $\varphi : G \to G'$, i. e., for all $e \in LG$ with $e = \{u, v\} \in E(G)$ one has $\varphi_1(e) = \{\varphi(u), \varphi(v)\}$.*

Proof. From Example 5.2.5, we know that $LK_3 = K_3 = LK_{1,3}$ but $K_3 \neq K_{1,3}$. Now suppose that $LG \cong LG'$ but $G \notin \{K_3, K_{1,3}\}$ or $G' \notin \{K_3, K_{1,3}\}$.

We consider all graphs with up to four vertices, except for the two graphs mentioned above. These are $K_2, K_4, P_2, P_3, C_4, K_4 \setminus \{e\}$ and K_3 amalgamated at one vertex with K_2. Consider the associated line graphs. It is clear that no two of them are isomorphic; cf. Example 5.2.5.

Now take G to be a graph with more than four vertices. We show that every isomorphism $\varphi_1 : LG \to LG'$ is induced by exactly one isomorphism $\varphi : G \to G'$, i. e., for every $e \in LG$, $e = \{u, v\} \in E(G)$ we have $\varphi_1(e) = \{\varphi(u), \varphi(v)\}$.

Uniqueness of φ: Assume that φ and ψ induce φ_1, i. e., for all $e = \{u, v\} \in E(G)$ we have $\varphi_1(e) = \{\varphi(u), \varphi(v)\} = \{\psi(u), \psi(v)\}$. Suppose that w is another vertex of G such that $\ell := \{v, w\} \in E(G)$, say. Then $\{e, \ell\} \in E(LG)$, and thus $\{\varphi_1(e), \varphi_1(\ell)\} \in E(LG')$ and $\varphi_1(\ell) = \{\varphi(v), \varphi(w)\} = \{\psi(v), \psi(w)\}$. Then $\varphi(v)$ and $\psi(v)$ are incident with the edges $\varphi_1(e)$ and $\varphi_1(\ell)$ in G'. Since two distinct edges cannot have two vertices in common, we get $\varphi(v) = \psi(v)$. And since $\varphi_1(e) \in E(G)$ contains only two vertices, $\varphi(v) = \psi(v)$ implies that $\varphi(u) = \psi(u)$.

Existence of φ: Now we have an isomorphism $\varphi_1 : LG \to LG'$.

(1) If $K_{1,3} = \{u\} + \{v_1, v_2, v_3\}$ with edges e_1, e_2, e_3 is contained in G, then the three edges $\varphi_1(e_1), \varphi_1(e_2), \varphi_1(e_3)$ of $\varphi_1(K_{1,3})$ in G' also form a $K_{1,3}$. To see this, we proceed as follows.

As G is connected and has at least five vertices, there exists $\ell = \{v_1, w\}$ or $\ell = \{u, w\}$ as an edge in G. In LG, the vertices e_1, e_2, e_3 form a K_3, and ℓ is adjacent only to e_1 or to all three vertices of the K_3. In $LG' = \varphi_1(LG)$, we have the same situation. Then $\ell' := \varphi_1(\ell)$ is adjacent to $\varphi_1(e_1) =: e'_1$, say, or to all of the $\varphi_1(e_i) =: e'_i$, $i = 1, 2, 3$. These vertices are edges in G'; that is e'_1, e'_2, e'_3 form a K_3 or a $K_{1,3}$ in G'.

Suppose that they form K_3. Then ℓ' has to be incident with all three edges, which is not possible in K_3. Otherwise, ℓ' has to be adjacent only to e'_1, which is also impossible in K_3. Thus e'_1, e'_2, e'_3 form $K_{1,3}$ in G', and this proves (1).

(2) Set $\text{in}(v) := \{e \in E \mid v \in e\}$ for $v \in G$, compare Definition 1.1.9. We consider two cases and show that in both cases $\varphi_1(\text{in}(v)) = \text{in}(v')$.

(a) If $\deg_G(v) \geq 2$, there exists exactly one $v' \in \bigcap_{e \in \text{in}(v)} \varphi_1(e)$. To see this, suppose that v in G is the common vertex of the edges e_1 and e_2. Then $\varphi_1(e_1) \neq \varphi_1(e_2)$ in G' and $\varphi_1(e_1) \bigcap \varphi_1(e_2) \neq \emptyset$, since φ_1 is a graph isomorphism and so $\{\varphi_1(e_1), \varphi_1(e_2)\} \in E(LG')$. In G', there exists exactly one $v' \in \varphi_1(e_1) \bigcap \varphi_1(e_2)$, since G and G' are simple and two edges can have only one common vertex. Since this is the case for any two edges in $\text{in}(v)$, we get the unique $v' \in \bigcap_{e \in \text{in}(v)} \varphi_1(e)$.

It remains to show that $\varphi_1(\text{in}(v)) = \text{in}(v')$. Take $e \in \text{in}(v)$, i. e., $v \in e$.
- If $\deg_G(v) > 2$, then we have the three edges e, e_1 and e_2 with common vertex v and, therefore, the three edges $\varphi_1(e), \varphi_1(e_1), \varphi_1(e_2)$ have the common vertex v' in G' because of (1). Consequently, $\varphi_1(\text{in}(v)) \subseteq \text{in}(v')$.
- If $\deg_G(v) = 2$, then $\text{in}(v) = \{e_1, e_2\}$ and thus again $\varphi_1(\text{in}(v)) \subseteq \text{in}(v')$ as $v' \in \bigcap_{e \in \text{in}(v)} \varphi_1(e)$.

Conversely, take $e' \in \text{in}(v')$ in G'. Then we get the reverse inclusion when considering φ_1^{-1}.

(b) If $\deg_G(v) = 1$, there exists exactly one $v' \in \varphi_1(e)$. Suppose that $e = \{v, u\}$ in G. Then $\deg_G(u) \geq 2$, as G is connected and has more than two vertices. As in (a), we get $\varphi_1(\text{in}(u)) = \text{in}(u')$, where $u' \in G'$ is unique in having this property. But since $e' := \varphi_1(e)$ in G' has exactly two end vertices, we again obtain that there exists exactly one $v' \in G'$ with $e' = \{u', v'\}$. It remains to show that $v' \in \varphi_1(e)$. This follows once we show that $\varphi_1(\text{in}(v)) = \text{in}(v')$. So suppose $e' \neq \ell'$ are both in $\text{in}(v')$ in G'. In LG', we get $\{e', \ell'\} \in E(LG')$, and as φ_1^{-1} is an isomorphism we have $\{e, \varphi_1^{-1}(\ell')\} \in$

$E(LG)$, i. e., $e \bigcap \varphi_1^{-1}(\ell') \neq \emptyset$ in G. As $\deg_G(v) = 1$, it follows that $u \in e \bigcap \varphi_1^{-1}(\ell')$. Then $\varphi_1^{-1}(\ell') \in \mathrm{in}(u)$ implies $\varphi_1\varphi_1^{-1}(\ell') \in \varphi_1(\mathrm{in}(u)) \bigcap \mathrm{in}(v') = \mathrm{in}(u') \bigcap \mathrm{in}(v')$, which contradicts the simplicity of G'. Thus $\deg_{G'}(v') = 1$, i. e., $|\mathrm{in}(v)| = |\mathrm{in}(v')| = 1$, and as $\varphi_1(e) \in \mathrm{in}(v')$, we get that in this case $\varphi_1(\mathrm{in}(v)) = \mathrm{in}(v')$, too.

This proves (2).

Now we can prove the rest of the theorem. Define $\varphi : G \to G'$ by $\varphi(v) := v'$ according to (2), which then is well-defined. It is apparent that $\varphi_1(e) = \{\varphi(v), \varphi(u)\}$ for $e = \{v, u\}$, since $\{e\} = \mathrm{in}(v) \bigcap \mathrm{in}(u)$. Thus $\{\varphi_1(e)\} = \mathrm{in}(v') \bigcap \mathrm{in}(u')$. Therefore, φ induces φ_1.

Moreover, $\varphi_1(\mathrm{in}(v)) = \mathrm{in}(\varphi(v)) = \mathrm{in}(\varphi(w)) = \varphi_1(\mathrm{in}(w))$ if $\varphi(v) = \varphi(w)$, and thus $\mathrm{in}(v) = \mathrm{in}(w)$, since φ_1 is an isomorphism. Now $\varphi(v) = \varphi(w)$ implies $v = w$, i. e., φ is injective; since G is simple, connected and has at least two edges, not both of v and w have degree 1. So both have degree at least 2 as $\mathrm{in}(v) = \mathrm{in}(w)$, and thus $v = w$ by 2(a).

Now φ is also surjective, as for $v' \in G'$ there exists $e' \in E(G')$ with $v' \in e'$. Upon setting $\{u, v\} = \varphi_1^{-1}(e')$, the definition of φ implies that $\varphi(u) = v'$ or $\varphi(v) = v'$.

Finally, φ is a graph homomorphism, as $\{\varphi(u), \varphi(v)\} = \varphi_1(e) \in E(G')$ for $e = \{u, v\} \in E(G)$ since φ_1 is a mapping, and analogously for φ^{-1}. □

5.3 Spectra of line graphs

After Proposition 5.3.1, we consider only line graph of undirected graphs in this section.

Proposition 5.3.1. *Take G to be without loops, simple, with $|E| = m$, and with the adjacency matrix $A(G)$. Let LG be the line graph of G. Denote by I_m the $m \times m$ identity matrix. Then*

$$^tB(G)\,B(G) = 2I_m + A(LG)$$

and

$$B(G)\,{}^tB(G) = D(G) - (A(G) + {}^tA(G))G$$

if G is directed, while

$$B\,{}^tB = D(G) + A(G)$$

if G is undirected.

Here, ^tB denotes the transpose of B, and we use the so-called directed/undirected vertex valency matrix $D(G) := (\mathrm{degree}(x_i)\delta_{ij})_{i,j=1,\dots,n} \in M(n \times n; \mathbb{N}_0)$, *where* $\mathrm{degree}(x_i) := \mathrm{indeg}(x_i) + \mathrm{outdeg}(x_i)$ *for directed graphs and* $\mathrm{degree}(x_i) := \deg(x_i)$ *for undirected graphs.*

Proof. Take G to be without loops and simple, with $|V| = n$. Then consider the (k, l)th entry

$$({}^t\!B\,B)_{kl} = \sum_{i=1}^{n} b_{ik}\, b_{il},$$

which is the standard scalar product of the kth and lth columns of B. For $k = l$, every column contributes 2. For $k \neq l$, the product is 1 if and only if the edges k and l are incident in the vertex i. This can happen at most once since G is simple. This is the value of the (k, l)th entry of $A(LG)$.

To prove the second equality, we consider the (i, j)th entry of the matrix:

$$(B\,{}^t\!B)_{ij} = \sum_{l=1}^{m} b_{il}\, b_{jl}.$$

For $i = j$, we get

$$\sum_{l=1}^{m} b_{il}\, b_{il} = \text{degree}(x_i).$$

The sum is taken over all edges that are incident with the vertex i, as $A(G)$ has zeros on the diagonal.

For $i \neq j$, we get that the standard scalar product of row i with row j contributes a nonzero value if and only if the two rows have a nonzero entry at the same place, which gives -1 as the product. This column corresponds to an edge between the vertices i and j, so we have the (i, j)th entry of $-(A(G) + {}^t\!A(G))$.

If G is undirected, we get only the entries of $A(G)$ and no negative numbers. □

Theorem 5.3.2 (Sachs). *If G is a simple d-regular graph without loops and with n vertices and $m = \frac{1}{2}nd$ edges, then for $m \geq n$ we have*

$$\text{chapo}(LG; t) = (t + 2)^{m-n}\,\text{chapo}(G; t + 2 - d).$$

Proof. Define two square matrices with $n + m$ rows and columns:

$$U := \begin{pmatrix} tI_n & -B \\ 0 & I_m \end{pmatrix} \quad \text{and} \quad V := \begin{pmatrix} I_n & B \\ {}^t\!B & tI_m \end{pmatrix},$$

where B is the incidence matrix of G and ${}^t\!B$ its transpose; cf. Definition 2.2.1. Then

$$UV = \begin{pmatrix} tI_n - B\,{}^t\!B & 0 \\ {}^t\!B & tI_m \end{pmatrix}, \quad VU = \begin{pmatrix} tI_n & 0 \\ t\,{}^t\!B & tI_m - {}^t\!BB \end{pmatrix}.$$

As

$$\det(UV) = \det(U)\det(V) = \det(V)\det(U) = \det(VU),$$

we get

$$\det(UV) = |tI_n - B\,{}^tB|\,|tI_m| = \det(UV) = \det(VU) = |tI_n|\,|tI_m - {}^tBB|.$$

The equality of the determinants gives the equations

$$\begin{aligned} t^m\,|tI_n - B\,{}^tB| &= t^n\,|tI_m - {}^tBB| \\ \text{or equivalently,}\quad t^{m-n}\,|tI_n - B\,{}^tB| &\overset{(\spadesuit)}{=} |tI_m - {}^tBB|. \end{aligned}$$

With ${}^tBB \overset{(\heartsuit)}{=} A(LG) + 2I_m$ and $B\,{}^tB \overset{(\diamondsuit)}{=} D(G) + A(G)$ (see Proposition 5.3.1), we calculate that

$$\begin{aligned} \operatorname{chapo}(LG;t) &= \det(tI_m - A(LG)) \\ &\overset{(\heartsuit)}{=} \det((t+2)I_m - {}^tBB) \\ &\overset{(\spadesuit)}{=} (t+2)^{m-n}\det((t+2)I_n - B\,{}^tB) \\ &\overset{(\diamondsuit)}{=} (t+2)^{m-n}\det((t+2-d)I_n - A(G)) \\ &= (t+2)^{m-n}\operatorname{chapo}(G;t+2-d). \end{aligned}$$

□

Corollary 5.3.3. *Let G be a d-regular graph with $m \geq n$ and spectrum*

$$\operatorname{Spec}(G) = \begin{pmatrix} \lambda_1 & \cdots & \lambda_{p-1} & d \\ m(\lambda_1) & \cdots & m(\lambda_p - 1) & 1 \end{pmatrix}.$$

Then

$$\operatorname{Spec}(LG) = \begin{pmatrix} -2 & \lambda_1 + d - 2 & \cdots & \lambda_{p-1} + d - 2 & 2d-2 \\ m-n & m(\lambda_1) & \cdots & m(\lambda_{p-1}) & 1 \end{pmatrix}.$$

Example 5.3.4 (Spectra of line graphs). The line graph LK_n is sometimes called a *triangle graph* and is denoted by Δ_n. Its vertices correspond to $n(n-1)/2$ pairs of numbers from the set $\{1, \dots, n\}$. Two vertices are adjacent if the corresponding pairs have a common member. The known spectrum of K_n and Theorem 5.3.2 imply that

$$\operatorname{Spec}(\Delta_n) = \begin{pmatrix} -2 & n-4 & 2n-4 \\ \frac{1}{2}n(n-3) & n-1 & 1 \end{pmatrix}.$$

We observe that

$$\operatorname{Spec}(\Delta_5) = \begin{pmatrix} -2 & 1 & 6 \\ 5 & 4 & 1 \end{pmatrix}.$$

Application of Theorem 5.1.3, taking into account that the Petersen graph $K_{5:2}$ has $n(n-1)/2 = 10$ vertices, gives

$$\operatorname{Spec}(K_{5:2}) = \begin{pmatrix} -2 & 1 & 3 \\ 4 & 5 & 1 \end{pmatrix}.$$

Theorem 5.3.5. *We always have $\lambda(LG) \geq -2$.*

Proof. The matrix ${}^tBB$ is positive semidefinite, since for all matrices of this form one has for the norm of Bz that

$$ {}^tz\,{}^tB\,Bz =: \|Bz\|^2 \geq 0 $$

for all $z \in \mathbb{R}^n$. This means that the eigenvalues of ${}^tBB$ are nonnegative. Now $A(LG) = {}^tBB - 2I_m$ implies that all eigenvalues of this matrix are greater than or equal to -2, as $({}^tBB - 2I_m)v = {}^tBBv - 2v = \lambda v - 2v = (\lambda - 2)v$ if λ is an eigenvalue of ${}^tBB$. □

Which graphs are line graphs?

Using the preceding theorem, we can conclude that G is not a line graph if $\lambda(G) < -2$. There also exist graphs with $\lambda(G) = -2$ which are not line graphs—one example is the Petersen graph; cf. [Biggs 1996], 3b on page 20, see also Exerceorem 8.5.3.

More generally, there is a characterization of line graphs by nine forbidden subgraphs with at most six vertices each; see L. W. Beineke [9].

Theorem 5.3.6. *A graph is a line graph if and only if it does not contain one of the following graphs as a strong subgraph.*

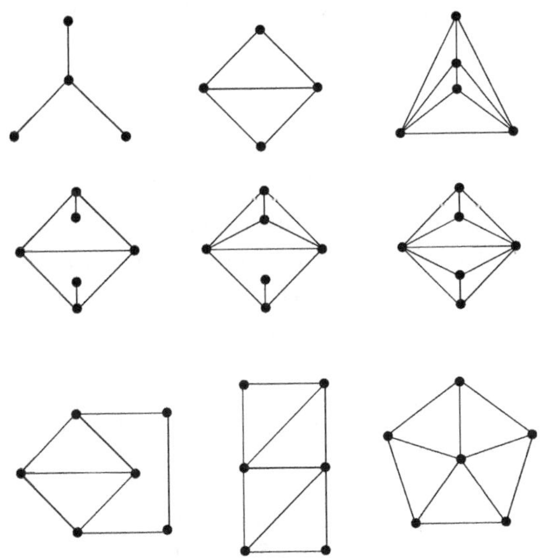

Remark 5.3.7. A connected, d-regular graph G with $d \geq 17$ and $\lambda(G) = -2$ is either a line graph or $K_{2,\ldots,2}$; cf. [Behzad et al. 1979], who point to Hofmann and Ray-Chaudhuri without giving a reference. According to [Biggs 1996], page 21, there are seven d-regular graphs with $d < 17$ and smallest eigenvalue -2 which are not line

graphs: the Petersen graph, the four exceptions from Theorem 5.3.8, a 5-regular graph with 16 vertices, and a 16-regular graph with 37 vertices.

Theorem 5.3.8.

(1) *If* $\operatorname{Spec}(G) = \operatorname{Spec}(LK_p)$ *for* $p \neq 8$, *then* $G \cong LK_p$.
For $p = 8$, *there exist three exceptional graphs.*

(2) *If* $\operatorname{Spec}(G) = \operatorname{Spec}(LK_{p,p})$ *for* $p \neq 4$, *then* $G \cong LK_{p,p}$.
For $p = 4$, *there exists one exceptional graph.*

For (1), see also J. Hoffman [39].
For (2) compare the following example.

Example 5.3.9 (The exceptional graph with $p = 4$ in Theorem 5.3.8, Shrikhande graph). In this graph, figure below, the first vertex in the upper row is identified with the first vertex in the bottom row, and so on; also, every vertex in the slanted line on the right is identified with the corresponding vertex in the left slanted line. This graph G then has 16 vertices and the same spectrum as $LK_{4,4}$. It is clear that the two are not isomorphic since $LK_{4,4} \cong K_4 \square K_4$, which has several copies of C_4; this is not the case in G.

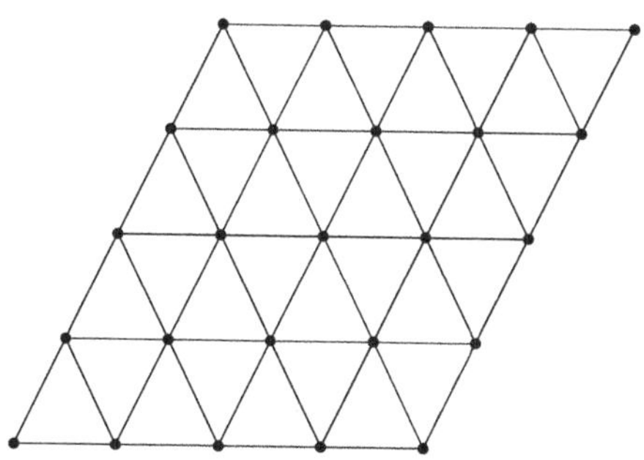

$\operatorname{Spec}(G) = \operatorname{Spec}(LK_{4,4})$

$$\operatorname{Spec}(K_{4,4}) \overset{\text{Example 2.5.9}}{=} \begin{pmatrix} -4 & 0 & 4 \\ 1 & 6 & 1 \end{pmatrix},$$

$$\begin{aligned} \operatorname{Spec}(LK_{4,4}) &\overset{\text{Theorem 5.3.2}}{=} \begin{pmatrix} -2 & -4+4-2 & 0+4-2 & 2-4-2 \\ 16-8 & 1 & 6 & 1 \end{pmatrix} \\ &= \begin{pmatrix} -2 & -2 & 2 & 6 \\ 8 & 1 & 6 & 1 \end{pmatrix} \\ &= \begin{pmatrix} -2 & 2 & 6 \\ 9 & 6 & 1 \end{pmatrix}. \end{aligned}$$

Compare with [Biggs 1996], page 21, and S. S. Shrikhande [85].

Remark 5.3.10. Let G be a connected, d-regular multigraph with n vertices and m edges, and let (λ, μ) be a pair of corresponding eigenvalues of G and LG. Then the incidence matrix $B(G)$ maps the eigenspace $\mathrm{Eig}(LG,\mu)$ onto the eigenspace $\mathrm{Eig}(G,\lambda)$ and ${}^tB(G)$ maps $\mathrm{Eig}(G,\lambda)$ onto $\mathrm{Eig}(LG,\mu)$; cf. [Cvetković et al. 1979] Theorem 3.36.

5.4 The total graph

This unary construction is based on the construction of the line graph. The total graph is a combination of the graph G and the line graph LG seen from the vertex set and from the edge set, plus some additional edges which form the third set in the edge set of the following definition.

Definition 5.4.1. The graph $TG = (V \bigcup E, E \bigcup E(LG) \bigcup \{\{v,e\} \mid v \in e\})$ is called the ***total graph*** of G.

Remark 5.4.2 (Total graphs of directed graphs). First, take into account the various possibilities for the line graph of directed graphs; see Remark 5.2.3. For the existence of an edge (v,e) or (e,v), we can now require that $v = t(e)$ or $v = o(e)$.

Exerceorem 5.4.3 (The matrix of a total graph). For a graph G, one has

$$A(TG) = \begin{pmatrix} A(G) & B(G) \\ {}^tB(G) & A(LG) \end{pmatrix}.$$

For a d-regular graph G (where $d > 1$) with n vertices, m edges and eigenvalues λ_i, $i = 1,\dots,n$, this implies that TG has $m-n$ eigenvalues -2 and the $2n$ eigenvalues $\frac{1}{2}(2\lambda_i + d - 2 \pm \sqrt{4\lambda_i + d^2 + 4})$, $i = 1,\dots,n$ (cf. [Cvetković et al. 1979] Theorem 2.20).

Example 5.4.4 (Total graph).

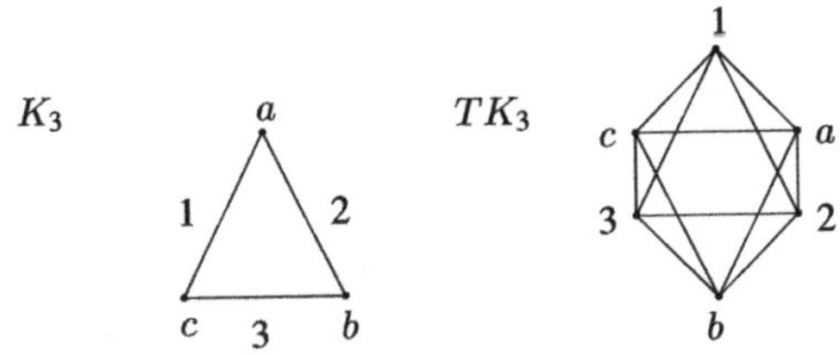

Note that TK_3 is the octahedron graph.

It is clear that TG always contains G and LG as subgraphs.

Exercise 5.4.5 (The total functor). Convince yourself that T becomes a covariant functor from the category ***Gra*** into the category ***EGra*** upon defining $Tf : TG \to TG'$ for $f : G \to G'$ by $Tf((v,w)) = (f(v), f(w))$ for $v, w \in V(G)$.

Question. Which properties shown for the line graph in the previous section can be generalized to the total graph?

5.5 The tree graph

This final unary construction gives a "graph from certain subgraphs of a graph."

Definition 5.5.1. Let $T_1, \dots, T_\ell$ denote all spanning trees of graph G on n-vertices. The ***(spanning) tree graph*** $\operatorname{Tr} G$ is defined by $V(\operatorname{Tr} G) = \{T_1, \dots, T_\ell\}$ and $E(\operatorname{Tr} G) = \{\{T_i, T_j\} \mid |E(T_i) \cap E(T_j)| = n-2\}$, i. e., two trees are adjacent if they coincide except for one edge. See Figure 5.3 for an example.

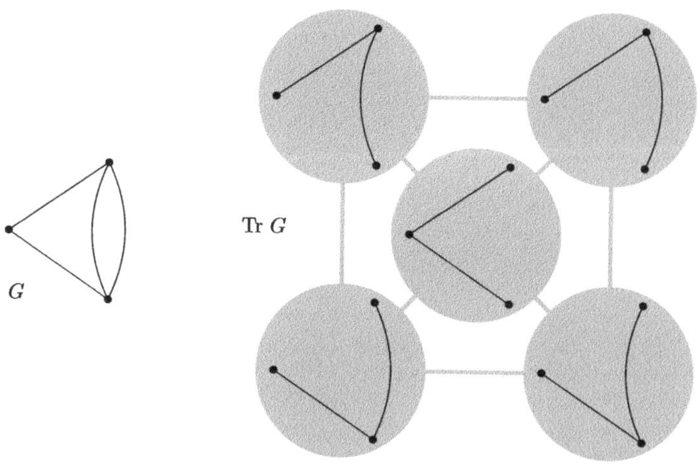

Figure 5.3: A graph G and its tree graph $\operatorname{Tr} G$.

Exercise 5.5.2 (The spanning tree functor). Interpret Tr as a functor from the category ***Gra*** into the category ***EGra*** by defining $\operatorname{Tr} f$. If, in the above example, we consider the mapping f which takes G onto K_3, then this implies that $\operatorname{Tr} G$ is mapped onto K_3. This means that $\operatorname{Tr} f$ is in ***EGra***. Note that for the homomorphisms f in this case, where G has multiple edges, we need a homomorphism concept which also takes care of edges like in Definition 1.4.1.

In general, under the functor Tr, different graph homomorphisms do not stay different, i. e., $\operatorname{Tr} f = \operatorname{Tr} g$ in ***EGra*** is possible even though $f \neq g$ in ***Gra***. This means that the functor is not faithful. Moreover, this functor does not preserve different objects, i. e., it is not injective on objects.

5.6 Comments

As mentioned earlier, it might be interesting to study unary operations as functors. In certain cases it will require some effort to define the appropriate categories; but apart from that, preservation and reflection of properties can be investigated.

On the non-categorical level, it could be interesting to study how properties of the total graph depend on the respective properties of the original graph. There is a monograph devoted to coloring questions in this context; see [Yap 1996].

After investigating determinants and permanents for graphs as mentioned in the Comments section of Chapter 2, it would be interesting to then examine these concepts for line graphs and total graphs.

6 Graphs and vector spaces

In this chapter, we use linear algebra to construct vector spaces from graphs and connect them by linear mappings. In the last four sections of this chapter, we give some applications to voltage and current problems.

Take a field F and a directed graph $G = (V, E)$ with $|V| = n$ and $|E| = m$. As for sets, we define

$$F^V := \{f : V \to F \mid f \text{ is a mapping}\}.$$

Since F is a field, addition and multiplication in F induce an addition and a scalar multiplication on F^V: for $g, h \in F^V$, we set $(f + g)(v) := f(v) + g(v)$ and $(kf)(v) := kf(v)$ for all $v \in V$ and $k \in F$. In this way, F^V becomes an F-vector space.

We denote by δ_{ij} the ***Kronecker symbol***, such that $\delta_{ij} = 0$ if $i \neq j$ and $\delta_{jj} = 1$.

6.1 Vertex space and edge space

We start with two vector spaces associated with every graph—the cycle space and the cocycle space. For undirected graphs, these vector spaces are considered over the two-element field $F_2 = \{0, 1\}$, where $1 + 1 = 0$. For directed graphs, we choose an arbitrary field F with characteristic zero, e. g., the real numbers $\mathbb{R}$.

Definition 6.1.1. The ***vertex space*** of $G = (V, E)$ over F is defined as $C_0(G) := F^V$ with operations induced by F. An element of $C_0(G)$ is called a 0-***chain*** (0-***simplex***).

The ***edge space*** of G over F is defined as $C_1(G) := F^E$, again with operations induced by F. An element of $C_1(G)$ is called a 1-***chain*** (1-***simplex***).

An arbitrary element of the vertex space is a formal linear combination of the vertices. For a vertex set $U \subseteq V$, the corresponding element in F^V is the indicator function $V \to F$, which assigns 1 to the vertices of U and 0 to the other vertices. The neutral element of $C_0(G)$ is the empty vertex set $\emptyset$.

Theorem and Definition 6.1.2. A basis of $C_0(G)$ is $(f_i)_{i=1,\dots,n}$ where $f_i \in C_0(G)$ with $f_i(x_j) = \delta_{ij}$ for $x_j \in V$, $i, j = 1, \dots, n$, and $\dim_F(C_0(G)) = |V| = n$. This basis is called the ***standard vertex basis***.

In an analogous way, we define the ***standard edge basis*** $(g_j)_{j=1,\dots,m}$, and we have $\dim_F(C_1(G)) = |E| = m$.

Proof. It is clear that we have minimal generating systems. □

https://doi.org/10.1515/9783110617368-006

Notation 6.1.3. For $V(G) = \{x_1,\ldots,x_n\}$ and $E(G) = \{e_1,\ldots,e_m\}$, we can write the elements of $f \in C_0(G)$ and $g \in C_1(G)$ as follows:

$$f = \sum_{i=1}^{n} \lambda_i f_i \quad \text{or} \quad f = (\lambda_1,\ldots,\lambda_n) \quad \text{with } \lambda_i = f(x_i) \in F \text{ for } x_i \in V(G),$$

$$g = \sum_{j=1}^{m} \mu_i e_j \quad \text{or} \quad g = (\mu_1,\ldots,\mu_m) \quad \text{with } \mu_j = g(e_j) \in F \text{ for } e_j \in E(G).$$

The boundary and co.

The following two linear mappings relate the vertex and edge spaces. Moreover, they have a representation by matrices already introduced.

Definition 6.1.4. The ***boundary operator*** $\partial : C_1(G) \to C_0(G)$ is defined by linear extension of

$$\partial(e) = o(e) - t(e) \text{ for } e \in E \quad \text{to} \quad C_1(G).$$

We call $\partial(g) := \sum_{e_j \in E} \mu_j \partial(e_j)$ the ***boundary*** of $g = \sum_{e_j \in E} \mu_j e_j$ in C_1.

The ***coboundary operator*** $\partial^* : C_0(G) \to C_1(G)$ is defined by linear extension of

$$\partial^*(x) := \sum_{j=1}^{m} \epsilon_j e_j \text{ where, for } x \in V,\ \epsilon_j = \begin{cases} 1 & \text{if } x = o(e_j) \\ -1 & \text{if } x = t(e_j) \\ 0 & \text{otherwise} \end{cases} \quad \text{to} \quad C_0(G).$$

We call $\partial^*(f) := \sum_{x_i \in V} \lambda_i \partial^*(x_i)$ the ***coboundary*** of $f = \sum_{x_i \in V} \lambda_i x_i$ in C_0.

The boundary operator takes 1-chains to 0-chains; the coboundary operator takes 0-chains to 1-chains.

Example 6.1.5 (Standard bases, boundary, and coboundary, directed).

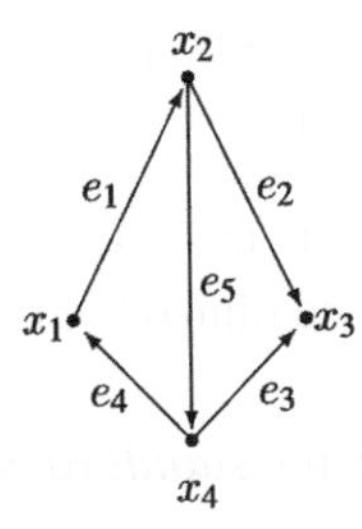

Standard vertex basis: (1, 0, 0, 0), (0, 1, 0, 0), (0, 0, 1, 0), (0, 0, 0, 1).
Standard edge basis: (1, 0, 0, 0, 0), …, (0, 0, 0, 0, 1).

We have

$$\begin{aligned}\partial(e_5) &= x_2 - x_4 = (0,1,0,-1),\\ \partial(e_2+e_3+e_5) &= 2x_2 - 2x_3 = (0,2,-2,0),\\ \partial(-e_2+e_3+e_5) &= 0.\end{aligned}$$

The kernel of ∂ corresponds to the directed cycles.
The image of ∂^* corresponds to the coboundaries.
We have

$$\begin{aligned}\partial^*(x_1) &= e_1 - e_4 = (1,0,0,-1,0),\\ \partial^*(x_2) &= -e_1 + e_2 + e_5 = (-1,1,0,0,1),\\ \partial^*(x_1+x_2) &= e_2 - e_4 + e_5 = (0,1,0,-1,1).\end{aligned}$$

Example 6.1.6 (Boundary and coboundary, over F_2).

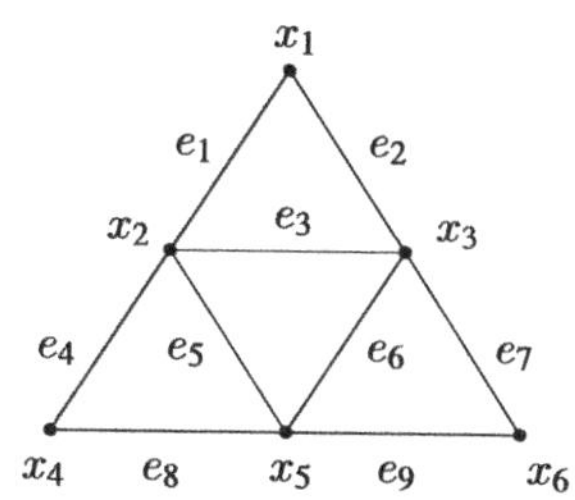

The 1-chain $\sigma_1 = e_1 + e_2 + e_4 + e_9$ has the boundary

$$\partial(\sigma_1) = (x_1+x_2) + (x_1+x_3) + (x_2+x_4) + (x_5+x_6) = x_3 + x_4 + x_5 + x_6.$$

The 0-chain $\sigma_0 = x_3 + x_4 + x_5 + x_6$ has the coboundary

$$\begin{aligned}\partial^*(\sigma_0) &= (e_2+e_3+e_6+e_7) + (e_4+e_8) + (e_5+e_6+e_8+e_9) + (e_7+e_9)\\ &= e_2 + e_3 + e_4 + e_5.\end{aligned}$$

Matrix representation

As is usual in linear algebra, we define the matrix of a linear mapping with respect to given bases.

Theorem 6.1.7. *Let B_0 and B_1 denote the standard bases of $C_0(G)$ and $C_1(G)$. Then the representing matrix of ∂ is the incidence matrix $B(G)$, and the representing matrix of ∂^* is ${}^tB(G)$, the transpose of $B(G)$; i. e., in the usual linear algebra notation,*

$$M_{B_0}^{B_1}(\partial) = B(G) \quad \textit{and} \quad M_{B_1}^{B_0}(\partial^*) = {}^tB(G).$$

Proof. The column i of $B(G)$ indicates the start and end vertices of the edge i. This means that it contains the coordinates with respect to B_0 of the image of the ith basis vector of B_1 under ∂.

The row j of $B(G)$, which is the same as the column j of ${}^tB(G)$, represents the edges of G which start from the vertex j by $+1$ and those which end in this vertex by -1. This means that it contains the coordinates with respect to B_1 of the image of the jth basis vector of B_0 under ∂^*. □

Example 6.1.8 (Matrix representation of ∂ for the graph in Example 6.1.5).

$$M_{B_0}^{B_1}(\partial) = B(G) = \begin{pmatrix} 1 & 0 & 0 & -1 & 0 \\ -1 & 1 & 0 & 0 & 1 \\ 0 & -1 & -1 & 0 & 0 \\ 0 & 0 & 1 & 1 & -1 \end{pmatrix}.$$

6.2 Cycle spaces, bases & Co.

The following definitions of the cycle space and the cocycle space are based on the possibility of using edges opposite in direction. We describe this using the notion of orientation.

Definition 6.2.1. Let $G = (V, E, o, t)$ be a directed graph, and let $E' \subseteq E$. A mapping $\operatorname{dir} : E' \to V \times V$ is called an ***orientation*** of E' if for $e \in E'$ we set

$$\operatorname{dir}(e) = (o(e), t(e)) \quad \text{or} \quad \operatorname{dir}(e) = (t(e), o(e)).$$

The cycle space

Let $L = \{e_{i_1}, \dots, e_{i_p}\}$ be a semicycle in G. Choose an orientation dir on L such that $\operatorname{dir}(L) = \{\operatorname{dir}(e_{i_1}), \dots, \operatorname{dir}(e_{i_p})\}$ is a cycle, and define

$$z_{\operatorname{dir}(L)} : \begin{cases} E \to F \\ e \quad \mapsto \begin{cases} 1 & \text{if } e \in L,\ e = \operatorname{dir}(e), \\ -1 & \text{if } e \in L,\ e \neq \operatorname{dir}(e), \\ 0 & \text{if } e \notin L. \end{cases} \end{cases}$$

The subspace generated,

$$Z(G) := \operatorname{span}\{z_{\operatorname{dir}(L)} \mid \operatorname{dir}(L) \text{ is an oriented semicycle in } G\} \subseteq C_1(G),$$

is called the ***cycle space*** of G.

For an orientation dir on E' and $e \in E'$, one has

$$\partial(e) = \partial(\mathrm{dir}(e)) \quad \text{or} \quad \partial(e) = -\partial(\mathrm{dir}(e)).$$

Corollary 6.2.14 will imply that the elements of $Z(G)$ are exactly the semicycles.

Lemma 6.2.2. *A semicircuit Z_1 is not a linear combination of other semicircuits in $Z(G)$ if it contains an edge which does not appear in any other semicircuit (and not only in this case). There always exists a basis of semicircuits for $Z(G)$.*

Proof. There exists a generating system for $Z(G)$ of semicircuits; as every semicycle is the union of semicircuits by Lemma 1.1.4, i. e., it belongs to $Z(G)$, it must be the sum of the corresponding elements in $Z(G)$. In Example 6.2.4, the semicircuit consisting of the edges e_3, e_6, e_9 is not a linear combination of the others, even though each of its edges appears also in another semicircuit. □

Definition 6.2.3. The ***cycle rank*** (***cyclomatic number, Betti number***) $\xi(G)$ of G is defined to be $\xi(G) := \dim_F(Z(G))$.

Example 6.2.4 (Cycles, cycle rank).

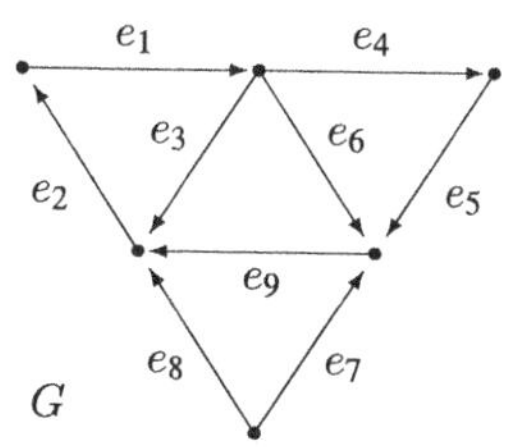

$$L_1 = (e_3, -e_8, e_7, -e_6)$$
$$L_2 = (-e_4, e_3, e_9, -e_5)$$

$$\begin{aligned}
z_{\mathrm{dir}(L_1)} &= (0,0,1,\ \ 0,\ \ 0,-1,1,-1,\ \ 0)\\
z_{\mathrm{dir}(L_2)} &= (0,0,1,-1,-1,\ \ 0,0,\ \ 0,\ \ 1)\\
z_{\mathrm{dir}(L_1)} + z_{\mathrm{dir}(L_2)} &= (0,0,2,-1,-1,-1,1,-1,-1)\\
z_{\mathrm{dir}(L_1)} - z_{\mathrm{dir}(L_2)} &= (0,0,0,\ \ 1,\ \ 1,-1,1,-1,\ \ 1)
\end{aligned}$$

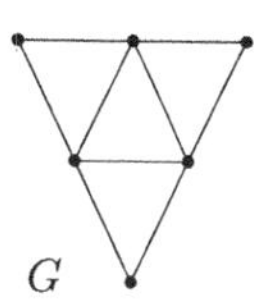
G

Basis of $Z(G)$

Cycle basis of $Z(G)$

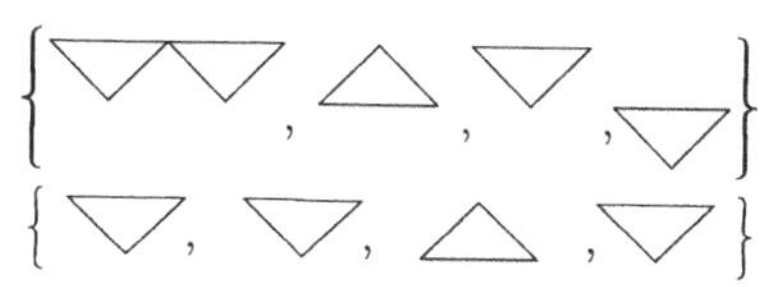

$$\xi(G) = 4.$$

Proposition 6.2.5. *Take $G = (V, E, o, t)$ with k weak components. Then*

$$\xi(G) \geq |E| - |V| + k.$$

In Corollary 6.2.14, it will be seen that we even have equality.

Proof. Every spanning forest has $|V| - k$ edges. Take one spanning forest. Adjoining one edge of G which does not belong to this forest gives exactly one semicircuit. For this we have $|E| - (|V| - k)$ possibilities. All of these are linearly independent by Lemma 6.2.2. (A spanning forest is the union of the spanning trees of the weak components of G.) □

The cocycle space

Definition 6.2.6. Let $G = (V, E, o, t)$ be a graph. The set of all edges of G connecting V_1 and V_2, for a given partition $V = V_1 \bigcup V_2$, is called a ***semicocycle (separating edge set, cut)*** of G. A minimal semicocycle is called a ***semicocircuit*** of G.

So a semicocycle (or separating edge set) S of a connected graph G is a set such that $G \setminus S$ is not connected. A semicocircuit is a minimal separating edge set.

Example 6.2.7 (Cut).

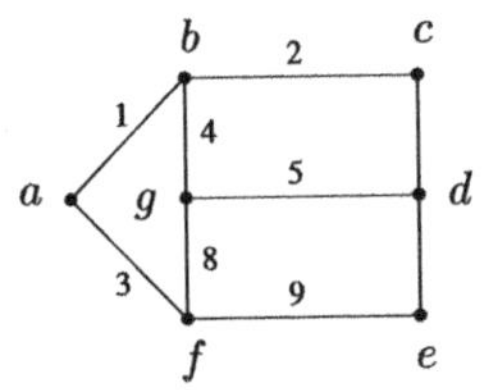

Take $V_1 = \{a\}$, V_2 to be the rest; then $\{1, 3\}$ is a cut.

Take $V_1 = \{b, f\}$, V_2 to be the rest; then $\{1, 2, 3, 4, 8, 9\}$ is a non-minimal cut.

Definition 6.2.8. Let $U = \{e_{i_1}, \ldots, e_{i_r}\}$ be a semicocycle in G with partition $V_1 \bigcup V_2 = V$. Choose an orientation dir on U such that in $\mathrm{dir}(U)$ all edges have the same direction (from V_1 to V_2, say). Define

$$s_{\mathrm{dir}(U)} : \begin{cases} E \to F \\ e \quad \mapsto \begin{cases} 1 & \text{if } e \in U,\ e = \mathrm{dir}(e), \\ -1 & \text{if } e \in U,\ e \neq \mathrm{dir}(e), \\ 0 & \text{if } e \notin U. \end{cases} \end{cases}$$

The subspace generated,

$$S(G) := \mathrm{span}\{s_{\mathrm{dir}(U)} \mid \mathrm{dir}(U) \text{ is an oriented semicocycle in } G\} \subseteq C_1(G),$$

is called the ***cocycle space*** of G.

Lemma 6.2.9. *A semicocircuit is not a linear combination of other semicocircuits (in $S(G)$) if it contains an edge which lies in no other semicocircuit (and not only in this case). There always exists a basis for $S(G)$ of semicocircuits.*

Definition 6.2.10. The ***cocycle rank*** (***cocyclomatic number***) $\xi^*(G)$ is defined by

$$\xi^*(G) := \dim_F S(G).$$

Proposition 6.2.11. *Let $G = (V, E, p)$ be a graph with k weak components. Then*

$$\xi^*(G) \geq |V| - k.$$

Proof. Every spanning forest has $|V| - k$ edges. Each of these (together with suitable other edges) defines a cut. By Lemma 6.2.9, they are all linearly independent. □

Again we even have equality, as we shall see in Corollary 6.2.14.

Example 6.2.12 (Cocycles, cocycle rank). Consider again Example 6.1.5 and one graph from Example 6.2.4.

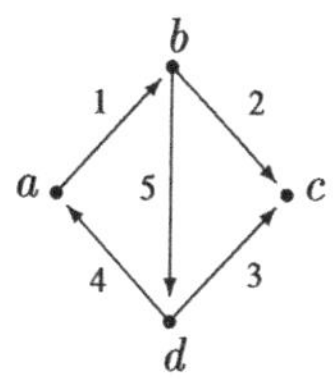

$V_1 = \{a\}, \quad U_1 = \{1, 4\}, \quad S_{\mathrm{dir}(U_1)} = (1, 0, 0, -1, 0)$

$V_1 = \{a, d\}, \quad U_2 = \{1, 3, 5\}, \quad S_{\mathrm{dir}(U_2)} = (1, 0, 1, 0, -1)$

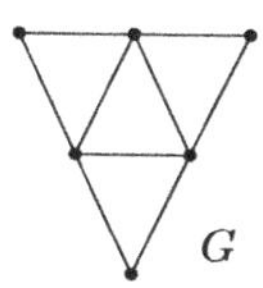

$\xi^*(G) \geq 6 - 1 = 5$

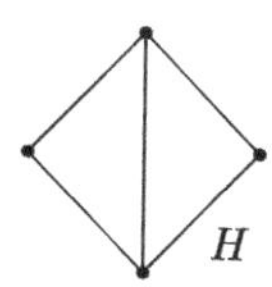

$\xi^*(H) \geq 4 - 1 = 3$

Orthogonality

Now we need the concept of orthogonality in the usual sense. Recall that for two coordinate vectors $v = (v_1, \ldots, v_n)$ and $w = (w_1, \ldots, w_n)$ in $\mathbb{R}^n$, the standard scalar product is defined by $\langle v, w \rangle := v_1 w_1 + \cdots + v_n w_n$.

Two vectors $v, w \in \mathbb{R}^n$ are said to be *orthogonal*, written as $v \perp w$, if $\langle v, w \rangle = 0$, where $\langle \, , \, \rangle$ denotes the standard scalar product in $\mathbb{R}^n$. For $U \subseteq \mathbb{R}^n$, we call $U^\perp := \{w \in \mathbb{R}^n \mid \langle u, w \rangle = 0, \ u \in U\}$ the *orthogonal complement* of U in $\mathbb{R}^n$. The zero vector is thus orthogonal to every vector.

Note that if we now consider vector spaces over $\mathbb{R}$, we get

$$C_1(G) = \mathrm{span}\{e_1, \ldots, e_m\} \cong \mathbb{R}^m.$$

Theorem 6.2.13. *With respect to the standard scalar product in $C_1(G)$, one has*

$$Z(G)^\perp = S(G).$$

Proof. We show that $Z(G)^{\perp} \supseteq S(G)$, i. e., for all $z_L \in Z(G)$ one has $\langle z_L, s_U \rangle = 0$ for all $s_U \in S(G)$, where L is a semicycle and U a semicocycle. Only those edges that belong to L and U contribute nonzero summands to the scalar product. We consider the following situation, where e_1 lies in U and L, and U separates V_1 and V_2.

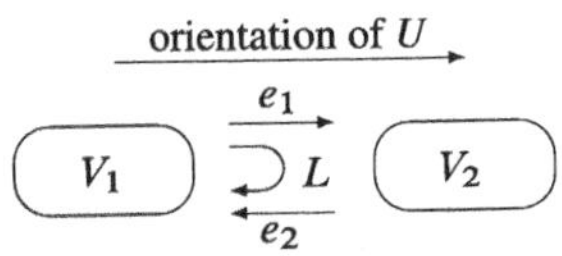

As L is a semicycle, there exists an edge e_2 with the given orientation. Otherwise, e_1 would have to be used twice and then the e_1th coordinate in the vector of L would have the value 0 (once with 1 and once with −1). Because of the orientation of U, the edge e_2 contributes to the scalar product $\langle z_L, z_U \rangle$ the summand −1 if e_1 gives the summand 1 (and vice versa). The same is true of all edges between V_1 and V_2, i. e., for every edge e_1 in U and L with summand 1 in the scalar product there exists an edge in the scalar product with summand −1. Thus $S(G) \subseteq Z(G)^{\perp}$.

To prove $S(G)^{\perp} \subseteq Z(G)$, we proceed as follows. As $C_1(G)$ is finite-dimensional, we get $S(G) \coprod S(G)^{\perp} \cong C_1(G)$. Thus

$$\begin{aligned} \dim S(G) + \dim S(G)^{\perp} &= m = |E|\,, \\ \dim S(G) + \dim Z(G) &\leq |E|\,. \end{aligned}$$

Lemmas 6.2.2 and 6.2.9 imply that

$$\dim S(G) + \dim Z(G) \geq |V| - k + |E| - |V| + k = |E|\,.$$

Therefore, we have equality! Consequently,

$$\begin{aligned} \dim Z(G) &= |E| - |V| + k, \\ \dim S(G) &= |V| - k, \\ S(G)^{\perp} &= Z(G). \end{aligned}$$

□

Corollary 6.2.14. *For graphs G with k components, we have*
(1) $C_1(G) \cong Z(G) \coprod S(G)$;
(2) $\xi(G) = |E| - |V| + k$;
(3) $\xi^*(G) = |V| - k$.

Definition 6.2.15. Let $G = (V, E, p)$ be a graph with k components, and let T be a spanning forest of G. Each of the $|V| - k$ edges of T defines a cocircuit. We call this a ***fundamental cocycle***. These cocycles form a basis of $S(G)$, a so-called ***cocycle basis***. Each of the $|E| - |V| + k$ edges of G which do not lie on T define a circuit, which is called a ***fundamental cycle***. These cycles form a basis of $Z(G)$, a so-called ***cycle basis*** in G.

The boundary operator & Co.

According to the next lemma, the elements of $Z(G)$ are closed semipaths.

Lemma 6.2.16. *The elements of $Z(G)$ are 1-chains with boundary 0; i. e.,*

$$Z(G) \subseteq \ker \partial = \{z \in Z(G) \mid \partial(z) = 0\}.$$

Proof. For oriented semicycles $z \in Z(G)$, one has $\partial(z) = 0$; similarly for linear combinations. □

Lemma 6.2.17. *The elements of $S(G)$ are coboundaries of 0-chains; i. e.,*

$$S(G) \subseteq \operatorname{Im} \partial^* = \{\partial^*(x) \mid x \in C_0(G)\} =: \operatorname{coker} \partial^*.$$

Proof. Let U be a fundamental cocircuit which separates V_1 and V_2, i. e., an element of a basis of $S(G)$, and let dir be an orientation. Consider

$$\partial^*\Big(\sum_{x_i \in V_1} x_i\Big) = \sum_{x_i \in V_1} \partial^*(x_i) = \sum_{e \in E} \mu_e e = s_{\operatorname{dir}(U)},$$

where

$$\mu_e = \begin{cases} +1 & \text{if } e \in U \text{ starts in } V_1, \\ -1 & \text{if } e \in U \text{ ends in } V_1, \\ 0 & \text{if } e \notin U. \end{cases}$$

□

Theorem 6.2.18. *The elements of $Z(G)$ are exactly the 1-chains with boundary 0; i. e.,*

$$Z(G) = \ker \partial.$$

The elements of $S(G)$ are exactly the coboundaries of 0-chains of G; i. e.,

$$S(G) = \operatorname{Im} \partial^* = \operatorname{coker} \partial^*.$$

Proof. As always in vector spaces, we have $\dim(\ker \partial) + \dim(\operatorname{Im} \partial) = \dim(C_1(G))$ for $\partial : C_1(G) \to C_0(G)$. For the ranks of the matrices, we have

$$\operatorname{rank}(B) = \operatorname{rank}({}^tB) = \dim(\operatorname{Im} \partial^*).$$

Thus $\dim(\ker \partial) + \dim(\operatorname{Im} \partial^*) = \dim(C_1(G))$. By virtue of Corollary 6.2.14, we have

$$Z(G) \coprod S(G) \cong C_1(G).$$

This implies the statement with Lemmas 6.2.16 and 6.2.17. □

Example 6.2.19 (Cycle rank).

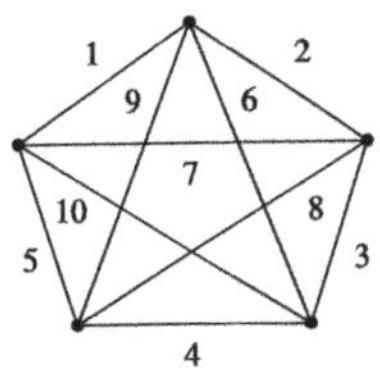

$$\dim(C_1(K_5)) = 10,$$

$$\dim(Z(K_5)) = 6 = \xi(K_5),$$

$$\dim(S(K_5)) = 4.$$

Exercise 6.2.20. Prove that $\xi(K_{3,3}) = 4$.

6.3 Application: MacLane's planarity criterion

In 1937, Saunders MacLane gave an algebraic characterization of planar graphs, which relies on an algebraic analysis of the boundary circuits of the regions in a *plane* graph. Plane graphs are graphs embedded in the plane such that edges intersect only in vertices. Graphs having such an embedding are said to be planar.

We recall that a graph is *planar* if and only if it does not contain a subgraph that is a subdivision of K_5 or $K_{3,3}$ (Kuratowski), or it does not contain K_5 or $K_{3,3}$ as a minor (Wagner); cf. Theorem 1.8.2. We will also use Euler's formula $|V| - |E| + |F| = 2$ for plane graphs $G = (V, E)$, where $|F|$ denotes the number of regions/faces of G, including the unbounded region; see Theorem 1.8.4.

Definition 6.3.1. A basis $\{C_1, \ldots, C_r\}$ of $Z(G)$ is called a ***two-cycle basis*** if every $e \in E$ appears in at most two of the C_i.

Example 6.3.2 (No two-cycle basis). The circuits $(1, 6, 19)$, $(2, 8, 9)$, $(7, 3, 10)$, $(8, 6, 4)$, $(5, 7, 8)$ in Example 6.2.19 are linearly independent but do not form a basis as $\dim(K_5) = 6$. A sixth circuit for a two-cycle basis must contain the edges $1, \ldots, 5$. This then has to be a linear combination of the above five circuits. Thus there does not exist a two-cycle basis.

Example 6.3.3 (Two-cycle basis). We show K_4 and a two-cycle basis of it:

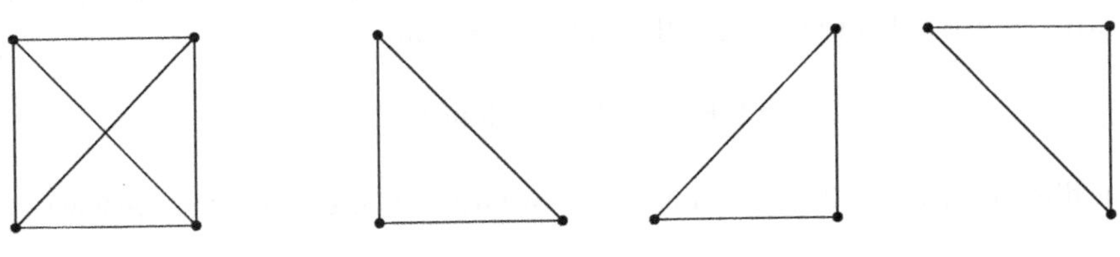

Lemma 6.3.4. *Let $\{D_1, \ldots, D_r\}$ be a basis of $Z(G)$ over $\mathbb{Z}_2$. Then there exist circuits $C_i \subseteq D_i$, $i \in \{1, \ldots r\}$, such that $\{C_1, \ldots, C_r\}$ is again a cycle basis of $Z(G)$.*

Proof. Let D_1 be an edge disjoint union of circuits, say $D_1 = C_1' \cup \ldots \cup C_t'$. Not all of the C_i' can be represented using $D_2, \ldots, D_r$, so there exists a circuit, say C_1', which is not a linear combination of $D_2, \ldots, D_r$. We form a new basis $C_1', D_2, \ldots, D_r$. By continuing in this way, we obtain a basis as desired. □

Lemma 6.3.5. *Take G_1 and G_2 with $|V_1 \cap V_2| \leq 1$, and let $\mathcal{B}_1$ and $\mathcal{B}_2$ be two bases of $Z(G_i)$. Then $\mathcal{B}_1 \cup \mathcal{B}_2$ is a basis of $Z(G)$, where $G = G_1 \coprod_{V_1 \cap V_2} G_2$ is an amalgam, and we have $Z(G) = Z(G_1) \coprod Z(G_2)$.*

Proof. We have the following situation:

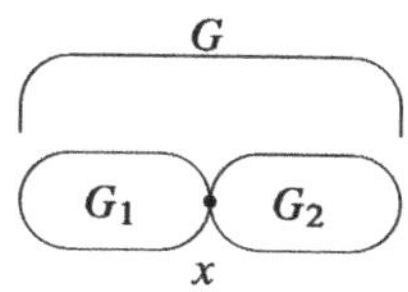

The statement about G is now clear; the statement about $Z(G)$ comes from linear algebra. □

Theorem 6.3.6 (MacLane). *The graph $G = (V, E)$ is planar if and only if $Z(G)$ as a vector space over $\mathbb{Z}_2$ has a two-cycle basis.*

Proof. By Lemma 6.3.4, we may assume that G is at least 2-connected.

For "⇒", let $R_1, \ldots, R_r$ be inner regions of a plane embedding of G and let $C_1, \ldots, C_r$ be the corresponding boundary circuits. Each $e \in E$ appears in at most two of the C_i. By Corollary 6.2.14, we get

$$\xi(G) = |E| - |V| + 1.$$

We have $|R| = r + 1$, taking into account also the exterior region. From Euler's formula, we get

$$|E| - |V| + 1 = r = \xi(G).$$

It remains to show that $C_1, \ldots, C_r$ is a generating system in $Z(G)$. For an arbitrary $C \in Z(G)$, suppose that $C_{i_1}, \ldots, C_{i_s}$ are those of the $C_j \in \{C_1, \ldots, C_r\}$, which lie in the inner region of C (possibly including C itself). Then

$$C = \sum_{l=1}^{s} C_{i_l},$$

since the edges of a circuit C_i belong to two of these C_{i_l} exactly if they do not lie on C. So the $C_1, \ldots, C_r$ generate $Z(G)$. Putting the above facts together, we see that $C_1, \ldots, C_r$ is a two-cycle basis of $Z(G)$.

For "⇐," let $C_1, \ldots, C_r$ be a two-cycle basis.

We show in two steps that for every $e \in E$, $Z(G \backslash e)$ also has a two-cycle basis.

1. If e is contained in two of the C_i, say C_1 and C_2, then $C_1+C_2, C_3, \dots, C_r$ is a two-cycle basis of $Z(G\backslash e)$.
2. If e is contained in only one of the C_i, say C_1, then $C_2, \dots, C_r$ is a two-cycle basis of $Z(G\backslash e)$.

In the first case, every circuit $C \subseteq G\backslash e$ in its linear representation by $C_1, \dots, C_r$ contains either none or both of the circuits C_1 and C_2. In the second case, the representation of C does not contain C_1. For both cases, 2-connectivity of G is used.

If G were not planar, then G would contain a subdivision of K_5 or of $K_{3,3}$. Using the fact that $Z(G\backslash e)$ has a two-cycle basis for all $e \in E$, this leads to a contradiction as follows. Observe that if a subdivision of a graph has a 2-cycle basis, than the graph itself has one, too. Hence, let $C_1, \dots, C_r$ be a two-cycle basis for $K_{3,3}$ or K_5. Consider

$$C_0 := \sum_{i=1}^{r} C_i.$$

Then $C_0 \subseteq Z(G)$, $C_0 \neq \emptyset$, is the set of edges which lie in exactly one of the C_i, for $1 \leq i \leq r$. Moreover, C_0 is itself a union of cycles. But for K_5, we have $|C_0| \geq 3$ and for $K_{3,3}$ we have $|C_0| \geq 4$.

By Lemma 6.2.2, we have

$$\xi(K_5) = 6, \quad \xi(K_{3,3}) = 4$$

(see also Example 6.2.19). This implies the following contradictions. For K_5, we have

$$6 \cdot 3 \leq \sum_{i=1}^{6} |C_i| = 2\,|E| - |C_0| = 20 - |C_0| \leq 17$$

(in the first place we have equality if all the C_i are triangles), and for $K_{3,3}$ we have

$$4 \cdot 4 \leq \sum_{i=1}^{4} |C_i| = 2\,|E| - |C_0| = 18 - |C_0| \leq 14. \qquad \square$$

6.4 Homology of graphs

We now take one more step toward abstraction in the direction of algebraic topology. We do this to obtain another view on direct decompositions of the edge space and vertex space of a graph. This section leads away from graphs; it can safely be skipped and returned to later as needed.

First, we recall the situation for arbitrary vector spaces over a field F.

Exact sequences of vector spaces

Definition 6.4.1. Consider the F-vector spaces $V_0, \dots, V_r$ and the linear mappings $f_1, \dots, f_r$ such that

$$V_r \xrightarrow{f_r} V_{r-1} \xrightarrow{f_{r-1}} \cdots \longrightarrow V_2 \xrightarrow{f_2} V_1 \xrightarrow{f_1} V_0.$$

This sequence is called an ***exact sequence*** if for all $i = 1, \dots, r-1$ one has $\operatorname{Im} f_{i+1} = \ker f_i$.

Now let $V \xrightarrow{f} W \longrightarrow 0$ (or $0 \longrightarrow V \xrightarrow{f} W$) be an exact sequence such that there exists a linear mapping $V \xleftarrow{g} W$ with $f \circ g = \mathrm{id}_W$ (or $g \circ f = \mathrm{id}_V$); then the sequence is said to be ***exact direct (split exact).***

Exercise 6.4.2. Let V and W be F-vector spaces. The sequence $0 \longrightarrow V \xrightarrow{f} W$ is exact if and only if f is injective. The sequence $W \xrightarrow{f} V \longrightarrow 0$ is exact if and only if f is surjective.

The next result explains the name "split exact." The proof follows directly from the definition of split exact.

Exercise 6.4.3. Let V and W be F-vector spaces. The sequence $0 \longrightarrow V \xrightarrow{f} W$ is split exact if and only if V is a direct summand of W, and the sequence $W \xrightarrow{f} V \longrightarrow 0$ is split exact if and only if V is a direct summand of W. The sequence $0 \longrightarrow V \xrightarrow{f} W \xrightarrow{f} V' \longrightarrow 0$ is split exact if and only if W is the direct sum of V and V', i. e., if and only if $W \cong V \coprod V' = V \oplus V'$.

Chain complexes and homology groups of graphs

We apply this bit of theory to the spaces associated with a graph.

Definition 6.4.4. Let G be a connected graph. The homomorphism

$$\varepsilon : C_0(G) \to F$$
$$\sum_{i=1}^{n} x_i v_i \mapsto \sum_{i=1}^{n} x_i$$

is called an ***augmentation mapping***.

The (in general not exact) sequence

$$0 \to C_1(G) \xrightarrow{\partial} C_0(G) \to 0$$

with boundary operator ∂ is called the ***chain complex*** of G.

If G is connected, we call

$$0 \to C_1(G) \overset{\partial}{\to} C_0(G) \overset{\varepsilon}{\to} F \to 0$$

the ***augmented chain complex*** of G.

All theorems in this section are merely reformulations of the results about vertex and edge spaces in a different language. They may be considered as "exerceorems."

Theorem 6.4.5. *Let G be a graph with k weak components, and let $Z(G)$, $C_1(G)$, and $C_0(G)$ be the F-vector spaces generated by G. Then*

$$0 \longrightarrow Z(G) \overset{\iota}{\longrightarrow} C_1(G) \overset{\partial}{\longrightarrow} C_0(G) \overset{\nu}{\longrightarrow} C_0(G)/\operatorname{Im}\partial \longrightarrow 0$$

is an exact sequence. Here, ι is the embedding, ∂ is the boundary operator and $\nu : C_0(G) \longrightarrow C_0(G)/\operatorname{Im}\partial$ is the natural surjection.

Furthermore, we have $C_0(G)/\operatorname{Im}\partial \cong F^k$, where now $\nu : C_0(G) \longrightarrow F^k$ on the components of G is the augmentation mapping into the respective component of F^k.

Definition 6.4.6. The factor group $H_0(G) := C_0(G)/\operatorname{Im}\partial$ is called the 0th ***homology group*** of G, and $H_1(G) := C_1(G)/\ker\partial \cong C_1(G)/Z(G) \cong S(G)$ is called the 1st ***homology group*** of G.

Theorem 6.4.7. *Let G be a graph with k weak components, and let $Z(G)$, $S(G)$, $C_1(G)$ and $C_0(G)$ be the F-vector spaces generated by G. Then*

$$0 \longleftarrow Z(G) \overset{\mu}{\longleftarrow} C_1(G) \overset{\partial^*}{\longleftarrow} C_0(G) \overset{\nu^*}{\longleftarrow} F^k \longleftarrow 0$$

is an exact sequence. Here, $\mu : C_1(G) \longrightarrow C_1(G)/S(G)$ is the natural homomorphism, ∂^* is the coboundary operator, ν^* is the embedding for which $\nu^*(b_j) = \sum_{\ell=1}^{n_j} v_{j_\ell}$, where b_j is the jth basis vector of F^k and v_{j_ℓ}, $\ell = 1, \dots, n_j$, are the vertices of the jth component of G. Furthermore, we have $C_1(G)/S(G) \cong Z(G)$.

Corollary 6.4.8. *We have $C_0(G) \cong \operatorname{Im}\partial \coprod \ker\partial^*$.*

Theorem 6.4.9. *Let G be a graph with k weak components, and let $Z(G)$, $S(G)$, $C_1(G)$, $C_0(G)$ be the F-vector spaces generated by G. Then in*

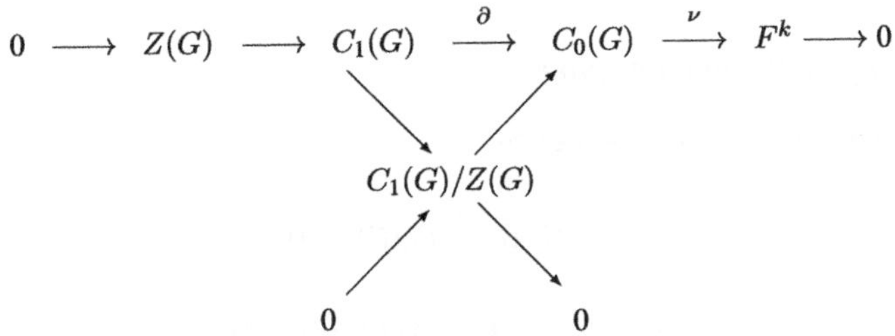

all sequences are exact, and the triangle is a commutative diagram with

$$C_1(G)/Z(G) \cong S(G) \cong \operatorname{Im}\partial.$$

By reversing all arrows, we get the diagram

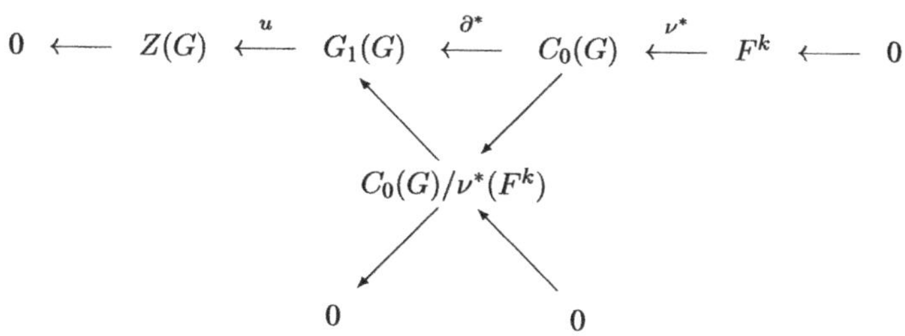

which has the same properties, with $C_0(G)/\nu^*(F^k) \cong \operatorname{Im}\partial^*$.

In both cases, the sequences from upper left to lower right, from lower left to upper right and conversely are exact direct.

The diagrams show that

$$\begin{aligned} C_1(G) &\cong C_1(G)/Z(G) \coprod Z(G) \cong S(G) \coprod Z(G) \\ &\cong C_0(G)/\nu^*(F^k) \coprod Z(G) \cong \operatorname{Im}\partial^* \coprod \ker\partial \end{aligned}$$

and

$$\begin{aligned} C_0(G) &\cong C_0(G)/\nu^*(F^k) \coprod F^k \cong C_1(G)/Z(G) \coprod F^k \\ &\cong S(G) \coprod F^k \cong \operatorname{Im}\partial \coprod \ker\partial^*. \end{aligned}$$

In particular,

$$C_1(G)/Z(G) \cong C_0(G)/\nu^*(F^k) \cong S(G) \cong \operatorname{Im}\partial \cong \operatorname{Im}\partial^*.$$

6.5 Application: number of spanning trees

In this section, we start with the first application of the theory developed earlier in this chapter.

Let $G = (V, E)$ be a directed, connected graph, with $|V| = n$ and $|E| = m$.

Lemma 6.5.1. *Let $\widetilde{B}$ and C be cocycle and cycle matrices of G, i. e., basis matrices of $S(G)$ and $Z(G)$, and take $L \subseteq E$. Denote by $\widetilde{B}|L$ and $C|L$ the submatrices which contain only elements belonging to L. Then the columns of $\widetilde{B}|L$ are linearly independent if and only if L has no semicircuit, and the rows of $C|L$ are linearly independent if and only if L has no semicocircuits.*

Example 6.5.2. We take the graph

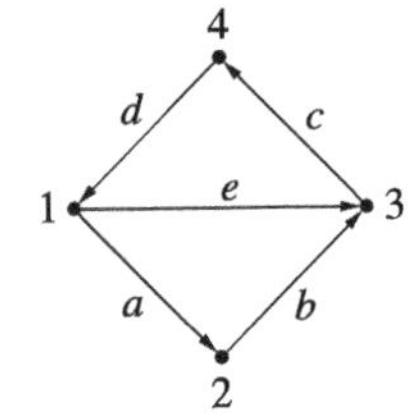

(a) Here, $L = \{a, b, c\}$ contains no semicircuit, and the columns of

$$\tilde{B}|L = \begin{pmatrix} 1 & 0 & 0 \\ -1 & 1 & 0 \\ 0 & 0 & -1 \end{pmatrix}$$

are linearly independent.
But $L = \{c, d, e\}$ contains a semicircuit, and the columns of

$$\tilde{B}|L = \begin{pmatrix} 0 & -1 & 1 \\ 0 & 0 & 0 \\ -1 & 1 & 0 \end{pmatrix}$$

are linearly dependent.

(b) Now $L = \{a, b\}$ contains a semicocircuit, and the rows of

$$C|L = \begin{pmatrix} 1 & -1 \\ 1 & -1 \end{pmatrix}$$

are linearly dependent. But $L = \{b, c\}$ contains no semicocircuit, and the rows of

$$C|L = \begin{pmatrix} 1 & -1 \\ 1 & 0 \end{pmatrix}$$

are linearly independent.

Proposition 6.5.3. *Let B be the incidence matrix of G; let $\tilde{B}$ be obtained from it by deleting one row. Then $\tilde{B}$ is a cocycle matrix of G, and this matrix is invertible.*

Proof. The row vectors of the incidence matrix B of G are cocircuits. For one row z, select those edges which do not have 0 at the entry z. Call this set U, which is a cocircuit (see Definition 6.2.6 ff.). Deletion of these edges isolates the vertex v. For $e \in U$, we have $s_{\mathrm{dir}(U)}(e) = z(e)$. Therefore, the rows are the elements of the cocycle space. Now B has rank $n - 1$ by Theorem 2.2.3, and any $n - 1$ rows are linearly independent. So deletion of one row gives a cocycle matrix, which clearly is invertible. □

Example 6.5.4. The incidence matrix of the graph from Example 6.5.2 is

$$\begin{pmatrix} 1 & 0 & 0 & -1 & 1 \\ -1 & 1 & 0 & 0 & 0 \\ 0 & -1 & 1 & 0 & -1 \\ 0 & 0 & -1 & 1 & 0 \end{pmatrix}.$$

Deletion of the third row gives

$$\begin{pmatrix} 1 & 0 & 0 & -1 & 1 \\ -1 & 1 & 0 & 0 & 0 \\ 0 & 0 & -1 & 1 & 0 \end{pmatrix},$$

which is a cocycle matrix.

Corollary 6.5.5. *The number of spanning trees of G is equal to the number of nonsingular $(n-1)\times(n-1)$ submatrices of $\widetilde{B}$.*

Proof. By Proposition 6.5.3, we know that $\widetilde{B}$ corresponds to the incidence matrix with one row deleted. The $(n-1)\times(n-1)$ submatrices of $\widetilde{B}$ therefore correspond to the incidence matrices of subgraphs of G with $(n-1)$ edges. By Lemma 6.5.1 and Proposition 6.5.3, these incidence matrices correspond to trees exactly when they are nonsingular. As they contain all vertices, the trees are spanning. □

Proposition 6.5.6. *The incidence matrix B of a directed graph is* ***totally unimodular****, i. e., every square submatrix has determinant* 0, 1 *or* −1.

Proof. We use Poincaré's Lemma (see, e. g., [Biggs 1996] p. 32).

Let S be a square submatrix of B. If every column of S has two nonzero entries, they must be +1 and −1. Then every column has sum 0. Therefore, S is singular and $\det S = 0$. Analogously, $\det S = 0$ if all entries are zero. The remaining case is where one column of S has exactly one nonzero entry. We expand the determinant with respect to this row: $\det S = \pm \det S'$, where S' contains one row and one column fewer than S. Continuing in this way, we get total unimodularity as the determinant is either 0 or a single entry of S. □

Example 6.5.7. We show that Proposition 6.5.6 is not valid for undirected graphs. Take K_3 with incidence matrix

$$\begin{pmatrix} 0 & 1 & 1 \\ 1 & 0 & 1 \\ 1 & 1 & 0 \end{pmatrix},$$

which has determinant equal to 2.

Theorem 6.5.8 (Matrix tree theorem). *Let G be a directed graph, and let $\widetilde{B}$ be its incidence matrix with one row deleted. The number of spanning trees is $\tau(G) = \det(\widetilde{B}\,{}^t\widetilde{B})$.*

Proof. By the determinant multiplication theorem, for a $p \times q$ matrix K and a $q \times p$ matrix L with $p \leq q$, we get that $\det KL = \sum_P \det K_P L_P = \sum_P \det K_P \det L_P$. Here, P denotes all p-element subsets of $\{1, \ldots, q\}$; K_P is the $p \times p$ submatrix of K that uses only the columns from P, and L_P is defined similarly. We apply this to $\widetilde{B}$ and get

$$\begin{aligned}\det \widetilde{B}\,{}^t\widetilde{B} &= \sum_P \det \widetilde{B}_P \det {}^t\widetilde{B}_P \\ &= \sum_{P_{\text{tree}}} \det \widetilde{B}_{P_{\text{tree}}} \det {}^t\widetilde{B}_{P_{\text{tree}}} + \sum_{P_{\text{non-tree}}} \det \widetilde{B}_{P_{\text{non-tree}}} \det {}^t\widetilde{B}_{P_{\text{non-tree}}}.\end{aligned}$$

Here, P_{tree} consists of the elements in P which form the spanning trees of G according to Lemma 6.5.1 and Corollary 6.5.5, and $P_{\text{non-tree}}$ is made up of the other elements. Now Lemma 6.5.1 and Proposition 6.5.6 imply that the determinant of a submatrix representing a tree is either 1 or −1, and the determinant of other $(n-1)\times(n-1)$ submatrices is 0. Thus $\det \widetilde{B}\,{}^t\widetilde{B} = \sum_{P_{\text{tree}}} 1 + \sum_{P_{\text{non-tree}}} 0$. □

Example 6.5.9. Consider again the graph from Example 6.5.2. Its spanning trees are as follows:

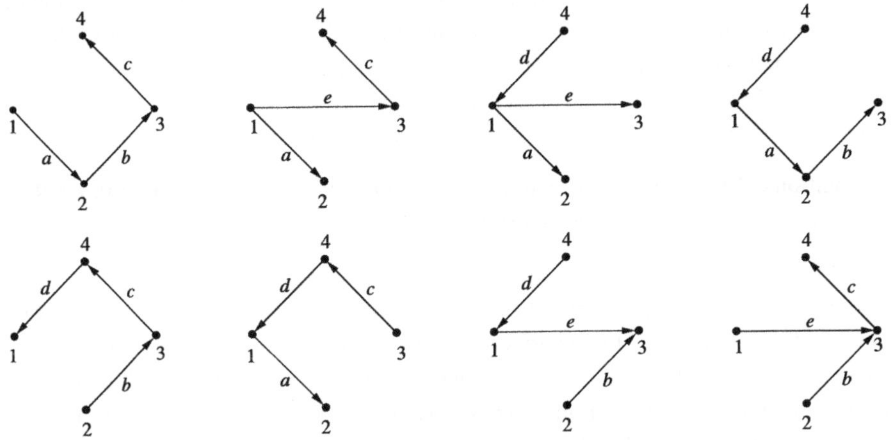

Take $\widetilde{B}$ from Example 6.5.4. It follows that

$$\begin{aligned}\det(\widetilde{B}^t\widetilde{B}) &= \det\left(\begin{pmatrix} 1 & 0 & 0 & -1 & 1 \\ -1 & 1 & 0 & 0 & 0 \\ 0 & 0 & -1 & 1 & 0 \end{pmatrix}\begin{pmatrix} 1 & -1 & 0 \\ 0 & 1 & 0 \\ 0 & 0 & -1 \\ -1 & 0 & 1 \\ 1 & 0 & 0 \end{pmatrix}\right) \\ &= \det\begin{pmatrix} 3 & -1 & -1 \\ -1 & 2 & 0 \\ -1 & 0 & 2 \end{pmatrix} = 8.\end{aligned}$$

Definition 6.5.10. Take $G = (V, E)$ with $|V| = n$ and $|E| = m$. An $m \times (m - n + 1)$ matrix C whose jth column is the jth basis vector of $Z(G)$ with respect to the standard basis

$e_1, \dots, e_m$ of $C_1(G)$ is called a ***cycle matrix*** of G. An $(n-1) \times m$ matrix $\widetilde{B}$ whose jth row is the jth basis vector $S(G)$ with respect to the standard basis $e_1, \dots, e_m$ in $C_1(G)$ is called a ***cocycle matrix*** of G.

Corollary 6.5.11. *Let $e_1, \dots, e_{n-1}$ be the edges of a spanning tree of G, and denote by $e_n, \dots, e_m$ the other edges (the cords with respect to the tree). Let C_i denote the circuit generated by e_{n-1+i} with the edges of the spanning tree oriented as e_{n-1+i}. The cycle matrix C formed with the cycle basis of $Z(G)$ obtained in this way has the form*

$$C = \begin{pmatrix} C_T \\ I_N \end{pmatrix},$$

where I_N denotes the $(m-n+1) \times (m-n+1)$ unit matrix and C_T the rest.

Proof. According to Definition 6.5.10, the jth column contains the jth basis vector, which contains 1 in the row of e_{n-1+j} and 0 in the rows from n to m; we get the $(m-n+1) \times (m-n+1)$ unit matrix I_N. Note that C_T is $(n-1) \times (m-n+1)$. □

Exercise 6.5.12. The number of spanning trees of G is $\tau(G) = |\det \binom{\widetilde{B}}{{}^tC}|$, where C is the cycle matrix of G from Corollary 6.5.11 and $\widetilde{B}$ is the incidence matrix with one row deleted. This means that $|\det \binom{\widetilde{B}}{{}^tC}| = \det(\widetilde{B}\,{}^t\widetilde{B})$.

Example 6.5.13. Select the edges a, b and c as the spanning tree of the graph in Example 6.5.2. Then C has the following form:

$$\begin{pmatrix} 1 & -1 \\ 1 & -1 \\ 1 & 0 \\ \hline 1 & 0 \\ 0 & 1 \end{pmatrix}.$$

Consider the spanning trees of this graph as in Example 6.5.9. With $\widetilde{B}$ from Example 6.5.4 and C as above, we get

$$\det \begin{pmatrix} \widetilde{B} \\ {}^tC \end{pmatrix} = \det \begin{pmatrix} 1 & 0 & 0 & -1 & 1 \\ -1 & 1 & 0 & 0 & 0 \\ 0 & 0 & -1 & 1 & 0 \\ 1 & 1 & 1 & 1 & 0 \\ -1 & -1 & 0 & 0 & 1 \end{pmatrix} = -8.$$

6.6 Application: electrical networks

Here, we come to the so-called Kirchhoff laws, well known in physics. The kernel is the law $U = IR$ as written in physics, which is hidden in Theorem 6.6.10. Here, U denotes the voltage, I the current and R the resistance of an electrical network.

Take $G = (V, E)$ to be a directed, connected graph, with $|V| = n$, $|E| = m$ and the $\mathbb{R}$-vector spaces $C_0(G)$ and $C_1(G)$. Note that $m \geq n - 1$ as G is connected.

Definition 6.6.1. A mapping $\mathrm{pot} : V \to \mathbb{R}$ is called a ***potential on*** G. Given a potential pot on G, the mapping $u : E \to \mathbb{R}$ defined by $u(e) = \mathrm{pot}(o(e)) - \mathrm{pot}(t(e))$ is called a ***voltage*** or ***tension*** on G. A mapping $r : E \to \mathbb{R}$, $r(e_i) = r_i$, is called an ***edge resistance***, for $i = 1, \ldots, m$.

Remark 6.6.2. The potentials on $G = (V, E)$ are exactly the elements of the $\mathbb{R}$-vector space $C_0(G) = \mathbb{R}^V$. Voltages and resistances are elements of $C_1(G)$. An element of $C_1(G)$ will sometimes be called a ***voltage generator***. The voltage of an edge is the potential difference between its endpoints, with the additional property seen in the next example and formulated in the next theorem. We will see in Definition 6.6.5 that currents are also elements of $C_1(G)$ with (another) additional property.

Example 6.6.3. Consider the following graph:

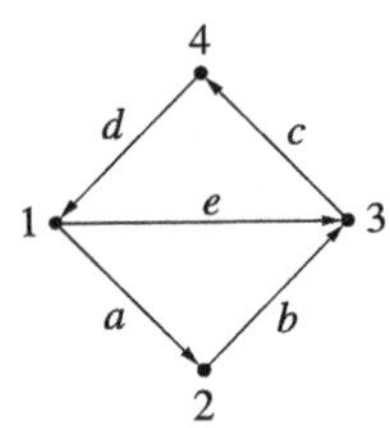

We define the potential $\mathrm{pot} : V \to \mathbb{R}$ by $\mathrm{pot}(1) = 1$, $\mathrm{pot}(2) = 3$, $\mathrm{pot}(3) = 4$, and $\mathrm{pot}(4) = 5$, and get the voltage $u : E \to \mathbb{R}$ with $u(a) = -2$, $u(b) = -1$, $u(c) = -1$, $u(d) = 4$, and $u(e) = -3$, as given in the following figure:

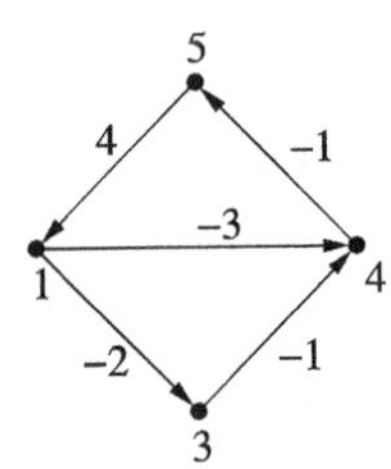

Consider the semicycles (c, d, e) and (a, b, e). Then

$$^t(0,0,1,1,1)\,^t(-2,-1,-1,4,-3) = 0 = \,^t(1,1,0,0,-1)\,^t(-2,-1,-1,4,-3).$$

This leads to the so-called Kirchhoff's voltage law: the voltage along circles is always 0 – otherwise one would get a "short-circuit" (Kurzschluss).

Theorem 6.6.4 (Kirchhoff's voltage law, mesh law). *An element $u \in C_1(G)$ is a voltage on G if and only if $\langle z, u \rangle = 0$ for all $z \in Z(G)$, i. e., if and only if $u \in S(G)$.*

Proof. For "$\Rightarrow$," let $u \in C_1(G)$ be a voltage on G and take $z \in Z(G)$. By Lemma 6.2.2, there exists a semicircuit $z_1, \ldots, z_n$ and factors $\lambda_1, \ldots, \lambda_n \in \mathbb{R}$ such that $z = \lambda_1 z_1 + \cdots + \lambda_n z_n$. We show that $\langle z_i, u \rangle = 0$ for $i \in \{1, \ldots, n\}$, since then we would have $\langle z, u \rangle = \langle \lambda_1 z_1 + \cdots + \lambda_n z_n, u \rangle = \lambda_1 \langle z_1, u \rangle + \cdots + \lambda_n \langle z_n, u \rangle = 0 + \cdots + 0 = 0$. We will prove by induction on n that for all semipaths with simple edges $L = (v_1, e_1, \ldots, e_{n-1}, v_n)$ with $n \in \mathbb{N} \setminus \{0, 1\}$, one has $\langle z_{\mathrm{dir}(L)}, u \rangle = \mathrm{pot}(v_1) - \mathrm{pot}(v_n)$.

Base step for induction: for $L = (v_1, e_1, v_2)$ suppose that v_1 is the starting point and v_2 the end of e_1, or vice versa. Then $z_{\mathrm{dir}(L)}(e_1) = 1$ or $z_{\mathrm{dir}(L)}(e_1) = -1$. In both cases, we have $z_{\mathrm{dir}(L)}(e_1)u(e_1) = \mathrm{pot}(v_1) - \mathrm{pot}(v_2)$. For edges $e \neq e_1$ of the graph, one has $z_{\mathrm{dir}(L)}(e) = 0$ and thus $z_{\mathrm{dir}(L)}(e)u(e) = 0$. This gives $\langle z_{\mathrm{dir}(L)}, u \rangle = \sum_{e \in E} z_{\mathrm{dir}(L)}(e)u(e) = \mathrm{pot}(v_1) - \mathrm{pot}(v_2)$.

Induction hypothesis: for $n \geq 2$, i. e., for all semipaths $L' = (v_1, e_1, \ldots, e_{n-1}, v_n)$, one has $\langle z_{\mathrm{dir}(L)}, u \rangle = \mathrm{pot}(v_1) - \mathrm{pot}(v_n)$.

Induction step: now take $L = (v_1, e_1, \ldots, e_{n-1}, v_n, e_n, v_{n+1})$; then we have $\langle z_{\mathrm{dir}(L')}, u \rangle = \mathrm{pot}(v_1) - \mathrm{pot}(v_n)$. Then, again, $z_{\mathrm{dir}(L)}(e_n) = 1$ or $z_{\mathrm{dir}(L)}(e_n) = -1$, and in both cases $z_{\mathrm{dir}(L)}(e_n)u(e_n) = \mathrm{pot}(v_n) - \mathrm{pot}(v_{n+1})$. With the definition of the standard scalar product, we get $\langle z_{\mathrm{dir}(L)}, u \rangle = \langle z_{\mathrm{dir}(L')}, u \rangle + z_{\mathrm{dir}(L)}(e_n)u(e_n) = \mathrm{pot}(v_1) - \mathrm{pot}(v_n) + \mathrm{pot}(v_n) - \mathrm{pot}(v_{n+1}) = \mathrm{pot}(v_1) - \mathrm{pot}(v_{n+1})$. This completes the induction proof.

If we now consider a semicircuit $L = (v_1, e_1, \ldots, e_{n-1}, v_n)$, then $v_1 = v_n$, and thus $\langle z_{\mathrm{dir}(L)}, u \rangle = \mathrm{pot}(v_1) - \mathrm{pot}(v_1) = 0$. Consequently $u \in S(G)$.

For "$\Leftarrow$", take $u \in S(G)$, i. e., $\langle z, u \rangle = 0$ for all $z \in Z(G)$. We define a potential $\mathrm{pot} : V \to \mathbb{R}$ by $\mathrm{pot}(v) := a$ for $v \in V$, with any $a \in \mathbb{R}$, e. g., $a = 0$. For $e \in \mathrm{out}(v)$ define $\mathrm{pot}(t(e)) := \mathrm{pot}(v) - u(e)$, and for $e' \in \mathrm{in}(v)$ define $\mathrm{pot}(o(e)) := u(e) + \mathrm{pot}(v)$. By continuing this procedure, we get a correctly defined mapping pot such that $u(e) = \mathrm{pot}(o(e)) - \mathrm{pot}(t(e))$ is a voltage, $u \in S(G)$. □

Definition 6.6.5. A ***current*** on G is a mapping $w : E \to \mathbb{R}$ with

$$\sum_{t(e)=v} w(e) - \sum_{o(e)=v} w(e) = 0 \quad \text{for all } v \in V.$$

Example 6.6.6. The next figure shows a current on the graph of Example 6.6.3, where we define $w : E \to \mathbb{R}$ by $w(a) = 1$, $w(b) = 1$, $w(c) = 2$, $w(d) = 2$, and $w(e) = 1$. Here, we indeed have "flow in = flow out."

Upon multiplying the associated vector ${}^t(1, 1, 2, 2, 1)$ by the vector of the voltage given above, we get 0.

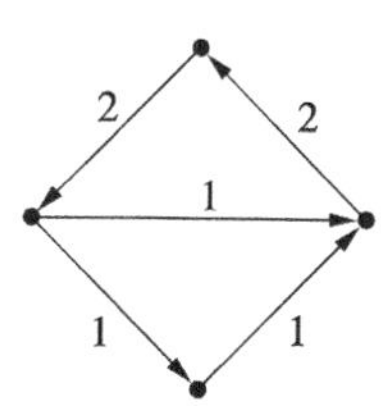

The reason is that the voltages are exactly the cocycles and the currents are exactly the cycles.

Theorem 6.6.7 (Kirchhoff's current law, vertex law). *An element $w \in C_1(G)$ is a current on G if and only if $\langle w, u \rangle = 0$ for all $u \in S(G)$, i. e., if and only if $w \in Z(G)$.*

Corollary 6.6.8. *Linear combinations of currents are currents; every current on a connected graph depends on $|E| - |V| + 1$ parameters and is determined completely by those parameters. Linear combinations of voltages are voltages; every voltage on a connected graph depends on $|V| - 1$ parameters and is determined completely by those parameters.*

Proof. It is clear that linear combinations of currents are currents, since linear combinations of cycles are disjoint unions of cycles. Linear combinations of voltages are voltages, since linear combinations of cocycles are disjoint unions of cocycles. Corollary 6.2.14 gives the rest, since $|E| - |V| + 1$ and $|V| - 1$ are the dimensions of the cycle and cocycle spaces, respectively. □

Corollary 6.6.9. *We have ${}^tCu = 0$ if and only if $u \in C_1(G)$ is a voltage on G, and $\tilde{B}w = 0$ if and only if $w \in C_1(G)$ is a current on G.*

Proof. We have that ${}^tCu = 0$ if and only if u is a voltage, as the multiplication of ${}^tC$ by u means that u is multiplied with vectors from $Z(G)$ and the results are then added. If this gives 0, we must have started from a voltage.

Conversely, multiplication of a voltage by a current gives 0.

We also have that $\tilde{B}w = 0$ if and only if w is a current, since the given multiplication means that $w \in C_1(G)$ is multiplied with basis vectors from $S(G)$ and the results are added. If this gives 0, we know that w was a current.

Conversely, multiplication of a current by a voltage gives 0. □

Theorem 6.6.10. *Let $G = (V, E)$ be a graph (an "electrical network") with a mapping $r : E \longrightarrow \mathbb{R}$, $r(e_i) = r_i$, for $i = 1, \ldots, m$ (the "edge resistances"). Take $g \in C_1(E)$ (a "voltage generator"), and set $R := (r_i\delta_{ij})_{i,j=1,\ldots,m}$. Then the current w with $u = Rw + g$ is given by*

$$w = -C({}^tCRC)^{-1}\,{}^tCg,$$

where C is the cycle matrix generated by a spanning tree of G according to Corollary 6.5.11 (w and g are written as column vectors).

Proof. We arrange the matrix B and the vectors w and u according to C in Corollary 6.5.11, i. e., $w = (w_T, w_N)$, $u = (u_T, u_N)$ and $B = (B_T, B_N)$. Then one part contains the information about the edges belonging to the spanning tree, and the other part contains the information about the other edges.

Corollary 6.6.9 now implies that $B_T w_T + B_N w_N = 0$, or $w_T = -B_T^{-1} B_N w_N = C_T w_N$. This implies $w = Cw_N$. Again by Corollary 6.6.9, we get ${}^tCu = 0$, as u is a voltage. Inserting $u = Rw + g$ gives ${}^tCRw + {}^tCg = 0$, and with $w = Cw_N$ we get $({}^tCRC)w_N = -{}^tCg$.

As $({}^tCRC)$ is invertible, multiplication by $C({}^tCRC)^{-1}$ from the left gives $-C({}^tCRC)^{-1}\,{}^tCg = Cw_N = w$. □

Example 6.6.11. Take C from Example 6.5.13, and let $r(a) = 2, r(b) = 1, r(c) = 3, r(d) = 1$ and $r(e) = 2$. Let u be the voltage from Example 6.6.3. Then

$$-w = C({}^tCRC)^{-1}\,{}^tCu =$$

$$\begin{pmatrix} 1 & -1 \\ 1 & -1 \\ 1 & 0 \\ 1 & 0 \\ 0 & 1 \end{pmatrix} \left(\begin{pmatrix} 1 & 1 & 1 & 1 & 0 \\ -1 & -1 & 0 & 0 & 1 \end{pmatrix} \begin{pmatrix} 2 & 0 & 0 & 0 & 0 \\ 0 & 1 & 0 & 0 & 0 \\ 0 & 0 & 3 & 0 & 0 \\ 0 & 0 & 0 & 1 & 0 \\ 0 & 0 & 0 & 0 & 2 \end{pmatrix} \begin{pmatrix} 1 & -1 \\ 1 & -1 \\ 1 & 0 \\ 1 & 0 \\ 0 & 1 \end{pmatrix} \right)^{-1} \begin{pmatrix} 1 & 1 & 1 & 1 & 0 \\ -1 & -1 & 0 & 0 & 1 \end{pmatrix} \begin{pmatrix} -2 \\ -1 \\ -1 \\ 4 \\ -3 \end{pmatrix}$$

$$= \begin{pmatrix} 1 & -1 \\ 1 & -1 \\ 1 & 0 \\ 1 & 0 \\ 0 & 1 \end{pmatrix} \begin{pmatrix} \frac{5}{26} & \frac{3}{26} \\ \frac{2}{26} & \frac{7}{26} \end{pmatrix} \begin{pmatrix} 1 & 1 & 1 & 1 & 0 \\ -1 & -1 & 0 & 0 & 1 \end{pmatrix} \begin{pmatrix} -2 \\ -1 \\ -1 \\ 4 \\ -3 \end{pmatrix}$$

$$= \frac{1}{26} \begin{pmatrix} 6 & 6 & 2 & 2 & -4 \\ 6 & 6 & 2 & 2 & -4 \\ 2 & 2 & 5 & 5 & 3 \\ 2 & 2 & 5 & 5 & 3 \\ -4 & -4 & 3 & 3 & 7 \end{pmatrix} \begin{pmatrix} -2 \\ -1 \\ -1 \\ 4 \\ -3 \end{pmatrix} = \begin{pmatrix} 0 \\ 0 \\ 0 \\ 0 \\ 0 \end{pmatrix}.$$

This is not surprising since voltages neutralize each other. Now select a voltage generator $g \in C_1(E), g \notin S(G)$, say $g = {}^t(1, 0, 0, 1, 0)$, and get

$$w = \frac{-1}{26} \begin{pmatrix} 6 & 6 & 2 & 2 & -4 \\ 6 & 6 & 2 & 2 & -4 \\ 2 & 2 & 5 & 5 & 3 \\ 2 & 2 & 5 & 5 & 3 \\ -4 & -4 & 3 & 3 & 7 \end{pmatrix} \begin{pmatrix} 1 \\ 0 \\ 0 \\ 1 \\ 0 \end{pmatrix} = \begin{pmatrix} \frac{-8}{26} \\ \frac{-8}{26} \\ \frac{-7}{26} \\ \frac{-7}{26} \\ \frac{1}{26} \end{pmatrix};$$

and as voltage u we get

$$u = Rw + g = \begin{pmatrix} \frac{10}{26} \\ \frac{-8}{26} \\ \frac{-21}{26} \\ \frac{19}{26} \\ \frac{2}{26} \end{pmatrix}.$$

The following figure collects the results:

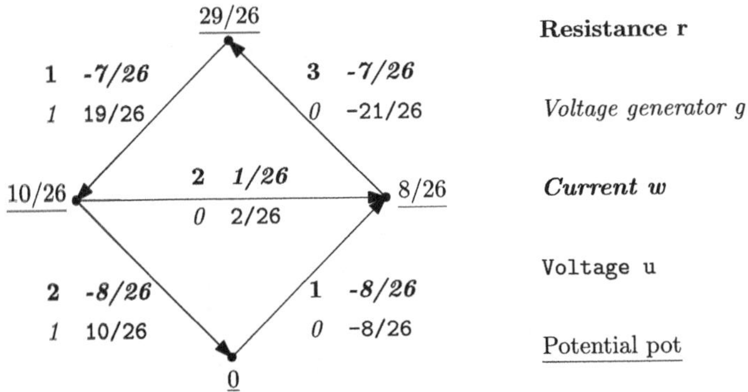

6.7 Application: squared rectangles

This application is quite surprising, as it is a kind of game. Much of the history of this problem can be found in the very personal book [Tutte 1998].

For various questions and results around this topic, you may check the internet, e. g., https://en.wikipedia.org/wiki/Squaring_the_square.

Definition 6.7.1. A ***squared rectangle*** is a rectangle which is decomposed into at least two squares. If all the squares making up a squared rectangle are of different sizes, one calls the rectangle a ***perfect rectangle***. The ***order*** of a squared rectangle is the number of constituent squares. A squared rectangle is said to be ***simple*** if it does not contain other squared rectangles.

The first perfect rectangle has order 9, i. e., it consists of nine different squares and has side length 32×33. It was found by Z. Moron and is depicted in Example 6.7.3. Now the search for squared rectangles has been computerized and the results listed up to order 21, according to [Tutte 1998]. The smallest perfect square has order 21.

The first perfect square, found by P. P. Sprague who published his result in 1939, has order 55. A smaller one of order 26 was composed of two perfect rectangles (377×608 and 231×377), and a square of length 231 was presented in 1940 by Tutte and coauthors.

Construction 6.7.2. A squared rectangle leads to a directed graph or electrical network as follows:

(a) Assign to each horizontal line segment a vertex.
(b) Put an edge between two vertices if the corresponding line segments contain segments which are borders of one square—top or bottom. The direction of the edge is "from top to bottom."

(c) Add the edge (y,x), where x is the "highest" and y the "lowest" vertex.
(d) Assigning to each vertex the distance to the lowest vertex gives a potential.
(e) Assigning resistance 1 to every edge makes Kirchhoff's current law (Definition 6.6.5) true for all vertices except x and y.

Example 6.7.3. We give an example of the construction of the graph from a squared rectangle. The diagram below is taken from [Tutte 1998], page 3.

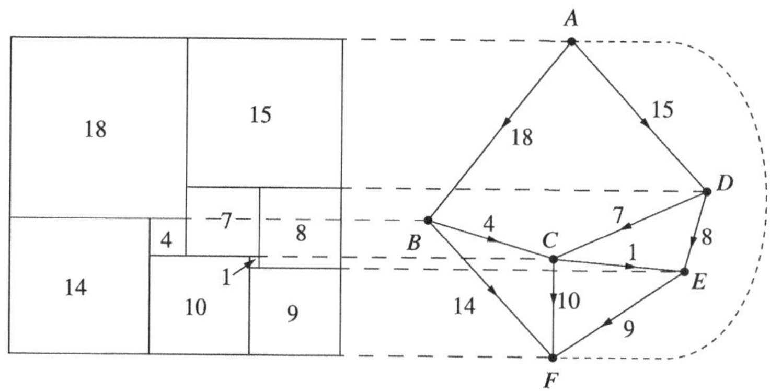

Theorem 6.7.4 ([13]). *Every graph of a simple squared rectangle (according to Construction 6.7.2) is 3-vertex connected and planar, with a current on the edges after adding one additional edge. Conversely, every current on a 3-vertex connected and planar graph gives a squared rectangle after deletion of (y,x).*

Proof. See R. L. Brooks, C. A. Smith, A. H. Stone, W. T. Tutte [13]. □

Construction 6.7.5 (to determine a simple squared rectangle).
(a) Start with a 3-vertex connected planar digraph $G' = (V', E')$ (see Definition 1.2.3), where $x, y \in V'$ are such that (y,x) is the only incoming edge of x and (y,x) is the only outgoing edge of y, with incidence matrix B'.
(b) Delete (y,x); here x is the first vertex (i. e., first row in B') and y is the last vertex corresponding to the row deleted from B'. Call the resulting graph G.
(c) Determine $\tau(G)$.
(d) Select a spanning tree in G.
(e) Form C.
(f) Solve

$$\begin{pmatrix} \tilde{B} \\ {}^tC \end{pmatrix} w = \begin{pmatrix} \tau(G) \\ 0 \\ \vdots \\ 0 \end{pmatrix} \quad \text{so that the components of } w \text{ are in } \mathbb{N}.$$

Proof. Consider the graph G' with n vertices and $m+1$ edges. Its incidence matrix after deleting the last row is of size $(n-1) \times (m+1)$ and has the form

$$\widetilde{B}' = \left(\begin{array}{c|c} & -1 \\ & 0 \\ \widetilde{B} & \vdots \\ & 0 \end{array} \right).$$

Here, the $(n-1) \times m$ matrix $\widetilde{B}$ is the incidence matrix of G after deletion of the last row of B corresponding to y, and x corresponds to the first row of B, where B is $n \times m$.

The cycle matrix C' of G' is of size $(m+1)\times(m-n+2)$ and has the form given below:

$$C' = \left(\begin{array}{c|c} C_T & c \\ \hline & 0 \\ I_N & \vdots \\ & 0 \\ \hline 0 \cdots 0 & 1 \end{array} \right).$$

Here, C_T is of size $(n-1) \times (m-n+1)$, where $m-n+1 = \xi(G)$, and I_N is the $(m-n+1) \times (m-n+1)$ unit matrix; both matrices are as in Corollary 6.5.11, i. e., $C = \binom{C_T}{I_N}$ is the cycle matrix of G, which is of size $m \times (m-n+1)$. The last row of C' corresponds to y and the last column of C' corresponds to the cycle of G' generated by the arc (y, x), so the vector c has length $n-1$.

Now we put a voltage on (y, x), i. e., we use the "voltage generator" $g = {}^t(0, \ldots, 0, s)$ of length $m+1$, so that $g((y, x)) = s$ and is 0 otherwise.

As in Construction 6.7.2, every edge gets assigned the resistance 1. So in the formula $u' = Rw' + g$, according to Theorem 6.6.10, we have $R = I$, the unit matrix. This implies that $u' = w' + g$, where we write $u' = (u'_1, \ldots, u'_{m+1})$ and similarly for w'. Corollary 6.6.9 implies that

$${}^tC'u' = 0 \quad \text{and} \quad B'w' = 0.$$

Deleting the arc (y, x) gives $u = w$, where $u = (u'_1, \ldots, u'_m)$ and similarly for w, since the difference between u' and w' was g. Moreover, the forms of C' and $\widetilde{B}'$ give

$$\widetilde{B}w = \begin{pmatrix} s \\ 0 \\ \vdots \\ 0 \end{pmatrix}, \qquad {}^tCu = 0,$$

and putting the above together we get

$$\begin{pmatrix} \widetilde{B} \\ {}^tC \end{pmatrix} w = \begin{pmatrix} s \\ 0 \\ \vdots \\ 0 \end{pmatrix}.$$

Now $\binom{\widetilde{B}}{'C}$, which is $m \times m$, is invertible by Exercise 6.5.12. Thus we have a unique solution. If we select $s = \tau(G)$, we get integer solutions. □

Exercise 6.7.6. Prove the last sentence in the proof, i. e., if we select $s = \tau(G)$, we get integer solutions. Do you remember from linear algebra why the system is solvable?

Example 6.7.7. We find the squared rectangle for the 3-connected planar graph G' drawn below. Note that this graph also gives a squared rectangle of order nine, but it is different from Example 6.7.3.

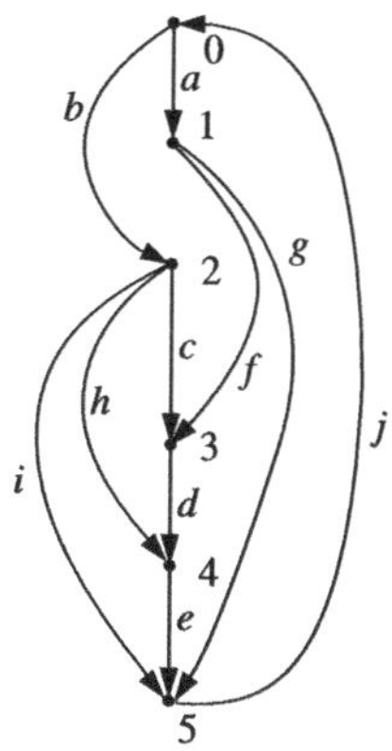

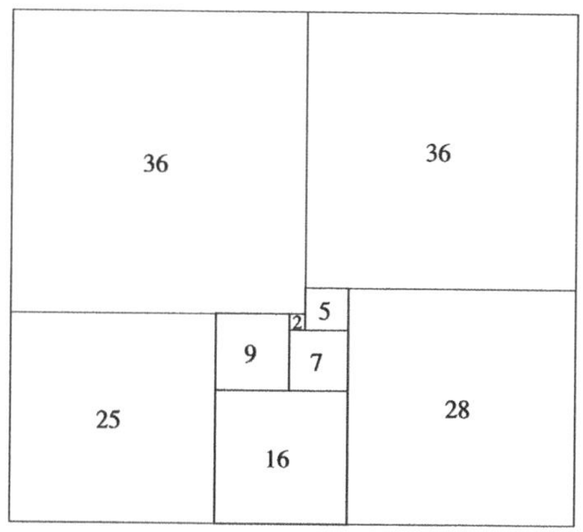

The above graph is G'. Deletion of the edge j gives the graph G which has the incidence matrix

$$B = \begin{pmatrix} 1 & 1 & 0 & 0 & 0 & 0 & 0 & 0 & 0 \\ -1 & 0 & 0 & 0 & 0 & 1 & 1 & 0 & 0 \\ 0 & -1 & 1 & 0 & 0 & 0 & 0 & 1 & 1 \\ 0 & 0 & -1 & 1 & 0 & -1 & 0 & 0 & 0 \\ 0 & 0 & 0 & -1 & 1 & 0 & 0 & -1 & 0 \\ 0 & 0 & 0 & 0 & -1 & 0 & -1 & 0 & -1 \end{pmatrix}.$$

Deletion of the last row gives

$$\widetilde{B} = \begin{pmatrix} 1 & 1 & 0 & 0 & 0 & 0 & 0 & 0 & 0 \\ -1 & 0 & 0 & 0 & 0 & 1 & 1 & 0 & 0 \\ 0 & -1 & 1 & 0 & 0 & 0 & 0 & 1 & 1 \\ 0 & 0 & -1 & 1 & 0 & -1 & 0 & 0 & 0 \\ 0 & 0 & 0 & -1 & 1 & 0 & 0 & -1 & 0 \end{pmatrix}.$$

Using Theorem 6.5.8, we get $\tau(G) = 69$.

With the spanning tree formed by a, b, c, d, and e, we obtain

$$C = \begin{pmatrix} C_T \\ I_N \end{pmatrix} = \begin{pmatrix} 1 & 1 & 0 & 0 \\ -1 & -1 & 0 & 0 \\ -1 & -1 & -1 & -1 \\ 0 & -1 & -1 & -1 \\ 0 & -1 & 0 & -1 \\ 1 & 0 & 0 & 0 \\ 0 & 1 & 0 & 0 \\ 0 & 0 & 1 & 0 \\ 0 & 0 & 0 & 1 \end{pmatrix}.$$

We have to solve the following linear system:

$$\left(\begin{array}{ccccccccc} 1 & 1 & 0 & 0 & 0 & 0 & 0 & 0 & 0 \\ -1 & 0 & 0 & 0 & 0 & 1 & 1 & 0 & 0 \\ 0 & -1 & 1 & 0 & 0 & 0 & 0 & 1 & 1 \\ 0 & 0 & -1 & 1 & 0 & -1 & 0 & 0 & 0 \\ 0 & 0 & 0 & -1 & 1 & 0 & 0 & -1 & 0 \\ \hline 1 & -1 & -1 & 0 & 0 & 1 & 0 & 0 & 0 \\ 1 & -1 & -1 & -1 & -1 & 0 & 1 & 0 & 0 \\ 0 & 0 & -1 & -1 & 0 & 0 & 0 & 1 & 0 \\ 0 & 0 & -1 & -1 & -1 & 0 & 0 & 0 & 1 \end{array}\right) w = \begin{pmatrix} 69 \\ 0 \\ 0 \\ 0 \\ 0 \\ 0 \\ 0 \\ 0 \\ 0 \end{pmatrix}.$$

The solution is $w = {}^t(33, 36, 2, 7, 16, 5, 28, 9, 25)$, which corresponds to the squared rectangle.

Exercise 6.7.8. Check the above example step by step: control C_T, and calculate B', C' (in particular the vector c), u', w', τ and w.

Find 3-vertex connected planar digraphs $G' = (V', E')$ where $x, y \in V'$ are such that (y, x) is the only incoming edge of x and (y, x) is the only outgoing edge of y. Apply Construction 6.7.5 to find simple squared rectangles.

6.8 Application: transport and shortest paths

Here, we show how to use currents and voltages to model applied problems such as transport and shortest paths. We start with transport.

Networks can be used to model the distribution of goods, data, etc. Suppose that the goods are produced at one point q, and as much as possible must be transported to some other point s. This means that all different paths from q to s should be used in a way that does not exceed their capacities.

The main idea is to use Kirchhoff's current law, which says that there are no positive or negative holes in the network, i. e., at intermediate points nothing is lost and nothing is added. This model makes sense only if the goods are being transported in single units, since flows may have to be split up differently at each vertex.

Definition 6.8.1. A directed, weakly connected graph $G = (V, E, q, s, k)$ without loops and multiple edges and with exactly one source q (named after the German word "Quelle") and one sink s (also after the German word "Senke") together with an edge valuation $k : E \to \mathbb{N} \bigcup \{\infty\}$ (after the German word "Kapazität") is called a ***transportation network***. For an edge $e \in E$, we call $k(e)$ the ***capacity*** of the edge e. The uniqueness requirement for the source and sink is sometimes relaxed.

Problems as the one formulated in the beginning of the section can now be formalized in the following way.

Transportation Problem 6.8.2. Let $G = (V, E, q, s, k)$ be a transportation network. Find a current $f : E \to \mathbb{R}$ on G, such that the potential $\sum_{(u,s)\in E} f(u) - \sum_{(s,u)\in E} f(u)$ is maximal, satisfies the capacity condition on the edges, i. e., $f(e) \leq k(e)$ for all $e \in E$, and Kirchhoff's current rule (Theorem 6.6.7) at every vertex other than q and s.

We now turn our attention to a different type of problem.

Potential Problem 6.8.3. Let $G = (V, E, q, s, k)$ be a transportation network. Find a potential $\mathrm{pot} : V \to \mathbb{R}$ on G such that the voltage satisfies $u(e) = \mathrm{pot}(t(e)) - \mathrm{pot}(o(e)) \leq k(e)$ for all $e \in E$, $\mathrm{pot}(q) = 0$ and $\mathrm{pot}(s)$ is maximal.

Theorem 6.8.4. *Let $G = (V, E, q, s, k)$ be a transportation network. The problem of finding a shortest q, s path is a potential problem.*

Proof. The problem of finding a shortest q, s path can be defined as follows. For all $e \in E$, we take $k(e)$ as the length of the edge e. Now we want to find a potential function $\mathrm{pot} : V \to \mathbb{R} \bigcup \{\infty\}$ on G that maximizes $\mathrm{pot}(s)$. One can show that $\mathrm{pot}(s)$ is maximized if $\mathrm{pot}(v)$ is the distance of v to the source q for all $v \in V$. Indeed a proof idea is as follows. Remove s and merge its neighbors into a new source s'. Then the claim holds for the smaller graph by induction. Now back in the original graph, maximizing the potential for s coincides with assigning to $\mathrm{pot}(s)$ the minimum $\mathrm{pot}(x) + k((x, s))$ over all neighbors of s. The latter coincides with the distance from q. □

Algorithm 6.8.5 (Shortest path). Determine a shortest path in a transportation network $G = (V, E, q, s, k)$ from q to any other vertex in G. Observe that, for the purpose of the algorithm, the uniqueness of the sink is not essential.

(1) (a) Set $V_1 := \{q\}$.
  (b) Set $\mathrm{pot}(q) := 0$.
(2) Now we have assigned $\mathrm{pot}(y)$ to $y \in V_k$, $k \geq 1$. Select $y \in V \setminus V_k$ and $x \in V_k$ such that $(x, y) \in E$ and $\mathrm{pot}(x) + k((x, y))$ is minimal.
  (a) Set $\mathrm{pot}(y) := \mathrm{pot}(x) + k((x, y))$.
  (b) Set $V_k := V_k \bigcup \{y\}$.

(3) If no y exists according to (2), then $V_k = V$ and the algorithm stops. Then pot(y) is the length of a shortest q, y path for all elements $y \in V$. The edges selected in (2) form a spanning tree which contains the shortest paths.

The following example illustrates the algorithm and suggests how to prove its correctness. Note that if y is not unique in (2), select any possible y. The other possible vertices will be selected in the next steps. Note, moreover, that the selection of $x \in V_k$ in (2) may also not be unique, specifically in the case where several vertices of V_k already have the same potential. A selection then implies deciding on one of several shortest paths. Step (3) is reached if $V_k = V$.

Example 6.8.6 (Shortest path algorithm). Consider the following graph:

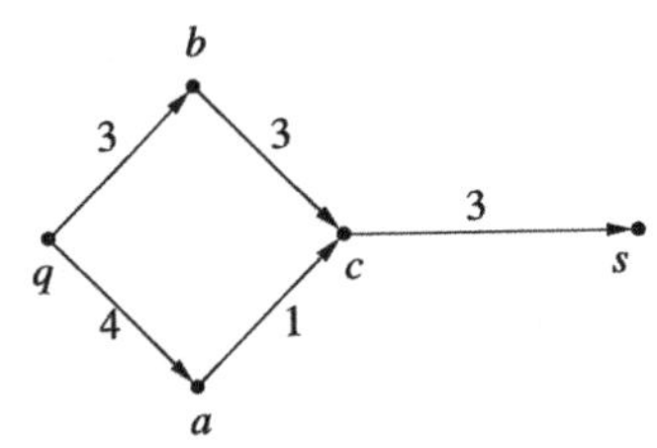

(1) (a) Set $V_1 := \{q\}$.
    (b) Set $\mathrm{pot}(q) := 0$.
(2) Select y and $q \in V_1$, with $y \notin V_1$, but $(q, y) \in E$, i. e., we get $y \in \{a, b\}$, and $\mathrm{pot}(q) + k((q, y))$ minimal implies $y = b$.
    (a) Set $\mathrm{pot}(b) := \mathrm{pot}(q) + k((q, b)) = 0 + 3 = 3$.
    (b) Set $V_2 := V_1 \bigcup \{b\} = \{q, b\}$.
(3) Select y and $x \in V_2$, with $y \notin V_2$, but $(x, y) \in E$, i. e. select (q, a) or (b, c); now $\mathrm{pot}(x) + k((x, y))$ minimal means select (q, a).
    (a) Set $\mathrm{pot}(a) := \mathrm{pot}(a) + k((q, a)) = 0 + 4 = 4$.
    (b) Set $V_3 := V_2 \bigcup \{a\} = \{q, a, b\}$.
(4) Select y and $x \in V_3$, with $y \notin V_3$ but $(x, y) \in E$; then $x \in \{a, b\}$ and $y = c$; now $\mathrm{pot}(x) + k((x, y))$ minimal implies $x = a$.
    (a) Set $\mathrm{pot}(c) := \mathrm{pot}(a) + k((a, c)) = 4 + 1 = 5$.
    (b) Set $V_4 := V_3 \bigcup \{c\} = \{q, a, b, c\}$.
(5) Select y and $x \in V_4$, with $y \notin V_4$ but $(x, y) \in E$, i. e., (c, s) is possible and $\mathrm{pot}(c) + k((c, s))$ is automatically minimal.
    (a) Set $\mathrm{pot}(s) := \mathrm{pot}(c) + k((c, s)) = 5 + 3 = 8$.
    (b) Set $V_5 := V_4 \bigcup \{s\} = \{q, a, b, c, s\}$.
(6) There are no further choices of y in step (2), so pot(y) is the length of a shortest q, y path. The spanning tree selected in this case contains all arcs except for (b, c).

Remark 6.8.7. There exist many algorithms for determining shortest paths, including the following:

(1) "Dantzig" (only for $k : K \to \mathbb{R}^{+}$) – gives shortest distances and one shortest path;
(2) "Warshall" – result as in (1);
(3) "Moore" – gives shortest distances and all shortest paths;
(4) "Dijkstra" – result as in (3);
(5) five other algorithms in [Marshall 1971].

See also [Kocay/Kreher 2005] and [Ebert 1985].

Finally, we also mention some applications of shortest path problems in other fields.

(1) "Kürzeste Wege beim Abbiegen und Umsteigen bzw. unter Belastungen"; see [Knödel 1969], pp. 46–47 and pp. 56–59, or search the internet for "shortest paths with delay."
(2) Critical paths in networks – CPM and PERT; see [Marshall 1971], pp. 98–104.
(3) "Graphentheoretisches Modell der menschlichen Niere" [Laue 1971], or A. Espinoza-Valdeza, R. Femata, F. C. Ordaz-Salazarb [16].

6.9 Comments

This chapter starts off very theoretically, but the concepts developed nevertheless have many applications. The applications we presented are on quite different levels; the shortest path and transportation problems do not really use the theory, while MacLane's planarity criterion and the other examples go deeper. The section on homology of graphs systematically synthesizes the results of the previous sections and does not contain much additional information about graphs and their connection to linear algebra. However, there is a natural way of defining the 2nd homology group of an embedded graph, that reflects the genus of the surface. This beautiful link between graph theory, topology and algebra is the subject of [Giblin 1977].

7 Graphs, groups, and monoids

The theory of groups is a powerful and effective tool for investigating symmetries of various objects with the help of their automorphisms. So it is not surprising that there is a fruitful correspondence between groups and graphs.

We recall that $(A, \cdot)$ is a *group* if A is closed with respect to the "multiplication" operation and the following three axioms are satisfied: associativity, existence of a unique identity element, and existence of an inverse for every element.

7.1 Groups of a graph

A bijective mapping of a finite set into itself is called a *permutation*. If a set of permutations is closed with respect to composition of mappings, then the above three axioms of a group are satisfied automatically and this set of permutations is called a *permutation group*.

An automorphism of a graph G is an isomorphism of G onto itself. So every automorphism α of G is a permutation of the vertex set which preserves the relation "is a neighbor of." Obviously, the bijection α takes a vertex to a vertex of the same degree.

It is also clear that the composition of two automorphisms is an automorphism; so the automorphisms of G form a permutation group on the vertex set of G. We call it the ***group*** of G and write $\operatorname{Aut}(G)$. Analogously, we talk about the monoid $\operatorname{End}(G)$ of the graph G.

We write permutations as mappings, cycles or lists as in the following example. We write transformations as mappings or as lists, as in the following example.

Example 7.1.1 (Automorphism group, endomorphism monoid).

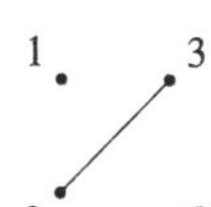

G

$$\operatorname{Aut}(G) = \left\{ \mathrm{id}\,, \begin{matrix} 1 \mapsto 1 \\ 2 \mapsto 3 \\ 3 \mapsto 2 \end{matrix} = (23) = \begin{pmatrix} 1 & 2 & 3 \\ 1 & 3 & 2 \end{pmatrix} \right\} \cong \mathbb{Z}_2,$$

$$\operatorname{End}(G) = \operatorname{Aut}(G) \bigcup \left\{ \begin{matrix} 1,2 \mapsto 2 \\ 3 \mapsto 3 \end{matrix}, \begin{matrix} 1,2 \mapsto 3 \\ 3 \mapsto 2 \end{matrix}, \begin{matrix} 1,3 \mapsto 3 \\ 3 \mapsto 2 \end{matrix}, \begin{matrix} 1,3 \mapsto 2 \\ 2 \mapsto 3 \end{matrix} = \begin{pmatrix} 1 & 2 & 3 \\ 2 & 3 & 2 \end{pmatrix} \right\}.$$

Exercise 7.1.2. We have $\operatorname{Aut}(G) \cong \operatorname{Aut}(\overline{G})$, where $\overline{G}$ denotes the complement graph of G.

https://doi.org/10.1515/9783110617368-007

Edge group

Definition 7.1.3. Let $G = (V, E)$ be undirected with $|E| \neq \emptyset$. An ***edge automorphism*** of a graph is a bijective mapping ψ of E to itself such that $\psi(e) \bigcap \psi(e') = \emptyset$ if and only if $e \cap e' = \emptyset$ for $e, e' \in E$. The ***edge group*** $\mathrm{Aut}_1(G)$ is the set of all edge automorphisms of G with composition.

An edge automorphism ψ of G is called an ***induced edge automorphism*** if there exists an automorphism φ of G such that for all $e \in E$ one has $\psi(e) = \{\varphi(o(e)), \varphi(t(e))\}$. The group of induced edge automorphisms is denoted by $\mathrm{Aut}_1^*(G)$.

Theorem 7.1.4. *For a connected graph G, we have $\mathrm{Aut}(G) \cong \mathrm{Aut}_1^*(G)$ if and only if $G \neq K_2$.*

Proof. It is clear that the statement is not true for K_2 since $\mathrm{Aut}(K_2) = \mathbb{Z}_2$ but $|\mathrm{Aut}_1(K_2)| = 1$, and thus $|\mathrm{Aut}_1^*(K_2)| = 1$. A proof of the positive part can be found in [Behzad et al. 1979], page 176 ff. It is not very complicated but quite long. Another proof is in [Harary 1969] on page 165. □

Corollary 7.1.5. *Let $G \neq K_2$ be connected. One has*

$$\mathrm{Aut}(G) \cong \mathrm{Aut}_1^*(G) \subseteq \mathrm{Aut}_1(G) \cong \mathrm{Aut}(LG),$$

where LG is the line graph of G.

This corollary raises one of those "natural questions" which the following theorem answers; see H. Whitney [90], or [Behzad et al. 1979].

Theorem 7.1.6 (Hemminger, Sabidussi, Whitney). *For a connected graph G, one has $\mathrm{Aut}_1^*(G) \cong \mathrm{Aut}_1(G)$ if and only if $G \neq$* 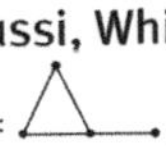, *K_4 or $K_4 \setminus e$.*

Exercise 7.1.7. Prove that there is no isomorphism for the three exceptional graphs.

Remark 7.1.8. It is quite obvious that induced edge endomorphisms will in general be egamorphisms. If we set $\mathrm{End}_1(G) := \mathrm{End}(LG)$, we have to take into account that the functor L goes into the category ***EGra***; cf. Remark 5.2.4.

Question. Can you find an analogue to Theorem 7.1.6 for endomorphisms?

7.2 Asymmetric graphs and rigid graphs

In this section, we deal with graphs that have small endomorphism monoids and automorphism groups. From Definition 1.7.1, we recall that a graph G is S unretractive if $\mathrm{SEnd}(G) = \mathrm{Aut}(G)$, and it is unretractive if $\mathrm{End}(G) = \mathrm{Aut}(G)$. More generally, G is said to be X-Y unretractive (or X-Y rigid) if $X(G) = Y(G)$ for $X, Y \in \{\text{End}, \text{HEnd}, \text{LEnd}, \text{QEnd}, \text{SEnd}, \text{Aut}\}$. Graphs G with $|\text{End}(G)| = 1$ are said to be rigid and graphs with $|\text{Aut}(G)| = 1$ are said to be asymmetric.

Recall that a graph G is said to be *k-vertex-color-critical* or simply *k-vertex-critical* if G can be vertex-colored with k colors, i. e., G has a k-coloring, and $G \setminus \{x\}$ can be colored with fewer than k colors for any vertex x. A vertex coloring assigns different colors to adjacent vertices.

Theorem 7.2.1. *If G is k-vertex-critical, then G is unretractive.*

Proof. If for an endomorphism f of G, one has $f(G) \subsetneq G$, then $f(G)$ can be colored with $h < k$ colors. But then we would get an h-coloring of G: color every preimage in G of a vertex in $f(G)$ with the same color as the image. This is a h-coloring of G since adjacent vertices do not have the same image under f. But then f is bijective and, therefore, $\text{End}(G) = \text{Aut}(G)$. □

Corollary 7.2.2. *The graphs C_{2n+1} are 3-vertex-critical, and the graphs K_n are n-vertex-critical for $n \in \mathbb{N}$. Therefore, they are unretractive.*

The first rigid graph was found by Z. Hedrlín and A. Pultr in [32]; see also Z. Hedrlín and A. Pultr [33].

Theorem 7.2.3. *The following graph G is rigid:*

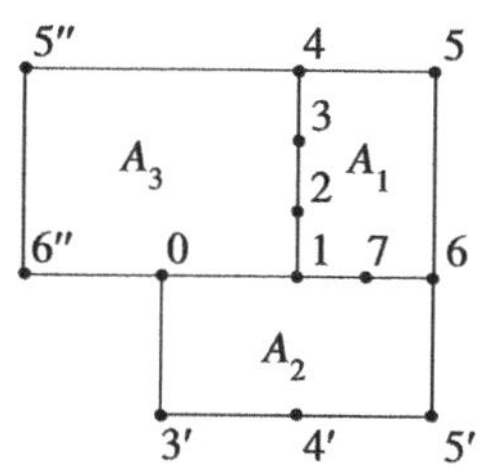

Proof. The graph consists of three copies of C_7, namely A_1, A_2 and A_3, which are unretractive by Corollary 7.2.2. Take $f \in \text{End}(G)$; then $f(A_i) = A_j$ for $i, j \in \{1, 2, 3\}$. Now, $f(A_1) = f(A_2)$ would imply $f(0) = f(2)$, since 1, 6, and 7 can only have one image each. But this is not possible since $f|_{A_3}$ is injective. By a similar argument, we get that the different cycles C_7 must stay different. Thus f is surjective, and hence bijective, i. e., it is an automorphism in this case. But then common points of at least two of the circuits must be fixed by f. Consequently, all points are fixed. Thus $f = \text{id}_G$. □

Theorem 7.2.4 (Vertex-minimal d-regular asymmetric graphs). *Let $\mu(d)$ be the minimal number of vertices of all asymmetric, d-regular graphs, i. e., graphs with vertex-*

degree d for all vertices. Then one has

d	0	1	2	3	4	5	6	d even	d odd	$(d > 6)$
$\mu(d)$	1	–	–	12	10	10	11	$d+4$	$d+5$	

Proof. It is clear that for $d = 1$ there is only K_2 while for $d = 2$ there are only circuits C_n, and both are not asymmetric. The following graphs are asymmetric with $d = 3$ and $d = 4$:

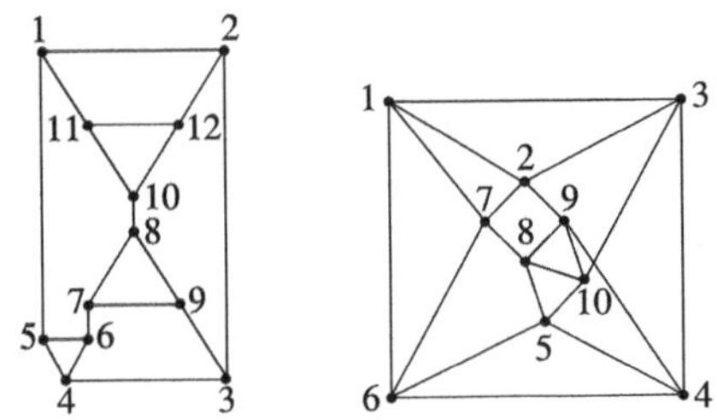

For the rest, see H. Whitney [90], or [Behzad et al. 1979]. □

Exercise 7.2.5. Observe that both graphs drawn in the proof of the previous theorem are not rigid since they can be mapped onto K_3, the first with congruence classes $\{1, 9, 12\}$, $\{2, 5, 8, 11\}$, and $\{3, 5, 7, 10\}$, and the second with congruence classes $\{1, 4, 10\}$, $\{2, 6, 8\}$, and $\{3, 5, 7, 9\}$.

Theorem 7.2.6. *For all $n \geq 8$, there exist rigid graphs with n vertices. There exist ten rigid graphs with eight vertices, and none with fewer than eight vertices.*

Proof. Pictures of these ten graphs can be found in U. Knauer [51]. They are reproduced below; under each graph, we give the number of edges, followed by the number of automorphisms in the next line.

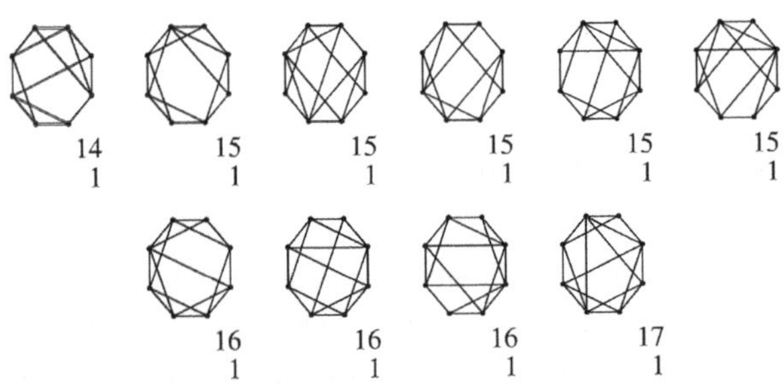

The first graph has the least possible number of edges, which is 14.

The infinite series can be constructed from the following graph with 8 vertices and 16 edges without using the vertex d; it is the third graph in the second row above:

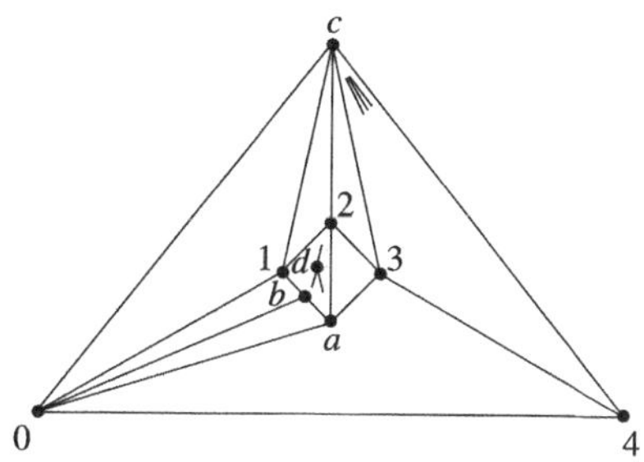

To get a graph with $10, 12, 14, \ldots$ vertices, insert two new edges starting from the vertex c to two new vertices on the edge joining 3 and 4, and so on. To get the graph with nine vertices, start with the eight-vertex graph and put in the vertex d with the three edges as indicated. The same procedure starting from the graph with 10 vertices gives a graph with 11 vertices and so on.

To prove this, we look at even n.

To see that $\operatorname{Aut}(G) = 1$ we look at the vertex c, which is the only vertex with all neighbors on an odd cycle. So it must be fixed under an automorphism. Because of a and b, the cycle cannot be reflected about c, so it must be fixed overall. It is clear that a, b, and also d —if it is used—cannot be permuted in this situation.

Now, since the neighbors of c form an odd cycle C, together with c they form a wheel. So a vertex coloring G needs four colors, and a and b can also be colored with these colors. The same is true for d – if it is used. So $G \setminus \{a, b\}$ and $G \setminus \{a, b, d\}$ are vertex-critical, and thus unretractive; cf. Theorem 7.2.1.

Next, we show that inserting a and b and possibly also d does not change the situation. We consider at the same time the possibly inserted pairs of points between 3 and 4 with the numbers up to $n - 4$ for $n \geq 4$. Suppose we have an endomorphism f such that $f(a) \in C = \{0, \ldots, n-4\}$ or $f(b) \in C = \{0, \ldots, n-4\}$. Since C is fixed, $f(a) = c$ and $f(b) \in C$ are impossible. So $f(a) = b$ implies $f(\{0, 2, 3\}) = f(N(a) \bigcap C) \subseteq N(b) \bigcap C = \{0, 1\}$, which is also impossible as C is fixed. Similarly, if $f(a) = d$ we get $f(d) = 0$, which is impossible since $\{d, 2\} \notin E$. Consequently, $f(a) = a$, $f(b) = b$ and possibly $f(d) = d$; that is, $\operatorname{End}(G) = \operatorname{Aut}(G)$. □

Definition 7.2.7. A family of rigid graphs $(G_i)_{i \in I}$, i. e., such that $|\operatorname{End}(G_i)| = 1$, is said to be ***mutually rigid*** if for $i, j \in I$, $\operatorname{Hom}(G_i, G_j) = \emptyset$ whenever $i \neq j$.

It can be checked that in the list of ten rigid graphs with eight vertices, the second up to the sixth are mutually rigid, as well as the seventh and the eighth.

Theorem 7.2.8. *The countably many graphs constructed in the proof of Theorem 7.2.6 are mutually rigid.*

Proof. (See [Hell 1974], pp. 291–301.) A homomorphism from a graph in the series from Theorem 7.2.6 to a smaller one cannot exist, as it would have to be a folding of the path in the middle from 3 to $n-4$, and this would mean shortening an odd cycle. A homomorphism from a graph in the series from Theorem 7.2.6 to a larger one would have to take the cycle C (with notation as in the proof of Theorem 7.2.6) to an odd cycle of the same length. Moreover, there has to be a vertex which is a neighbor to all vertices on this cycle. This can only be c, since the length of this cycle is at least five. So all these graphs are mutually rigid. □

Example 7.2.9. For illustration, below we present all unretractive graphs (i. e., graphs with End = Aut) with seven and on the next page with eight vertices. Again, under each graph we have in the first line the number of edges, and in the second line the number of automorphisms, which in this case is also the number of endomorphisms. The graphs are taken from M. Böttcher and U. Knauer [11].

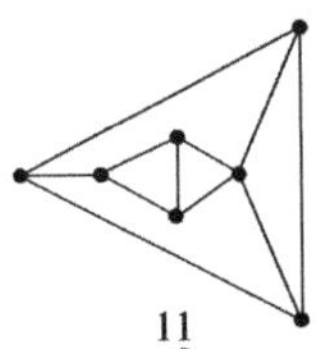

11
8

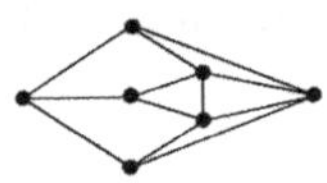

12
6

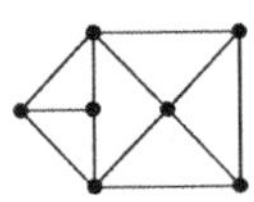

12
4

13
4

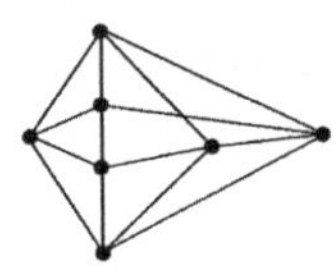

14
4

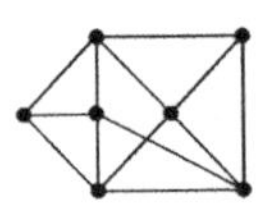

13
2

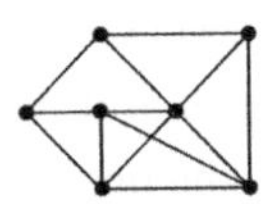

12
2

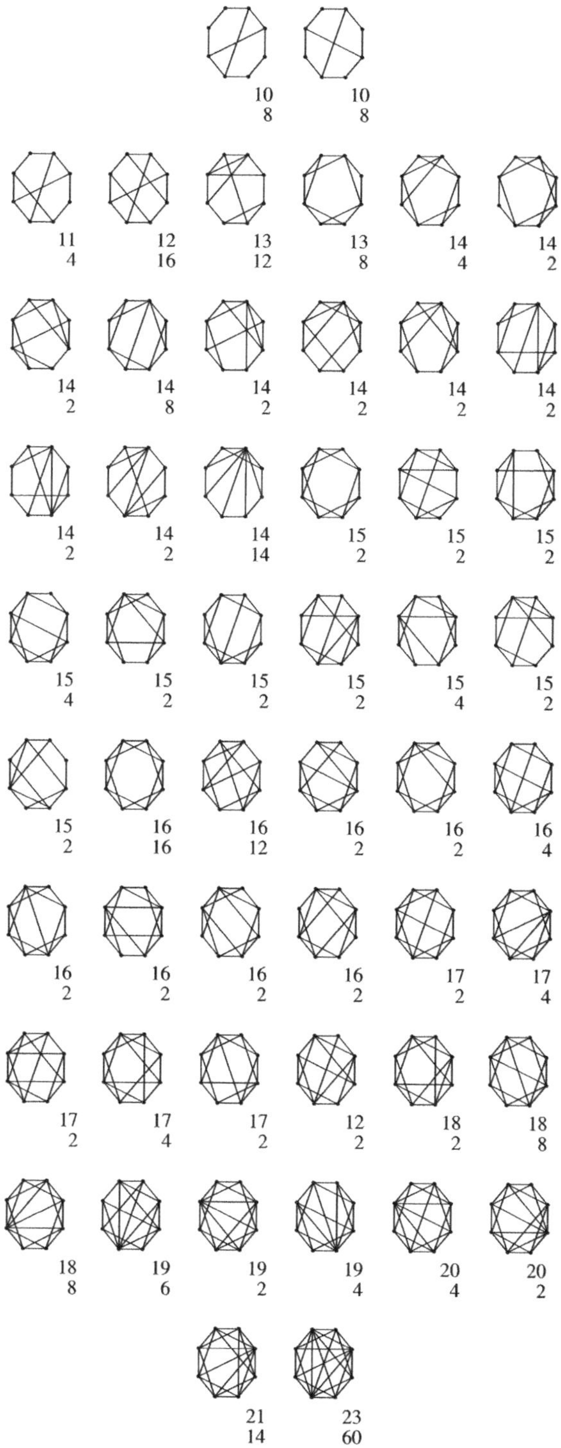
10 8
10 8
11 4
12 16
13 12
13 8
14 4
14 2
14 2
14 8
14 2
14 2
14 2
14 2
14 2
14 2
14 14
15 2
15 2
15 2
15 4
15 2
15 2
15 2
15 4
15 2
15 2
16 16
16 12
16 2
16 2
16 4
16 2
16 2
16 2
16 2
17 2
17 4
17 2
17 4
17 2
12 2
18 2
18 8
18 8
19 6
19 2
19 4
20 4
20 2
21 14
23 60

Exercise 7.2.10. Check that the endomorphism monoids (= automorphism groups) in the graphs with seven vertices are $\mathbb{Z}_2 \times \mathbb{Z}_2 \times \mathbb{Z}_2$, $\mathbb{Z}_6 \cong \mathbb{Z}_2 \times \mathbb{Z}_3$, $\mathbb{Z}_2 \times \mathbb{Z}_2$ and $\mathbb{Z}_2$.

Using a computer program, one can determine the following numbers. Note that in these computations the "isomorphism problem" plays an important role.

Theorem 7.2.11. *Let $\kappa(n)$ denote the number of nonisomorphic simple undirected graphs with n vertices, let $\alpha(n)$ denote the number of nonisomorphic simple undirected asymmetric graphs with n vertices, and let $\varrho(n)$ denote the number of non-isomorphic simple undirected rigid graphs with n vertices. Then we have the following:*

n	1	2	3	4	5	6	7	8	9	10
$\kappa(n)$	1	2	4	11	34	156	1044	12 346	274 668	12 005 168
$\alpha(n)$	0	0	0	0	0	9	152	3 697	126 148	7 971 623
$\varrho(n)$	0	0	0	0	0	0	0	10	682	52 905

The next theorem, which needs a new definition to make it precise, suggests an extrapolation of this list.

Definition 7.2.12. Denote by $G(n)$ the set of all simple graphs without loops and with n vertices, and denote by $GP(n)$ the set of all graphs from $G(n)$ with a certain property P. We say that ***almost all*** graphs have property P if

$$\lim_{n\to\infty} \frac{|GP(n)|}{|G(n)|} = 1.$$

Theorem 7.2.13. *Almost all graphs are asymmetric and almost all graphs are rigid.*

The first statement is almost folklore; it probably goes back to P. Erdős. For a proof, see [Godsil/Royle 2001], Corollary 2.3.3 on page 24. The second assertion is sometimes considered to be almost the same, but this is in fact not the case. A relatively short and independent proof of the second statement using the first is given in J. Koetters [57]; see also Theorem 4.7 in [Hell/Nešetřil 2004].

Remark 7.2.14. Similar "almost all" results can be found in A. D. Korschunov [58] and, e. g., under "Random graphs" in [Chartrand/Lesniak 2004], the fourth edition of *Graphs and Digraphs* or in [West 2001].

For example:

- Almost all graphs have a unique vertex of maximal (minimal) degree.
- Almost all graphs are connected. Almost all graphs have diameter 2.
- Almost all trees are cospectral (cf. Remark 2.7.3).

Project 7.2.15. Develop a suitable program to compute (all?) X-Y unretractive graphs with a small number of vertices, where $X, Y \in \{\mathrm{End}, \mathrm{HEnd}, \mathrm{LEnd}, \mathrm{QEnd}, \mathrm{SEnd}, \mathrm{Aut}\}$.

Find all small graphs for which these endomorphisms sets are monoids or not monoids.

7.3 Cayley graphs

We recall the following definitions.

A set M with a binary composition $M \times M \to M$, $(a,b) \mapsto ab$, is called a *groupoid*. A groupoid is called a *semigroup* if the composition is associative. A semigroup is called a *monoid* if there exists a neutral element $e \in M$, i. e., $a\,e = e\,a = a$ for all $a \in M$.

Denes König asked the following question: when is a given group isomorphic to the automorphism group of a simple undirected graph? Roberto Frucht answered this question by a construction. His proof that every group is isomorphic to the automorphism group of a graph uses the Cayley color graph. We define this a little more generally for groupoids, since it turns out that König's question has the same positive answer for certain groupoids as it has for groups.

Note that the Cayley graph for groupoids may have multiple arcs and loops.

Definition 7.3.1. Let A be a (finite) groupoid and $C = \{c_i \mid i = 1, \ldots, n\} \subseteq A$ a subset. The directed graph $\mathrm{Cay}(A, C) := (A, E(C))$ such that, for $x, y \in A$,

$$(x,y) \in E(C) \Leftrightarrow xc_i = y \quad \text{for some } c_i \in C$$

is called the ***Cayley (color) graph*** of A with ***connection set*** C. We say that the edge $e = (x,y)$ ***has color*** c_i. We use the same notation for the ***uncolored Cayley graph***, which is obtained by neglecting the colors. We use $\overline{\mathrm{Cay}}(A, C)$ for the corresponding undirected graph.

For groups—and possibly quasi-groups as well—one defines another variant of Cayley graphs, which will be used in Theorems 7.5.8, 7.7.12, and 7.7.13. We give the definition from [Klin et al. 1988], page 107.

Definition 7.3.2. Let A be a group, and let $\Omega \subseteq A$ be a system of generating elements with $1 \notin \Omega$ but such that $i \in \Omega$ implies $i^{-1} \in \Omega$. We denote by $\mathrm{Cay}(A, \Omega)$ the ***König graph*** of A with respect to Ω which is uncolored and undirected.

Both definitions can also be given using multiplication from the left by the elements of the connection set.

We observe that the requirement that $\Omega = \Omega^{-1}$ makes the graph undirected; see the middle graph in Example 7.3.3. Since this is usual for groups, most authors use the term Cayley graph instead of König graph for groups, too.

For the investigation of Cayley graphs of right groups, e. g., it will be important to consider generating sets C which give directed Cayley graphs of groups.

Example 7.3.3 (Cayley graphs and connection sets). We consider Cayley graphs of $\mathbb{Z}_5$ with respect to different connection sets in the figure below.

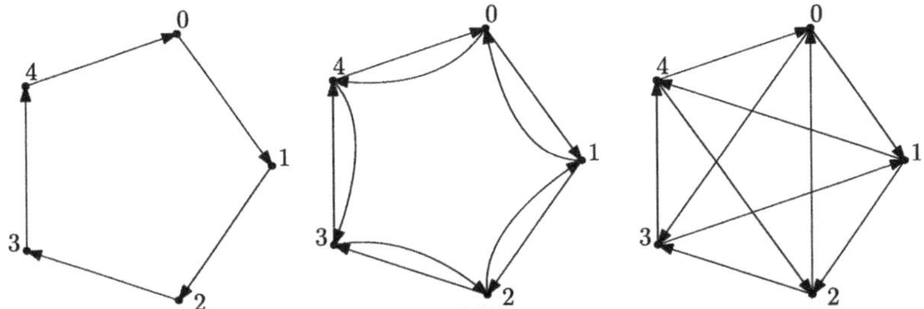

$\mathrm{Cay}(\mathbb{Z}_5, \{1\})$ $\mathrm{Cay}(\mathbb{Z}_5, \{1,4\})$ $\mathrm{Cay}(\mathbb{Z}_5, \{1,3\})$

Definition 7.3.4. A mapping $\varphi : A \to A$ with $\varphi(xi) = \varphi(x)i$ for all $x \in A$ and $i \in C$ is called a ***color endomorphism*** of $\mathrm{Cay}(A, C)$. The monoid formed by such mappings is denoted by $\mathrm{ColEnd}(\mathrm{Cay}(A, C))$. We define ***color automorphisms*** and $\mathrm{ColAut}(\mathrm{Cay}(A, C))$ analogously.

Proposition 7.3.5. *Let A be a groupoid and $C \subseteq A$. A mapping $\varphi : A \to A$ is an endomorphism of* $\mathrm{Cay}(A, C)$ *that preserves colors if and only if $\varphi(xi) = \varphi(x)\, i$ for all $x \in A$, $i \in C$.*

Proof. For "$\Leftarrow$," take $(x, y) \in E(\mathrm{Cay}(A, C))$, i. e., $xi = y$. Then $\varphi(y) = \varphi(xi) = \varphi(x)i$ and $(\varphi(x), \varphi(y)) \in E(\mathrm{Cay}(A, C))$.

The proof of "$\Rightarrow$" follows from the definition. □

Observe that color endomorphisms of $\mathrm{Cay}(A, C)$ are graph endomorphisms which are "linear" with respect to the operation of C on A.

Corollary 7.3.6. *If φ is a bijective color endomorphism, then φ is a color automorphism.*

Proof. Let φ be a bijective color endomorphism and consider the mapping φ^{-1}. Take $(\varphi(x), \varphi(y)) \in E(C)$, i. e., $\varphi(x)j = \varphi(y) = \varphi(xj)$. As φ is injective, we get $xj = y$, i. e., $(x, y) \in E(C)$, $(x, y) \in \varphi^{-1}(\varphi(x), \varphi(y))$. □

Recall that every element of a groupoid (or monoid) is a finite product of elements of a generating set C of the groupoid.

Theorem 7.3.7. *Let A be a monoid. For every generating set C of A, the mapping*

$$\begin{aligned} \Lambda : A &\to \mathrm{ColEnd}(\mathrm{Cay}(A, C)) \\ b &\mapsto \lambda_b, \end{aligned}$$

where λ_b is left translation by b, i. e., $\lambda_b(x) := b\,x$ for all $x \in A$, defines a monoid isomorphism.

Proof. First, we have that $\lambda_b \in \text{ColEnd}(\text{Cay}(A, C))$, since for $x \in A$ and $i \in C$ one has

$$\lambda_b(x\,i) = b(x\,i) = (bx)i \overset{\text{Proposition 7.3.5}}{=} \lambda_b(x)i.$$

Next, we prove that Λ is injective. Suppose that $b \neq b' \in A$. Then $\lambda_b \neq \lambda_{b'}$ as $1_A \in A$, since $\lambda_b(1_A) = b \neq b' = \lambda_{b'}(1_A)$.

Now, Λ is a monoid homomorphism. To show this, for $x \in A$ and $b_1, b_2 \in A$ calculate

$$\begin{aligned}\Lambda(b_1b_2)(x) &= \lambda_{b_1b_2}(x) = (b_1b_2)x = b_1(b_2x) = \lambda_{b_1}(b_2x) = \lambda_{b_1}(\lambda_{b_2}(x))\\ &= \Lambda(b_1)(\Lambda(b_2)(x)) = (\Lambda(b_1)\Lambda(b_2))(x),\end{aligned}$$

which means that

$$\Lambda(b_1b_2) = \Lambda(b_1)\Lambda(b_2) \quad \text{and} \quad \Lambda(1_A) = \text{id}_{\text{Cay}(A,C)}.$$

Finally, Λ is surjective. Take $\varphi \in \text{ColEnd}(\text{Cay}(A, C))$ with $\varphi(1_A) = b \in A$. We shall show that $\Lambda(b) = \varphi$. Take $a \in A$, i. e., $a = i_1 \cdots i_s$ with $i_1, \ldots, i_s \in C$. Then

$$\Lambda(b)(a) = \lambda_b(a) = \lambda_b(1_A\, i_1 \cdots i_s) = \lambda_b(1_A)\, i_1 \cdots i_s = \varphi(1_A) i_1 \cdots i_s = \varphi(a). \qquad \square$$

Remark 7.3.8. We have seen that for a monoid A we get $A \cong \text{ColAut}(\text{Cay}(A, C))$. In general, injectivity of Λ means that for any two elements $b \neq b' \in A$ there exists $x \in A$ such that $bx \neq b'x$. This is the case, e. g., if A is a monoid or left cancellable. If A does not have an identity, then surjectivity of Λ onto $\text{ColEnd}' = \text{ColEnd} \setminus \text{ColAut}$ turns out complicated.

Exercise 7.3.9. Take the group $\mathbb{Z}_2 \times \mathbb{Z}_2$ with two-element and three-element generating sets, the group $\mathbb{Z}_6$ with generating sets $\{1\}$ and $\{2, 3\}$, and the symmetric group S_3 with generating sets $\{(12), (23)\}$ and $\{(12), (123)\}$, permutations written as cycles, and draw the six Cayley graphs. Check that $A \cong \text{ColAut}(\text{Cay}(A, C))$ in each case. Find the connection sets for which $A \cong \text{Aut}(\text{Cay}(A, C))$.

Take the two-element right zero semigroup $R_2 = \{r_1, r_2\}$ with multiplication $rr' = r'$ for $r, r' \in R_2$. It is clear that the only generating set is R_2 itself. Now $\text{ColEnd}'(R_2, R_2) = \emptyset$, so R_2 and $\text{ColEnd}'(R_2, R_2)$ are not isomorphic.

Now take the left zero semigroup $L_2 = \{l_1, l_2\}$ with multiplication $ll' = l$ for $l, l' \in L_2$. Then $L_2 \cong \text{ColEnd}'(L_2, L_2)$.

7.4 Frucht-type results

A Frucht-type result is a construction of an undirected and uncolored graph with a prescribed automorphism group or a prescribed endomorphism monoid. We consider this type of problem in this section. The straightforward question of which graphs have a one-element group or a one-element monoid, i. e., which graphs are asymmetric or rigid, has already been discussed in the previous section.

Frucht's theorem and its generalization for monoids

Theorem 7.4.1 (Robert Frucht [21]). *For every finite group A, there exists a simple undirected graph G with $\mathrm{Aut}(G) \cong A$.*

Proof. Consider the Cayley color graph $\mathrm{Cay}(A, C)$ with its natural edge coloring, where C is a generating set which does not contain the identity 1_A of A. We then replace every edge of color k, e. g., (x, y), by a subgraph of the following form:

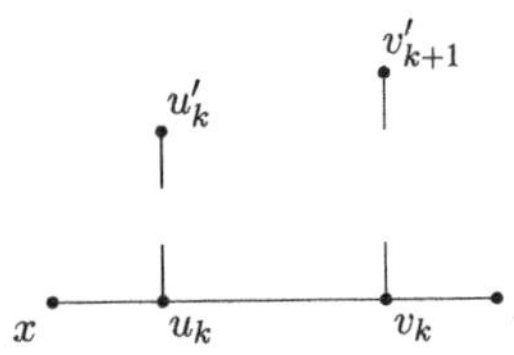

Subgraphs which take over this function are also known as *šip* (a Czech word, because the first papers on this subject were from Prague) or *gadgets* (e. g., in *Groups* by P. J. Cameron, in [Beineke/Wilson 1997]). If $y = xk$, say, we add new vertices u_k, u'_k, and v_k, v'_{k+1}, as well as a u_k, u'_k path of length $2k$ and a v_k, v'_{k+1} path of length $2k + 1$ as indicated in the figure.

We see that Frucht's construction replaces every directed edge (x, y) by an undirected graph with one starting vertex and one end vertex. The resulting graph has $n^2(2n - 1)$ vertices altogether. It is clear that every color automorphism of $\mathrm{Cay}(A, C)$ is an automorphism of G. Conversely, it is clear that G has no other automorphisms. (cf. [Harary 1969], page 177). □

Corollary 7.4.2. *For every group A, there exist infinitely many non-isomorphic graphs G with $\mathrm{Aut}(G) \cong A$.*

Remark 7.4.3. A similar result is valid for infinite groups; see G. Sabidussi [79]. If A has a countable generating system, one can use the same principle of construction. Otherwise, one has to find suitable families of graphs which can be "inserted."

For monoids A, this construction does not lead to the desired result, since "folding the tails" gives many new endomorphisms which do not correspond to elements of A. The situation can be repaired by inserting other graphs with the property that they do not have nontrivial endomorphisms or homomorphisms between each other, i. e., mutually rigid graphs. This idea goes back to Pavol Hell, therefore, we use the symbol H in the next theorem, which stands for *Hell graph*.

Theorem 7.4.4. *For every finite monoid A, there exists a simple undirected graph H with* $\mathrm{End}(H) \cong A$.

Proof. In Z. Hedrlín and A. Pultr [32], only one rigid graph (the graph from Theorem 7.2.3) is used for the construction of a suitable graph H.

We follow P. Hell and use the idea of the proof of Theorem 7.4.1; i. e., we construct the Cayley color graph for a generating set of the monoid. If there are loops, we replace those of color c by a 2-cycle colored with colors c_1 and c_2. In this case, we replace any directed edge of any color, say a, by a directed path of length two colored with the colors a_1 and a_2. If there are no loops, we omit this step. Now we insert different mutually rigid graphs for different colors, identifying endpoints of the original arc with two non-adjacent vertices of the respective rigid graph, say a and c, from the drawing of the respective family in the proof of Theorem 7.2.6. □

Corollary 7.4.5. *For every finite monoid A, there exists a graph H such that*

$$\mathrm{HEnd}(H) = \mathrm{End}(H) \cong A;$$

however, in general, $\mathrm{End}(H) \neq \mathrm{HEnd}(H)$.

Proof. In the original graph, we consider a situation where $f(1) = f(2)$ but there is no edge between the two preimage sets, like in Example 1.5.11. Then $f(1)$ is no longer adjacent to a vertex in the image graph of the Hell graph H constructed in the proof of Theorem 7.4.4. To check whether the argument stays true for connected graphs, consider the graph from Example 1.5.11 plus K_1.

For the second statement, it is clear that already for $\mathrm{Cay}(\mathbb{Z}_2^e, \{1, e\})$ we have End $\neq$ HEnd, where $\mathbb{Z}_2^e = \{e, 0, 1\}$ is the two-element group with an externally adjoint new identity $e = 1_{\mathbb{Z}_2^e}$. □

Exercise 7.4.6. Check both parts of the previous proof.

Question. For which monoids A, do there exist graphs G with $\mathrm{LEnd}(G) \cong A$, with $\mathrm{QEnd}(G) \cong A$, or with $\mathrm{SEnd}(G) \cong A$? A partial answer can be found in Suohai Fan [17].

7.5 Graph-theoretic requirements

In the previous section, we constructed graphs with a given monoid or a given group. Now we sharpen the requirements by imposing additional conditions on the graphs.

In this section, many results are not proved, or the proofs are only partial or sketched. Some of the proofs can be taken as extended exercises and could serve as starting points for theses at the Bachelor's or higher level.

Smallest graphs for given groups

Theorem 7.5.1 ([6]). *For a group A, denote by $\mu(A)$ the minimal number of vertices of all graphs G with $\mathrm{Aut}(G) \cong A$. Then one has*

A	$\mathbb{Z}_2$	$\mathbb{Z}_3$	$\mathbb{Z}_4$	$\mathbb{Z}_5$	$\mathbb{Z}_p$	A
$\mu(A)$	2	9	10	15	$2p$	$\leq 2\lvert A\rvert$

where p is a prime number greater than 5.

This theorem is due to L. Babai [6].

Example 7.5.2. The following graph is vertex-minimal with group $\mathbb{Z}_3$. We can interpret it as the Cayley graph $\mathrm{Cay}(\mathbb{Z}_3, \{1\})$ with gadget $K_4 \setminus \{e\}$.

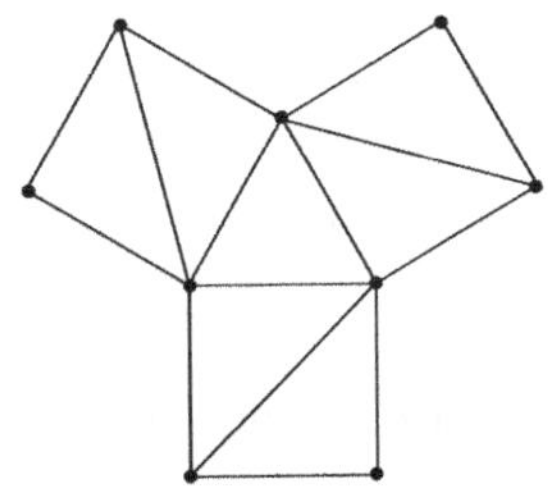

Exercise 7.5.3. This graph has (many) nontrivial endomorphisms, i. e., is not unretractive. Prove that it becomes unretractive, i. e., has endomorphism monoid $\mathbb{Z}_3$, by subdividing every edge of the inner triangle by an additional vertex.

Theorem 7.5.4. *Let $n \geq 3$. The only connected graph with groups S_n and*

$$\begin{array}{ll} n & \text{vertices is } K_n, \\ n+1 & \text{vertices is } K_{1,n}, \\ n+2 & \text{vertices is } K_1 + \overline{K}_{1,n}. \end{array}$$

Proof. See [26]. □

For further results on vertex-minimal graphs with a given group, see [Arlinghaus 1985].

Corollary 7.5.5. *Since, by Theorem 7.5.1, the smallest graph with group $\mathbb{Z}_4$ has ten vertices, we can conclude that in Example 7.2.9, among the first seven graphs with seven and eight vertices in the middle line we have the smallest graph with (group and) monoid $\mathbb{Z}_2 \times \mathbb{Z}_2$. Moreover, there the second graph in the first row is the smallest graph with (group and) monoid $\mathbb{Z}_6$.*

Question. Can you find smallest graphs with given monoids? To have a better chance of success, one should restrict the classes of monoids considered, for instance to groups themselves, right groups, Clifford semigroups (see Definition 9.1.1), etc. See also Example 7.5.2 and Corollary 7.5.5.

Additional properties of group-realizing graphs

We now describe some further properties of a graph, which it has in addition to a given automorphism group.

For the proof of the next theorem, we introduce the "type" of a vertex of an r-regular graph, and describe a graph by a quadratic form. Here, we define the type only for $r = 3$.

Definition 7.5.6. Let v be a vertex of a cubic graph G which is incident with the edges e_1, e_2, and e_3. For $i \neq j$, we denote by μ_{ij} the length of a shortest cycle containing e_i and e_j. We set $\mu_{ij} = 0$ if e_i and e_j do not lie on a cycle. Let μ_1, μ_2, μ_3 be the numbers μ_{ij} arranged in nondecreasing order. The triple (μ_1, μ_2, μ_3) is called the ***type*** of the vertex v in G.

For examples, see the proof of the next theorem.

Lemma 7.5.7. *Every (undirected) graph $G = (V, E)$, $V = \{x_1, \ldots, x_n\}$, with adjacency matrix A can be characterized by a quadratic form which defines—and is defined by—the upper triangle of A in the variables $\{x_1, \ldots, x_n\}$. This form is unchanged under automorphisms of G.*

Proof. Instead of giving a formal proof, we just look at the graph K_3 which, with the upper triangle of its adjacency matrix filled up with zeros, gives the quadratic form $x_1x_2 + x_2x_3 + x_1x_3 = (x_1, x_2, x_3)A(K_3)^t(x_1, x_2, x_3)$; this remains unchanged under permutation of the indices. □

Theorem 7.5.8. *For every finite group A with $|A| = n$ and generating set Ω, where $|\Omega| = m$, there exists a 3-regular graph G, i. e., a graph G such that all vertices have degree 3, $\mathrm{Aut}(G) \cong A$ and the number of vertices $|V(G)|$ is given as follows:*

A	$\{1\}$	$\mathbb{Z}_2, \mathbb{Z}_3$	$\mathbb{Z}_{n,n\geq 4}$	cyclic	non-cyclic
$\lvert V(G)\rvert$	12	10	$3n, 6n$	$2(m+2)n$	$2mn$

Proof. See Roberto Frucht [22]. First, we display graphs corresponding to the groups $\{1\}$ (which appeared already in Theorem 7.2.4), $\mathbb{Z}_2$ and $\mathbb{Z}_3$:

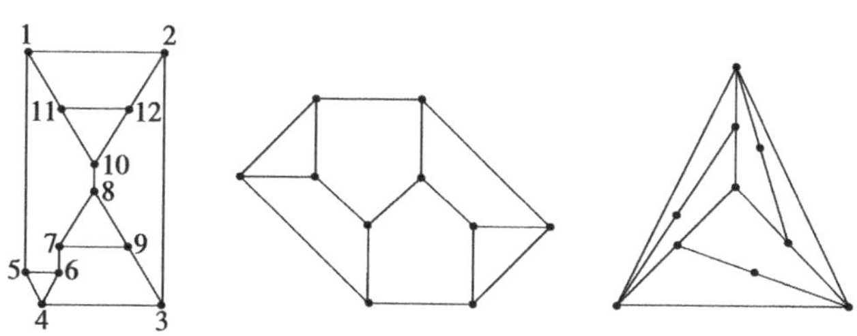

The following interesting proof for the case of $|V(G)| = 6n$, which uses the representation of a graph by a quadratic form, is taken from [Behzad et al. 1979], pages 184–185.

Consider the following quadratic form with $6n$ variables $a_i, \dots, f_i$, $i = 1, \dots, n$, modulo n:

$$\sum(a_ib_i + a_ie_i + a_if_i + b_ic_i + c_id_i + c_if_i + f_ie_i) + \sum(b_je_{j+1} + d_jd_{j+1}) + b_ne_1 + d_1d_n.$$

This form represents the following 3-regular graph, drawn for $n = 5$:

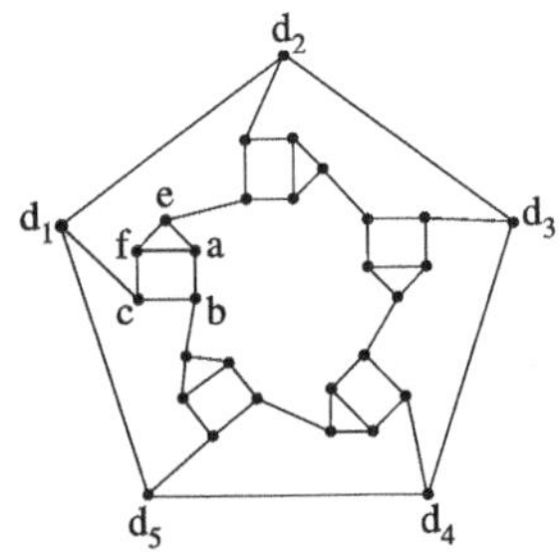

Clearly, any cyclic permutation of the indices leaves the quadratic from unchanged, so $\mathbb{Z}_n$ is a subgroup of the automorphism group. Using the type of a vertex, we show that there exist no other automorphisms of this graph. First, we list the vertices along with their types:

$$a_i, f_i : (3,4,5), \qquad b_i : (4,7,9), \qquad c_i : (4,7,7), \qquad e_i : (3,7,8),$$

$$d_i : \begin{cases} (n,7,7) & \text{if } n \leq 7, \\ (7,7,n) & \text{if } 7 < n < 11, \\ (7,7,11) & \text{if } 11 \leq n. \end{cases}$$

It is clear that only vertices of the same type can be permuted.

Now denote by α the turn by 1 of the 5-cycle, i. e., of the entire graph, so α is an automorphism. Let β be another automorphism.

Case 1. Suppose $\beta(b_1) = b_1$. Since the neighbors a_1, c_1, and e_2 of b_1 have different types, they cannot be permuted; so they must be fixed. The same argument applies to all vertices with index 1.

Case 2. Suppose $\beta(b_1) \neq b_1$. Then because of the type we get $\beta(b_1) = b_j$, with $j \neq 1$. But since $\alpha^{j-1}(b_1) = b_j$, we get that $(\alpha^{j-1})^{-1}(\beta(b_1)) = b_1$. Now by Case 1 we get $(\alpha^{j-1})^{-1}\beta = \alpha^n$ and thus $\beta = \alpha^{j-1}$.

So in both cases $\beta \in \mathbb{Z}_n$. □

The proof of the following theorem (see H. Izbicki [41], and G. Sabidussi [76]) uses the concept of fixed point-free graphs, which will be defined a little later in Definition 7.7.5.

Theorem 7.5.9. *For every finite group A and for all $c, d \in \mathbb{N}$ with $2 \leq c \leq d$ and $d \geq 3$, there exist infinitely many graphs G with* Aut(G) $\cong$ *A which are d-regular and have the chromatic number $\chi(G) = c$, that c is the minimal number of colors needed to color the vertices of G such that adjacent vertices have different colors.*

Proof. The proof goes as follows. Construct a connected graph G' which is fixed point-free, $\Box$-prime (see Theorem 10.5.5) and such that $\mathrm{Aut}(G') = A$. Construct a connected graph $G'' \not\cong G'$ which has the required properties, is $\Box$-prime and satisfies $|\mathrm{Aut}(G'')| = 1$. One then has to prove that $\mathrm{Aut}(G' \Box G'') = \mathrm{Aut}(G') \times \mathrm{Aut}(G'') = A$ (which was done in G. Sabidussi [75]) and that $\Box$ preserves the required properties in the following sense: if G' is fixed point-free, then $G' \Box G''$ is fixed point-free; if G' is m-regular and G'' is n-regular, then $G' \Box G''$ is $m+n$-regular; the chromatic number of $G' \Box G''$ is the maximum of the chromatic numbers of G' and G''; if G' is m-fold connected and G'' is n-fold connected, then $G' \Box G''$ is $m+n$-fold connected. The construction of G uses the Frucht principle. □

Theorem 7.5.10. *For every monoid M and every group A,,, there exists a graph G with a vertex or an edge x such that* $\mathrm{End}(G) \cong A$ *and* $\mathrm{End}(G \setminus \{x\}) \cong M$.

Proof. (See p. 101 ff in P. Hell [38].) The idea of the proof is as follows. Let G_A be a graph with $\mathrm{End}(G_A) = A$ and let G_M be a graph with $\mathrm{End}(G_M) = M$. Such graphs can be obtained by the method of Theorem 7.4.4. We then consider the graph G which is the union of G_M and $|M|$ copies of G_A. Now add edges from the identity element $1_M \in M$ to all vertices of the first G_A, and set $e = (1_M, 1_{G_A})$; also add edges from every other vertex of G_M to all vertices of the corresponding copy of G_A except for the vertex 1_A. It can be shown that any endomorphism fixes everything but the component G_A which is adjacent to 1_M. On this component any automorphism of G_A is possible. One can show analogously that after removing e, all endomorphisms of G_M are possible. Then $\mathrm{End}(G \setminus \{e\}) = M$ and $\mathrm{End}(G) = A$.

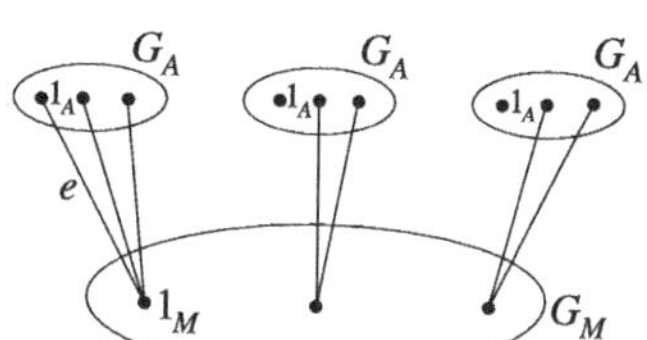

The statement about the vertex x is obtained upon replacing all vertical edges by a path of length 2, where the middle point of the path corresponding to the edge e is called x. Now deletion of the vertex x actually means that we delete e. □

Theorem 7.5.11. *Let H be an arbitrary (finite or infinite) graph and $B \subseteq \mathrm{End}(H)$ a subsemigroup. Then there exists a graph G such that:*

(a) *$H \subseteq G$ is a strong subgraph;*
(b) *for all $\varphi \in \mathrm{End}(G)$ one has $\varphi(H) \subseteq H$;*
(c) *for all $\varphi, \varphi' \in \mathrm{End}(G)$ one has that $\varphi|_H = \varphi'|_H$ implies $\varphi = \varphi'$;*
(d) $\mathrm{End}(G)|_H := \{\varphi|_H \mid \varphi \in \mathrm{End}(G)\} = B$.

Proof. See G. Foldes and G. Sabidussi [20]. □

Theorem 7.5.12. *Every graph with chromatic number k is a strong subgraph of a rigid graph with chromatic number* $k + 1$.

Proof. See L. Babai and J. Nešetřil [8]. □

Theorem 7.5.13. *For every monoid M, there exists a graph G with* $\mathrm{End}(G) \cong M$ *such that G has one of the following properties:*
(a) *G has no cycle shorter than k,* $k > 7$.
(b) *G has chromatic number* 3.
(c) *G is directed and has an arbitrary chromatic number greater than or equal to* 2.

Proof. See E. Mendelsohn [61]. □

Theorem 7.5.14. *For directed graphs, we have the following.*

Exactly those directed cycles $\vec{C}_{p^k}$, *with p prime and* $k \in \mathbb{N}$, *are the graphs G such that:*
(a) $\mathrm{End}(G) \cong \mathbb{Z}_{p^k}$;
(b) $|V(G)| = p^k$; *and*
(c) *no proper subgraph of G has property* (a).

Proof. See R. Goebel [25]. □

7.6 Transformation monoids and permutation groups

We know from earlier sections of this chapter that every group or monoid is the automorphism group or endomorphism monoid, respectively, of an uncolored undirected graph. Here, by "group" we mean what is sometimes called an abstract group, which historically is distinguished from a permutation group. A famous theorem due to A. Cayley shows that every group is a permutation group of some set. The proof goes as follows. Start with the group $A = \{a_1, \ldots, a_n\}$ and define for $a \in A$ the permutation p_a by left multiplication: $p_a(a_i) = aa_i$. Then $\{p_a \mid a \in A\}$ is a subgroup of S_n acting from the left on $\{1, \ldots, n\}$, and is hence what is known as a ***permutation group***.

Another "natural" question that arises is which permutation groups are automorphism groups of graphs.

To make the difference clear, we will use the terminology from representation theory of monoids; see, e. g., [Kilp et al. 2000], which formalizes what we might call nonadditive module theory.

Definition 7.6.1. Let X be a set and M a semigroup. We call (M, X) a ***left M-act*** if there exists a "scalar multiplication" $M \times X \to X$, $(m, x) \mapsto mx \in X$, such that $m'(mx) =$

$(m'm)x$ for $x \in X$ and $m, m' \in M$. If M is a monoid, we require in addition that $1_M x = x$. Analogously, one defines a ***right M-act*** and writes (X, M). In both cases we say that M ***acts*** on X, from the left or from the right, if more precision is needed.

Let (M, X) and (M, Y) be left M-acts. A "linear" mapping $\xi : (M,X) \to (M, Y)$ is called a left ***act (homo)morphism*** if $\xi(mx) = m\xi(x)$ for all $x \in X$ and $m \in M$.

Let (M, X) and (N, Y) be left acts for two semigroups M and N. A pair of mappings $(\mu, \xi) : (M,X) \to (N, Y)$ is called a ***semilinear morphism*** if μ is a semigroup homomorphism and $\xi(mx) = \mu(m)\xi(x)$ for $x \in X$, $m \in M$. If ξ and μ are bijective, we use the term ***semilinear isomorphism***.

We can think of an M-act X as a vector space without addition of vectors and with scalars taken from a semigroup or monoid instead of from the scalar field. Indeed, from the usual eight axioms characterizing vector spaces, only one remains if M is a semigroup and not a monoid, and two remain if we have an identity 1_M.

So every F-vector space V is a two-sided F-act. In this way, linear mappings become act morphisms. The concept of a semilinear mapping from an F-vector space to an F'-vector space, used in linear algebra, turns into the semilinear morphism just defined.

Other examples of left acts over monoids include $(\mathrm{Aut}(G), G)$, $(\mathrm{SEnd}(G), G)$, $(\mathrm{End}(G), G)$, $(\mathrm{Cnd}(G), G)$, $(\mathrm{EEnd}(G), G)$, and $(\mathrm{SEEnd}(G), G)$ if G is a graph and we write endomorphisms from the left.

Note that for a given semigroup M, the left M-acts together with the left act morphisms form a category, denoted by ***M-Act***. See Example 3.1.13.

Definition 7.6.2. Let the group A be a subgroup of S_n, i. e., there exists a semilinear morphism $(\mu, \mathrm{id}_{\{1,\dots,n\}}) : (A, \{1, \dots, n\}) \to (S_n, \{1, \dots, n\})$ where μ is injective. Then we call the left A-act $(A, \{1, \dots, n\})$ a ***permutation group***. For a connected graph $G = (V, E)$ with $|V| = n$, the permutation group $(A, \{1, \dots, n\})$ is the automorphism group of G if $(\mathrm{Aut}(G), V)$ and $(A, \{1, \dots, n\})$ are semilinearly isomorphic as left acts. In this case, we call $(A, \{1, \dots, n\})$ the ***permutation group of the graph*** G.

A monoid A is called the ***transformation monoid of the graph*** G if there exists a connected graph $G = (V, E)$ such that $(\mathrm{End}(G), V)$ and $(A, \{1, \dots, n\})$ are semilinearly isomorphic as left acts.

Question. Which groups are permutation groups and which monoids are transformation monoids of graphs?

We give some examples.

Example 7.6.3. It is clear that the permutation group $(\mathbb{Z}_3, \{1, 2, 3\})$ cannot be the permutation group of a graph G, since G must have three vertices and no such undirected graph has automorphism group $\mathbb{Z}_3$. This is a brute-force argument.

It is also easy to see that the permutation group $(A_4, \{1,2,3,4\})$ is not the permutation group of a graph, by checking all graphs with four vertices; see also *Groups* by P. J. Cameron, in [Beineke/Wilson 1997], page 130.

Now consider the permutation group $(\mathbb{Z}_2 \times \mathbb{Z}_2, \{1,2,3,4\})$. A graph with this group as its permutation group must have four vertices. By considering all graphs with four vertices, we wee that only $K_4 \setminus \{1,3\}$ has automorphism group $\mathbb{Z}_2 \times \mathbb{Z}_2$. Now $(\mu, \mathrm{id}_{\{1,2,3,4\}})$ with $\mu((1,0)) = (13)$ and $\mu((0,1)) = (24)$ is a group isomorphism and establishes the semilinear isomorphism.

Also, $(\mu, \xi) : (\mathbb{Z}_2 \times \mathbb{Z}_2, V(L(K_4 \setminus \{e\}))) \to (\mathrm{Aut}(L(K_4 \setminus \{e\})), V(L(K_4 \setminus \{e\})))$ where ξ is a bijective mapping.

It is clear that S_n is the permutation group and the transformation monoid of the complete graph K_n.

For later use, we define orbits and collect some information about them.

Definition 7.6.4. Let $G = (V, E)$ be a graph and $U \subseteq \mathrm{End}(G)$ a subsemigroup. For $x \in G$, we call $Ux := \{u(x) \mid u \in U\}$ the ***U*-orbit of *x* in *G***.

With this definition the following lemma is clear.

Lemma 7.6.5. *If U is a subgroup of* $\mathrm{Aut}(G)$, *then the U-orbits in G form a partition* $V = V_1 \bigcup \cdots \bigcup V_k$.

Lemma 7.6.6. *Let $V_1, \ldots, V_k$ be the U-orbits in G for a subgroup U of* $\mathrm{Aut}(G)$. *Then for all $i, j \leq k$ and $v, v' \in V_i$, we have $|N_G(v) \bigcap V_j| = |N_G(v') \bigcap V_j|$.*

Proof. Suppose that $\varphi(v) = v'$, i. e., v and v' are in one orbit V_i. As φ is an automorphism, it follows that $\varphi(N_G(v)) = N_G(\varphi(v)) = N_G(v')$. In particular, $|N_G(v')| = |N_G(v)|$ since φ is bijective; see Proposition 1.4.7. For every orbit V_j, one has $\varphi(V_j) = V_j$, and thus

$$|N_G(v') \bigcap V_j| = |N_G(\varphi(v)) \bigcap \varphi(V_j)| = |\varphi(N_G(v) \bigcap V_j)| = |N_G(v) \bigcap V_j|. \qquad \square$$

7.7 Actions on graphs

In this section, we relate automorphism groups and endomorphism monoids even more closely to the elements of a graph by considering the action of the group or monoid of the graph on the vertices of the graph. In particular, we consider transitive actions and fixed point-free actions. As main results, we show that graphs with a transitive action are retracts of Cayley graphs of groups, graphs with a fixed point-free action contract to Cayley graphs of groups, and Sabidussi's theorem saying that graphs with a regular action are Cayley graphs of groups.

Again, some of the results that are not proved can be starting points for further research.

Transitive actions on graphs

The various concepts of the transitive action of a group impose symmetry conditions on the graph. The following definitions can also be formulated for monoid action, in which case symmetry requirements are much weaker. Vertex transitivity is taken up again in Chapter 12.

Definition 7.7.1. A graph $G = (V, E)$ is said to be:
- ***vertex transitive (vertex symmetric)*** if for all $u, x \in V$ there exists $\varphi \in \mathrm{Aut}(G)$ with $\varphi(u) = x$;
- ***edge transitive (edge symmetric)*** if for all $(u, v), (x, y) \in E$ there exists $\varphi \in \mathrm{Aut}(G)$ with $(\varphi(u), \varphi(v)) = (x, y)$;
- ***transitive (symmetric)*** if it is both vertex transitive and edge transitive;
- ***s-transitive*** if for all $u, v, x, y \in V$ with $d(u, v) = d(x, y) = s$ there exists $\varphi \in \mathrm{Aut}(G)$ with $\varphi(u) = x$ and $\varphi(v) = y$;
- ***distance transitive*** if for all $v, u, x, y \in V$ with $d(v, u) = d(x, y)$ there exists $\varphi \in \mathrm{Aut}(G)$ with $\varphi(v) = x$ and $\varphi(u) = y$.

Each of these concepts can be considered for the action of any subset U of the endomorphism monoid $\mathrm{End}(G)$, in which case we will add the prefix U- and write, e. g., "U-vertex transitive."

We note that the one-element group is the permutation group of any asymmetric graph whose points are all fixed points and, (therefore), the group does not act vertex transitively.

The action of $\mathbb{Z}_2 \times \mathbb{Z}_2$ on $K_4 \setminus \{e\}$ is fixed point-free, but $K_4 \setminus \{e\}$ is not $\mathbb{Z}_2 \times \mathbb{Z}_2$-vertex transitive; cf. Example 7.6.3.

Remark 7.7.2. A vertex transitive action of U on G is a *(globally) surjective action* on the vertex set of G, i. e., every element of G is in the image of some $\varphi \in U$.

If G is connected, then 1-transitive implies transitive (i. e., 0-transitive). Note that 1-transitive is stronger than transitive, at least for undirected graphs.

It is clear that a graph is distance transitive if it is s-transitive for all $s \in \mathbb{N}$.

We will now proceed to prove that connected vertex-transitive graphs are retracts of connected Cayley graphs of groups.

Let $D = (V, E)$ be a digraph, $U \subseteq \mathrm{Aut}(D)$ a group and $x \in V$. The ***stabilizer*** U_x of x is $\{s \in U \mid s(x) = x\}$.

Lemma 7.7.3. *Let $D = (V, E)$ be a digraph, $U \subseteq \mathrm{Aut}(D)$ a group and $x, y \in V$. We have that $\{s \in U \mid s(x) = y\}$ is either empty or a left-coset of U_x.*

Proof. If $\{s \in U \mid s(x) = y\} \neq \emptyset$, let $s, s' \in U$ such that $s(x) = s'(x) = y$. Thus $s^{-1}s'(x) = x$, i. e., $s^{-1}s' \in U_x$ and $s' \in sU_x$. □

For the following, recall the definition of the lexicographic product; see Definition 4.4.1.

Theorem 7.7.4. *Let $D = (V,E)$ be a connected digraph without parallel multiple arcs, $U < \mathrm{Aut}(D)$ an inclusion-minimal group acting transitively on D, $x \in V$, and $k = |U_x|$. The digraph $D[\overline{K}_k]$ is a connected Cayley graph of U, and thus in particular, D is a retract of a connected Cayley graph of a group.*

Proof. Set $C = \{s \in U \mid (x, s(x)) \in A\}$. Let us summarize some properties of C:
- By Lemma 7.7.3, C is a disjoint union of left cosets of U_x.
- We have $(s(x), s'(y)) \in E$ if and only if $(x, s^{-1}s'(y)) \in E$ if and only if $s^{-1}s' \in C$.
- If $s \in C$ and $t, t' \in U_x$, then we have $x = t(x) = t't(x)$ and, therefore, $(t(x), st(x)) \in E$, which implies $(t't(x)), t'st(x)) \in E$, i. e., $t'st \in C$. Thus, $U_xCU_x \subseteq C$, while clearly $U_xCU_x \supseteq C$, and hence $U_xCU_x = C$.

By induction on the distance $d(x,y)$ it is easy to see that the group U' generated by C contains an element mapping x to y for all $x, y \in V$. Thus, U' acts transitively on D and $U' \subseteq U$. By the assumption of minimality in the choice of U, we have $U' = U$.

Consider now $\mathrm{Cay}(U, C)$. Since the left cosets of U_x partition U, we can express any element of U in the form st for some $t \in U_x$. If $t, t' \in U_x$, then vertices $(st, s't') \in E$ if and only if $t^{-1}s^{-1}s't' \in C$ which happens if and only if $s^{-1}s' \in C$. Thus, between any two left cosets of U_x either every element of one coset has an arc to all elements of the other or none has. Thus, $\mathrm{Cay}(U, C) \cong D[\overline{K}_k]$.

It is easy to see, that D is a retract of $D[\overline{K}_k]$. □

Regular actions

In order to define regular actions, we first need to define fixed point-free actions. We will come back to such actions with respect to general semigroups in the next section.

Definition 7.7.5. We say that a group $U < \mathrm{Aut}(G)$ acts ***strictly fixed point-free*** on G if for all $x \in G$ and all $\varphi \in U$, $\varphi \neq \mathrm{id}_G$, we have $\varphi(x) \neq x$. In other words, $|U_x| = 1$ for all $x \in V$.

We are ready to define regular actions.

Definition 7.7.6. Let $U < \mathrm{Aut}(G)$ be a subgroup which acts strictly fixed point-free on G. If, in addition, $G = (V, E)$ is U-vertex transitive, we say that we have a ***regular action*** of U on G.

Remark 7.7.7. The notion of a regular action of U on G is fundamental to the concept of a (graphical) regular representation of an abstract group (see, e. g., [Biggs 1996], Definition 16.4 on p. 124). A group which is the automorphism group of a graph G and acts regularly on G is said to have a ***graphical regular representation.***

It has been shown that the only groups which have no graphical regular representation are Abelian groups of exponent greater than 2, generalized dicyclic groups, and 13 exceptional groups, among them $\mathbb{Z}_2^2, \mathbb{Z}_2^3, \mathbb{Z}_2^4$, the dihedral groups D_6, D_8, D_{10}, and the alternating group A_4; see [Biggs 1996] Exercise 16g on page 128. These groups must all be solvable (for the definition check any book on group theory).

We have the following result.

Proposition 7.7.8. *If the subgroup $U < \text{Aut}(G)$ acts strictly fixed point-free on the finite graph G and if $|U| = |G|$, then G is U-vertex transitive. If G is U-vertex transitive, then $|U| \geq |G|$.*

Proof. For $x \in G$, we have $|Ux| = |U|$, since otherwise there would exist $\varphi \neq \varphi' \in U$ with $\varphi(x) = \varphi'(x)$, and thus $\varphi^{-1}\varphi'(x) = x$. Since $|U| = |G|$, we have $|Ux| = |G|$, and consequently $Ux = G$, as $Ux \subseteq G$ and everything is finite. But this means that for every $x, y \in G$ there exists $\varphi \in U$ with $\varphi(x) = y$. This implies the second statement. □

Theorem 7.7.9. *Exactly the following implications hold:*

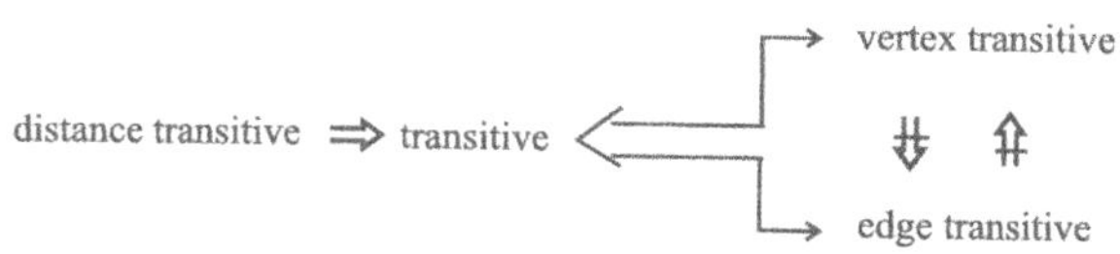

In particular, vertex transitive and edge transitive together imply transitive.

Proof. All implications follow directly from the definitions.

We proved the nonimplications by examples.

The graph $K_{1,2} \cong P_2$ is edge transitive, but it is not vertex transitive since no endpoint can go to the middle point via an automorphism.

The graph $K_3 \square K_2$ (the 3-prism) is vertex transitive but not edge transitive, since otherwise a C_4 would have to go onto a K_3.

The graph $K_4 \setminus \{e\}$ is not edge transitive and not vertex transitive.

The graph C_4 is distance transitive, and thus has all the other properties.

The graph $P(8, 3)$ is depicted below:

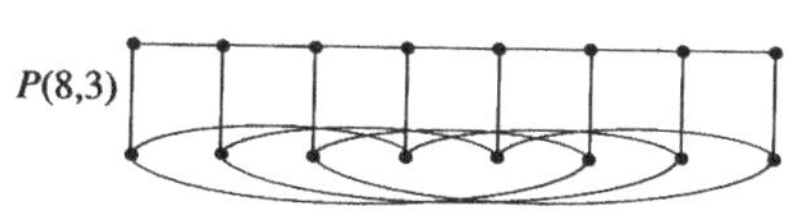

It can be shown to be transitive but not distance transitive; see [Biggs 1996], 15e on page 119. □

Theorem 7.7.10 ([Biggs 1996] p. 115). *If a connected graph is edge transitive but not vertex transitive, then it is bipartite.*

Cayley graphs of groups have been characterized by several authors, one of the first being G. Sabidussi [78]. Since in a regular action all stabilizers are of size 1 an immediate corollary of Theorem 7.7.4 (and Remark 7.3.8 for the backward direction) is, what is known as Sabidussi's theorem.

Corollary 7.7.11. *A connected digraph $D = (V, E)$ without parallel double-arcs is a Cayley graph of a group if and only if there exists a subgroup $U < \mathrm{Aut}(D)$ that acts regularly on D.*

The following theorem (see [Biggs 1996] 16.2, p. 123) concerns vertex transitivity of the König graph of a group A. In the second part, certain group automorphisms of a group A are identified as graph automorphisms of a suitable König graph.

Theorem 7.7.12. *Let A be a group with a generating set $\Omega = \Omega^{-1}$.*
(a) *The König graph $\mathrm{Cay}(A, \Omega)$ is $\mathrm{Aut}(\mathrm{Cay}(A, \Omega))$-vertex transitive.*
(b) *If π is a group automorphism of the group A with $\pi(\Omega) = \Omega$, then $\pi \in \mathrm{Aut}(\mathrm{Cay}(A, \Omega))$ and $\pi(1) = 1$.*

The next theorem (see [Biggs 1996] 16.3, p. 124) gives a criterion for a graph G to be a König graph. It is an immediate consequence of Corollary 7.7.11.

Theorem 7.7.13. *Let G be connected. There exists a subgroup $U \subseteq \mathrm{Aut}(G)$ which acts regularly on the graph G if and only if $G = \mathrm{Cay}(U, \Omega)$ for a suitable generating set $\Omega \subseteq U$.*

Note that the Petersen graph (which is $\overline{LK_5}$) is vertex transitive but not a König graph. With Theorem 7.7.4, a Cayley graph that retracts to the Petersen graph can be constructed.

Fixed point-free actions on graphs

A fixed point-free action is an action for which there are no one-element orbits, apart from orbits under the identity. We give the following definition in its general form for monoids. So far, it is mostly used for subgroups U of $\mathrm{Aut}(G)$.

Definition 7.7.14. We say that a subsemigroup $U \subseteq \mathrm{Aut}(G)$ acts ***strictly fixed point-free*** on G if for all $x \in G$ and all $\varphi \in U$, $\varphi \neq \mathrm{id}_G$, we have $\varphi(x) \neq x$. In other words, every element of U other than id_G does not fix any vertex of G. We say that U acts ***fixed point-free*** on G if for all $x \in G$ there exists $\varphi \in U$ with $\varphi(x) \neq x$.

Lemma 7.7.15. *A subgroup $U \subseteq \mathrm{End}(G)$ acts fixed point-free on G if and only if for all $x, y \in G$ there exists at most one $\varphi \in U$ with $\varphi(x) = y$.*

If the subgroup $U \subseteq \mathrm{End}(G)$ acts strictly fixed point-free on G, then we have $|U| \le |G|$.

Proof. To prove necessity, suppose $\varphi(x) = \psi(x) = y$; then $x = \varphi^{-1}\varphi(x) = \varphi^{-1}\psi(x)$ and $x = \psi^{-1}\varphi(x)$ imply that $\varphi = \psi$.

Sufficiency is obvious, since we can then assume that there exists one $x \in G$ such that $\varphi(x) = x$ for all $\varphi \in U$.

The last statement is clear since no $x \in G$ can have more than $|G|$ images. □

Example 7.7.16. The action of $U = \{0, 2, 4 = -2\}$ on $\mathrm{Cay}(\mathbb{Z}_6, \{1\})$ is strictly fixed point-free. Observe that $\mathrm{Cay}(\mathbb{Z}_6, \{1\})$ is not U-vertex transitive; cf. Definition 7.7.1.

Remark 7.7.17. The action of U on G being strictly fixed point-free is equivalent to saying that G is a *strongly faithful* U-act (i. e., $\varphi(x) = \varphi'(x)$ for some $x \in G$ implies $\varphi = \varphi'$).

The weaker property of U acting fixed point-free on G is equivalent to saying that G is a *faithful* U-act (i. e., $\varphi(x) = \varphi'(x)$ for all $x \in G$ implies $\varphi = \varphi'$).

While Theorem 7.7.4 treated the case of transitive actions of graphs and Corollary 7.7.11 specialized this to regular, i. e., transitive and fixed point-free, actions, the following lemma considers the case of only strictly fixed point-free actions. This lemma is due to L. Babai [5]. For the definition of a contraction, see Definition 1.4.6.

Lemma 7.7.18. *Let $D = (V, E)$ be a connected digraph and $U < \mathrm{Aut}(D)$ a subgroup acting strictly fixed point-free on G. Then D can be contracted to a connected Cayley graph of U.*

Proof. Denote the orbits of U by $M_1, \ldots M_r$. Since D is connected, there is a subset $S \subseteq V$ with $|S \cap M_i| = 1$ for all $1 \leq i \leq r$ such that S induces a connected subgraph of D. More precisely, if there is an edge between two orbits than there is an edge from every vertex of one of these orbits to a vertex of the other orbit. So a way to find S is to start with any vertex and in each round add a neighbor of one of the already taken vertices from an orbit that has not been taken yet.

Now, since for any two vertices $x, y \in V$ there exists at most one $s \in U$ such that $s(x) = y$, no $s \in U \setminus \{\mathrm{id}\}$ has a fixed point. Thus, if $s \in U \setminus \{\mathrm{id}\}$, then $S \cap s(S) = \emptyset$ and therefore $s(S) \cap s'(S) \neq \emptyset$ implies $s = s'$ for all $s, s' \in U$. Moreover, we clearly have $\bigcup_{s \in U} s(S) = V$. Thus, we can define a map $\varphi : X \to U$ by setting $\varphi(x)$ to be the unique $s \in U$ such that $x \in s(S)$. This mapping defines a contraction to a digraph D' on vertex set U, where the connected subgraph $s(S)$ of D is contracted to the vertex s. It is straightforward to check that the left action of U on D' is regular and D' is connected. Thus by Corollary 7.7.11, D' is a Cayley graph of U with respect to a generating set of U. □

7.8 Comments

Permutation groups of graphs have attracted much attention to date, but less attention has been paid to transformation monoids of graphs; see Definition 7.6.2 and the question before Example 7.6.3. The various transitivities may also hold some interest for further research.

Theorems like 7.5.4, 7.5.8, and 7.5.9 can serve as models for questions arising when Aut is replaced by SEnd or even bigger subsets of End such as HEnd, LEnd or QEnd.

Another line of thought would involve replacing the group A in these theorems by a semigroup or monoid which is close to groups, e. g., right or left groups or Clifford semigroups (see Chapter 9 and later).

8 The characteristic polynomial of graphs

We continue the discussion started in Sections 2.5 and 5.3 concerning eigenvalues and characteristic polynomials of graphs.

8.1 Eigenvectors of symmetric matrices

It is often difficult to determine the eigenvalues of graphs or matrices, so it is sometimes useful to obtain bounds for them. We use the so-called Rayleigh quotient of an eigenvector to achieve this aim. The next definition and the two subsequent theorems are valid for any symmetric matrices.

Definition 8.1.1. Take $A = (a_{ij})_{i,j}$, with $i, j \in \{1, \dots, n\}$, and $v = (v_1, \dots, v_n) \in \mathbb{R}^n$. We call

$$R(v) = \frac{\langle v , Av \rangle}{\langle v , v \rangle} = \frac{\sum_{i,j=1}^n a_{ij} v_i v_j}{\sum_{i=1}^n v_i^2}$$

the ***Rayleigh quotient*** of v with respect to A.

Theorem 8.1.2. *If A is symmetric, then for all $v \in \mathbb{R}^n$, $v \neq 0$, we have*

$$\lambda = \lambda(A) \le R(v) \le \Lambda(A) = \Lambda.$$

Moreover,

$$\lambda = R(v) \quad or \quad R(v) = \Lambda$$

if and only if v is an eigenvector for λ or for Λ, respectively.

Proof. Let $u_1, \dots, u_n$ be an orthonormal basis of eigenvectors for A. Choose an arbitrary linear combination $v = \sum_{i=1}^n \xi_i u_i$ and compute

$$\begin{aligned} R(v) &= \frac{\langle v , Av \rangle}{\langle v , v \rangle} = \frac{\sum_{i=1}^n \langle \xi_i u_i , A \xi_i u_i \rangle}{\sum_{i=1}^n \langle \xi_i u_i , \xi_i u_i \rangle} \\ &\overset{u_i \text{ is eigenvector}}{=} \frac{\sum_{i=1}^n \langle \xi_i u_i , \lambda_i u_i \xi_i \rangle}{\sum_{i=1}^n \langle \xi_i u_i , \xi_i u_i \rangle} = \frac{\sum_{i=1}^n \lambda_i \xi_i^2}{\sum_{i=1}^n \xi_i^2}. \end{aligned}$$

This implies that

$$\lambda = \frac{\sum_{i=1}^n \lambda \xi_i^2}{\sum_{i=1}^n \xi_i^2} \overset{\lambda \le \lambda_i}{\le} R(v) \overset{\lambda_i \le \Lambda}{\le} \Lambda.$$

Moreover, if v is an eigenvector for λ, then

$$R(v) = \frac{\langle v , Av \rangle}{\langle v , v \rangle} = \frac{\langle v , \lambda v \rangle}{\langle v , v \rangle} = \frac{\lambda \langle v , v \rangle}{\langle v , v \rangle} = \lambda,$$

and similarly for Λ.

https://doi.org/10.1515/9783110617368-008

Conversely, if v is not an eigenvector for λ, then A has at least two different eigenvalues; therefore, all inequalities are strict. □

Theorem 8.1.3. *Let A be a symmetric matrix with only nonnegative entries, and let $v = (v_1, \ldots, v_n) \in \mathbb{R}^n$ be an eigenvector corresponding to $\Lambda(A)$. Then $\tilde{v} = (|v_1|, \ldots, |v_n|)$ is an eigenvector for $\Lambda(A)$ and $|\lambda(A)| \leq \Lambda(A)$.*

Proof. Note that $a_{ij} \geq 0$ implies $R(\tilde{w}) \geq R(w)$ for all $w = (w_1, \ldots, w_n) \in \mathbb{R}^n$ by the definition of R. Also, Theorem 8.1.2 implies that

$$R(\tilde{w}) \leq \Lambda(A) \Leftrightarrow -R(\tilde{w}) \geq -\Lambda(A) \Rightarrow -R(\tilde{w}) \leq R(w).$$

If w is an eigenvector corresponding to $\lambda(A)$, then again using Theorem 8.1.2 we get

$$-\Lambda(A) \leq -R(\tilde{w}) \leq \lambda(A) \overset{w \text{ EV for } \lambda}{=} R(w) \leq R(\tilde{w}) \leq \Lambda(A), \quad \text{i.e., } |\lambda| \leq \Lambda.$$

If v is an eigenvector corresponding to $\Lambda(A)$, then $\Lambda(A) = R(v) \leq R(\tilde{v}) \leq \Lambda(A)$. Theorem 8.1.2 implies that $\tilde{v}$ is an eigenvector for $\Lambda(A)$. □

Eigenvalues and connectedness

Theorem 8.1.4. *If G is connected, then $\Lambda = \Lambda(G)$ is a simple eigenvalue and every eigenvector of Λ has only nonzero entries of the same sign.*

Proof. By assumption, $A(G)$ cannot be decomposed into blocks; see Theorem 2.1.8.

(a) We show that no entry of v is 0 if $Av = \Lambda v$.
Take $v = (v_1, \ldots, v_s, \underbrace{v_{s+1}, \ldots, v_n}_{=0})$. Then $A\tilde{v} = \Lambda\tilde{v}$ (with $\tilde{v}$ as in Theorem 8.1.3). This means that

$$\sum_{j=1}^{n} a_{ij} \, |v_j| = \Lambda \, |v_i| = 0 \quad \text{for all } i = s+1, \ldots, n.$$

Explicitly, this is saying that

$$\begin{pmatrix} a_{11} & \cdots & a_{1s} & \cdots & a_{1n} \\ \vdots & \ddots & \vdots & \ddots & \vdots \\ a_{s1} & \cdots & a_{ss} & \cdots & a_{sn} \\ a_{(s+1)1} & \cdots & a_{(s+1)s} & \cdots & a_{(s+1)n} \\ \vdots & \ddots & \vdots & \ddots & \vdots \\ a_{n1} & \cdots & a_{ns} & \cdots & a_{nn} \end{pmatrix} \begin{pmatrix} |v_1| \\ \vdots \\ |v_s| \\ 0 \\ \vdots \\ 0 \end{pmatrix} = \Lambda \begin{pmatrix} |v_1| \\ \vdots \\ |v_s| \\ 0 \\ \vdots \\ 0 \end{pmatrix}.$$

As all entries of A are nonnegative and all the $|v_i|$ are positive, we get that the lower left rectangle of the matrix consists entirely of zeros. As A is symmetric, the same is true for the corresponding upper right rectangle of the matrix.
But then A would be block decomposable, which is a contradiction.

(b) Now we show that all components of v have the same sign. Set

$$N_+(v) := \{i \mid v_i > 0\},$$
$$N_-(v) := \{j \mid v_j < 0\},$$

which implies, by (a), that

$$N_+(v) \bigcup N_-(v) = \{1, \ldots, n\}. \qquad (\spadesuit)$$

By Theorems 8.1.2 and 8.1.3, we get that $\Lambda = R(v) = R(\bar{v})$. For $i \in N_+(v)$ and $j \in N_-(v)$ it follows that

$$a_{ij} v_i v_j < 0 \quad \text{or} \quad a_{ij} = 0 \qquad (\diamond)$$

or

$$a_{ij} |v_i| |v_j| > 0 \quad \text{or} \quad a_{ij} = 0.$$

Now $\spadesuit$ and $\diamond$ imply $a_{ij} = 0$ for all such i, j. The symmetry of A would again give a block decomposition. Thus $N_+(v) = \emptyset$ or $N_-(v) = \emptyset$.

(c) Because of (b), there does not exist an eigenvector of Λ orthogonal to v. Otherwise, it would have components smaller than zero as well as components greater than zero to give the scalar product 0 with v. This implies that Λ is simple. □

Corollary 8.1.5. *Every eigenvector corresponding to an eigenvalue $\lambda_i \neq \Lambda(G)$ has at least one negative and at least one positive component.*

Regular graphs and eigenvalues

Theorem 8.1.6. *For a graph G with n vertices, the following statements are equivalent:*
(i) *G is d-regular.*
(ii) *$\Lambda(G) = d_G$, i.e., the sum of the vertex degrees divided by n.*
(iii) *G has $v = {}^t(1, \ldots, 1)$ as an eigenvector for $\Lambda(G)$.*

Moreover, if $m(\Lambda(G)) = r$, then G has exactly r components.

Proof. (i) ⇒ (ii). Owing to the d-regularity of G, all row sums of the adjacency matrix are equal to d, i. e., $Av = dv$ for $v := {}^t(1, \ldots, 1)$. Then d is an eigenvalue corresponding to the eigenvector v. By hypothesis, we have $d = d_G$. By Theorem 8.4.8, we get $d = d_G \leq \Lambda(G) \leq \Delta_G = d$.

(ii) ⇒ (iii). Again take $v := {}^t(1, \ldots, 1)$; then

$$d_G = \frac{1}{n} \sum a_{ij} \overset{\text{by hyp.}}{=} R(v).$$

From Theorem 8.1.2, we get that v is an eigenvector corresponding to $\Lambda(G) = d_G$.

(iii) $\Rightarrow$ (i): By hypothesis, we have $Av = \Lambda(G)v$. As $v = (1, \ldots, 1)$ is an eigenvector corresponding to $\Lambda(G)$, we get for every row i that $\sum_{j=1}^{n} a_{ij} = \Lambda$. Therefore, all vertex degrees are Λ.

Moreover, if G is connected, we get $m(\Lambda(G)) = 1$ by Theorem 8.1.4. If G is not connected, then in the case of d-regularity every component has the eigenvalue $\Lambda = d$ with multiplicity one, so in total we get $m(\Lambda(G)) = r$ if G has r components. □

Exercise 8.1.7. All connected regular graphs with largest eigenvalue 3 and exactly three different eigenvalues are known; see, for instance, J. J. Seidel [84].

8.2 Interpretation of the coefficients of chapo(G)

It turns out that the coefficients of the characteristic polynomial of a directed graph can be interpreted relatively easily. The interpretation for undirected graphs will follow from this.

Theorem 8.2.1. *Let $\vec{G}$ be a directed graph (possibly with loops and multiple edges). For the coefficients of the characteristic polynomial* chapo$(\vec{G}) = \sum_{i=0}^{n} a_i t^{n-i}$, *we have*

$$a_i = \sum_{\vec{L}_i \in \vec{\mathcal{L}}_i} (-1)^{k(\vec{L}_i)}$$

where

$$\begin{aligned} \vec{\mathcal{L}}_i &:= \{\vec{L}_i \subseteq \vec{G} : |\vec{L}_i| = i, \textit{ the components of } \vec{L}_i \textit{ are directed circuits,} \\ &\qquad \text{i. e., } \operatorname{indeg}(x) = \operatorname{outdeg}(x) = 1 \textit{ for all } x \in \vec{L}_i\}, \\ k(\vec{L}_i) &:= \textit{number of components of } \vec{L}_i, \end{aligned}$$

and $a_i = 0$ for $\vec{\mathcal{L}}_i = \emptyset$.

This means that every subgraph $\vec{L}_i$ with i vertices contributes $+1$ or -1 to a_i, depending on whether $\vec{L}_i$ has an even or odd number of directed circuits (cf. [Sachs 1972], pp. 119–134).

Proof. By the Leibniz formula for determinants, we get that the constant coefficient is

$$a_n = (-1)^n \det((a_{ij})_{i,j}) = \sum_{p=\cdots} (-1)^{n+l(p)} a_{1i_1} a_{2i_2} \cdots a_{ni_n},$$

where $l(p)$ is the number of inversions of the permutation $p : j \mapsto i_j$ for $j, i_j \in \{1, \ldots, n\}$. A summand of a_n is therefore not equal to zero if and only if all the $a_{1i_1}, \ldots, a_{ni_n}$ are nonzero, i. e., if and only if $(x_1, x_{i_1}), (x_2, x_{i_2}), \ldots, (x_n, x_{i_n})$ are edges in $\vec{G}$.

As p is a permutation, all second indices are again numbers from $1, \ldots, n$. Thus, in $\vec{G}$, we have a circuit of length n or a circuit of length 2, say x_1, x_2, x_1, and a circuit of length 3, say x_3, x_4, x_5, x_3, etc. For p, this implies $i_2 = 1$, $i_5 = 3$, etc. This is true if all $a_{ij} = 1$, i. e., for simple graphs. For $a_{ij} > 1$, the statement remains true, since in that case $|a_{ij}|$ edges go from x_i to x_j which generate the same number of circuits containing x_i and x_j in this sense.

We now compare this to the right-hand side of the formula. Take $\vec{L}_n \in \vec{\mathcal{L}}_n$ and consider the possible cases:

(a) $\vec{L}_n$ is an n-angle, i. e., $l(p) = n - 1$ and the summand is -1.
Geometric interpretation: we get the negative number of n-angles.

(b) $\vec{L}_n$ is an $(n-1)$-angle and a loop, i. e., $l(p) = n - 2$ and the summand is $+1$.
Geometric interpretation: we get the positive number of $(n-1)$-angles, such that the remaining vertex has a loop.

(c) $\vec{L}_n$ is an $(n-2)$-angle and a 2-circuit or an $(n-2)$-angle and two loops, i. e., $l(p) = n-2$ or $l(p) = n - 3$.
Geometric interpretation: we get the negative number of $(n-2)$-angles and 2-circuits, or the negative number of $(n-2)$-angles with one loop at each of the two other vertices.

(d) $\vec{L}_n$ is an $(n-3)$-angle and one triangle or an $(n-3)$-angle, one 2-circuit, and one loop, etc.

We now use that for the coefficients a_i with $i < n$ one has that $(-1)^i a_i$ is the sum of the principal ith row minors of $A = (a_{ij})$. Each of these corresponds uniquely to a subgraph of $\vec{G}$ on i vertices. □

Example 8.2.2 (corresponding to Theorem 8.2.1).

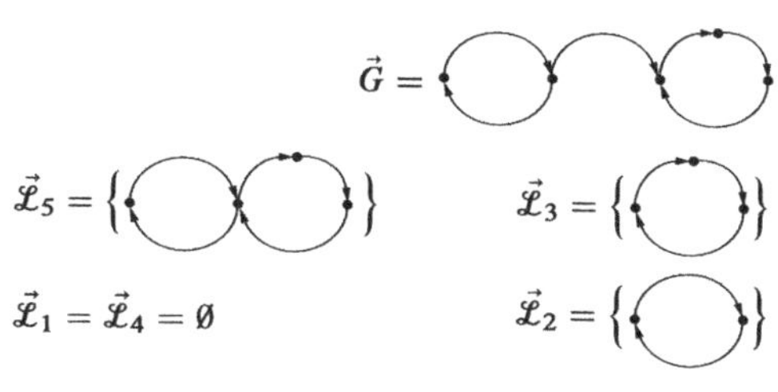

Interpretation of the coefficients for undirected graphs

Theorem 8.2.3. *Let G be without loops and multiple edges with the characteristic polynomial* chapo(G) $= \sum_{i=0}^{n} a_i t^{n-i}$. *Then $a_0 = 1$ and, for $1 \leq i \leq n$,*

$$a_i = \begin{cases} \sum_{H \in \mathcal{K}_i} (-1)^{k(H)} 2^{c(H)} & \text{for } \mathcal{K}_i \neq \emptyset, \\ 0 & \text{for } \mathcal{K}_i = \emptyset, \end{cases}$$

where

$$\mathcal{K}_i := \{H \subseteq G : |H| = i, \text{ components of } H \text{ are } K_2 \text{ or circuits}\},$$
$$k(H) := \text{number of components of } H,$$
$$c(H) := \text{number of circuits of } H.$$

Proof. The idea is to replace in G the edge $\{x_1, x_2\}$ by (x_1, x_2) and (x_2, x_1); call the result $\vec{G}$. Now we count the circuits of G in $\vec{G}$ twice; the edges become 2-circuits and are counted in $\vec{G}$ only once, all according to Theorem 8.2.1. □

Corollary 8.2.4. *Take G without loops and multiple edges. Then the coefficients of the characteristic polynomial are such that:*

$$\begin{aligned} a_0 &= 1; \\ a_1 &= 0; \\ -a_2 &= |E|; \\ -\frac{a_3}{2} &= \text{number of triangles in } G; \\ a_4 &= \text{number of pairs of disjoint edges} \\ &\quad - \text{twice the number of rectangles.} \end{aligned}$$

If G has loops, then $-a_1$ = *the number of loops, but the other coefficients of* chapo(*G*) *are more difficult to interpret.*

Corollary 8.2.5. *The length of the shortest odd circuits in G is the first odd index* $i \neq 1$ *with* $a_i \neq 0$*, and there are* $-a_i/2$ *shortest odd circuits.*

Exactly for the bipartite graphs, all coefficients with odd i are zero.

For trees, we have that the number of choices of r disjoint edges in the tree is $(-1)^r a_{2r}$.

Proof. The statements follow from the structure of the $\mathcal{K}_i$. For odd i, each $H \in \mathcal{K}_i$ does not consist of K_2 only. In $\mathcal{K}_5$, say, there is no C_5 but nevertheless $a_5 \neq 0$; so there exists at least one C_3, which then, however, appears in $\mathcal{K}_3$. The cardinality of this $\mathcal{K}_i$ is $-a_i/2$, since the smallest odd circuit is counted twice.

For bipartite graphs everything is clear, since they don not have odd circuits, i. e., $\mathcal{K}_i = \emptyset$ for odd i.

The statement for trees is also clear from Corollary 8.2.4. You may want to check this statement for some (small) trees. □

Example 8.2.6 (Coefficients for undirected graphs).

$G \qquad \text{chapo}(G) = t^4 - 5t^2 - 4t$

$i = 1: \quad \mathcal{K}_1 = \emptyset$

$a_1 = 0$

$i = 2:$ $\mathcal{K}_2 =$ 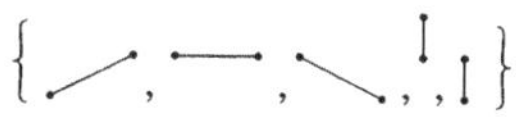

$a_2 = -5$

$k(H) = 1$ for all H, $c(H) = 0$ for all H

$i = 3:$ $\mathcal{K}_3 =$ 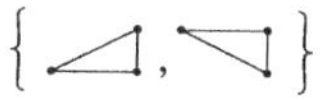

$a_3 = (-1) \cdot 2 + (-1) \cdot 2 = -4$

$k(H) = 1$ for all H $c(H) = 1$ for all H

$i = 4:$ $\mathcal{K}_4 =$ 

$a_4 = 0$

$k(H_1) = 1,\quad c(H_1) = 1$

$k(H_2) = 2,\quad c(H_2) = 0$

$k(H_3) = 2,\quad c(H_3) = 0$

8.3 Characteristic polynomials of trees

Theorem 8.3.1 (Recursion formula for trees). *Let $G = (V, E)$ be a tree with $|E| > 2$, and take $x, x' \in V$ with $\deg(x) = 1$ and $\{x, x'\} \in E$. Then*

$$\mathrm{chapo}(G; t) = t \cdot \mathrm{chapo}(G \backslash x) - \mathrm{chapo}((G \backslash x) \setminus x').$$

See E. Heilbronner [35].

Example 8.3.2 (Characteristic polynomials of trees).

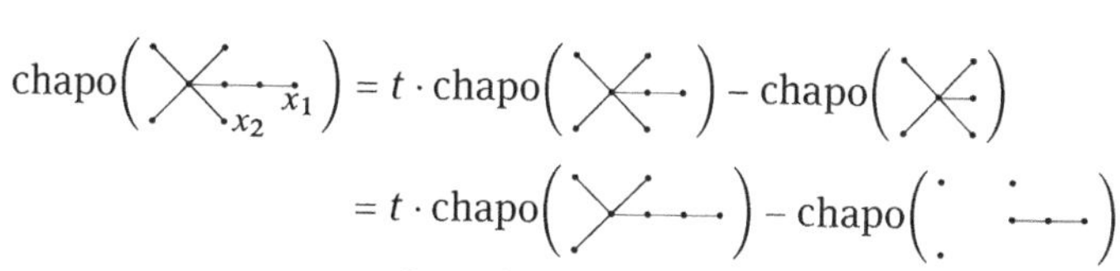

$$= t^8 - 7t^6 + 9t^4.$$

Corollary 8.3.3. *For paths P_n with $n - 1$ edges, where $n \geq 2$, we get*

$$\mathrm{chapo}(P_n; t) = t \cdot \mathrm{chapo}(P_{n-1}; t) - \mathrm{chapo}(P_{n-2}; t).$$

Exerceorem 8.3.4. If G is composed from G_1 and G_2 such that there is exactly one joining in-between, say $x_1 \in V(G_1)$ is joined by an edge to $v_2 \in V(G_2)$, then

$$\mathrm{chapo}(G) = \mathrm{chapo}(G_1)\,\mathrm{chapo}(G_2) - \mathrm{chapo}(G_1 \setminus x_1)\,\mathrm{chapo}(G_2 \setminus x_2).$$

Example 8.3.5 (Characteristic polynomials of undirected paths).

$$\begin{aligned}
\text{chapo}(P_0) &= t,\\
\text{chapo}(P_1) &= t^2 - 1,\\
\text{chapo}(P_2) &= t^3 - 2t,\\
\text{chapo}(P_3) &= t^4 - 3t^2 + 1,\\
\text{chapo}(P_4) &= t^5 - 4t^3 + 3t,\\
\text{chapo}(P_5) &= t^6 - 5t^4 + 6t^2 - 1,\\
\text{chapo}(P_6) &= t^7 - 6t^5 + 10t^3 - 4t,\\
\text{chapo}(P_7) &= t^8 - 7t^6 + 15t^4 - 10t^2 + 1,\\
\text{chapo}(P_8) &= t^9 - 8t^7 + 21t^5 - 20t^3 + 5t.
\end{aligned}$$

See F. Harary, Clarence King and A. Mowshowitz [31].

8.4 The spectral radius of undirected graphs

Let $G = (V, E)$ be a graph with $|V| = n$ and $|E| = m$. Denote by $\lambda(G)$ the smallest eigenvalue and by $\Lambda(G)$ the largest eigenvalue of G.

Subgraphs

Theorem 8.4.1. *If $G' \subseteq G$ is a subgraph, then $\Lambda(G') \leq \Lambda(G)$.*

If G' is a strong subgraph, then, in addition, $\lambda(G) \leq \lambda(G')$.

Proof. We prove the first statement for $G' = G \setminus e$, with $e = \{x_i, x_j\}$. Let $v = (v_1, \dots, v_n)$ be an eigenvector corresponding to the eigenvalue $\Lambda(G')$, with norm $\|v\| = 1$. By Theorem 8.1.4, we have $v_l \geq 0$ for all $l = 1, \dots, n$. Then, using Theorem 8.1.2, the definition of R, $v_i, v_j > 0$ and again Theorem 8.1.2 in this order, we get

$$\Lambda(G \setminus e) = R_{G\setminus e}(v) = R_G(v) - 2v_i v_j \leq R_G(v) \leq \Lambda(G).$$

By induction, we get the statement about $\Lambda(G)$.

Now let G' be a strong subgraph. We prove both assertions for $G' = G \backslash x_i$. We obtain $A(G \backslash x_i)$ from $A(G)$ by deletion of the ith row and column. For $v \in \mathbb{R}^{n-1}$, we denote by $\hat{v} \in \mathbb{R}^n$ the vector obtained from v by inserting 0 at the position i.

For all $v \in \mathbb{R}^{n-1}$, $v \neq 0$, we have $R_{G\backslash x}(v) = R_G(\hat{v})$ by the definition of R.

Let v be an eigenvector of $G \setminus x$ for $\Lambda(G \setminus x)$. By Theorem 8.1.2, we have

$$\Lambda(G \backslash x) = R_{G\backslash x}(v) = R_G(\hat{v}) \leq \Lambda(G).$$

Moreover, we have that for an eigenvector v corresponding to $\lambda(G \setminus x)$,

$$\lambda(G) \le R_G(\hat{v}) = R_{G\setminus x}(v) = \lambda(G\setminus x).$$

The statement follows again by induction. □

Example 8.4.2. To illustrate the situation, we look at the path P_2 (see Example 8.3.5). It has smallest eigenvalue $-\sqrt{2}$. It is not a strong subgraph of K_3. So we have $-1 = \lambda(K_3) > \lambda(P_2) = -\sqrt{2}$.

The following theorem goes back to Cauchy, although we are not sure whether its name (which to some extent describes how the eigenvalues are arranged) is due to Cauchy too.

Theorem 8.4.3 (Interlacing Theorem). *Let $\lambda_1 \le \cdots \le \lambda_n$ be the spectrum of G and $\mu_1 \le \cdots \le \mu_{n-1}$ the spectrum of $G \setminus x$. Then*

$$\lambda_1 \le \mu_1 \le \lambda_2 \le \mu_2 \le \cdots \le \mu_{n-1} \le \lambda_n.$$

Upper bounds

As spectra and spectral radii are often not easy to determine, we now derive some upper bounds. In the following, n denotes the number of vertices and m the number of edges of G.

Theorem 8.4.4. *We have $\Lambda(G) \le \sqrt{\frac{2m(n-1)}{n}}$.*

Proof. For $n = 1$ everything is clear since $\Lambda(K_1) = 0$.

If $n \ge 2$, then in $\mathbb{R}^p$ we use the Cauchy–Schwarz inequality

$$\Big(\sum_{i=1}^{p} a_i b_i\Big)^2 \le \Big(\sum_{i=1}^{p} a_i^2\Big)\Big(\sum_{i=1}^{p} b_i^2\Big).$$

Setting $p = n-1$, $a_i = 1$ and $b_i = \lambda_i$ for all $1 \le i \le n-1$, we get

$$\Big(\sum_{i=1}^{n-1} 1\lambda_i\Big)^2 \le (n-1)\Big(\sum_{i=1}^{n-1} \lambda_i^2\Big).$$

As $\sum_{i=1}^{n} \lambda_i = 0$ (Theorem 2.5.6), we get

$$\sum_{i=1}^{n-1} \lambda_i = -\lambda_n \quad \text{and thus} \Big(\sum_{i=1}^{n-1} \lambda_i\Big)^2 = \lambda_n^2 \quad \text{with } \Lambda(G) = \lambda_n.$$

Consequently,

$$\lambda_n^2 \le (n-1)\Big(\sum_{i=1}^{n-1} \lambda_i^2\Big) + (n-1)\lambda_n^2$$

$$\Leftrightarrow n\lambda_n^2 \le (n-1)\underbrace{\sum_{i=1}^{n}\lambda_i^2}_{2|E|} \overset{\text{Theorem 2.5.6}}{=} (n-1)\cdot 2\cdot |E| \overset{|K|=m}{=} (n-1)\cdot 2\cdot m$$

$$\Rightarrow \Lambda(G) \le \sqrt{\frac{2m(n-1)}{n}}. \qquad \square$$

We state some more results without giving proofs here.

Theorem 8.4.5 (Schwenk, unpublished according to [Behzad et al. 1979]).

$$\Lambda(G) \le \sqrt{2m-n+1}.$$

Remark 8.4.6. Theorem 8.4.5 gives a better bound than does Theorem 8.4.4, since

$$\frac{2m}{n} = \frac{1}{n}\sum_{x\in G}\deg(x) =: d_G \quad \text{and, by Theorem 8.4.4,} \quad \Lambda(G) \le \sqrt{2m-d_G}.$$

As $d_G \le n-1$ (with equality for K_n), it follows that $\sqrt{2m-n+1} \le \sqrt{2m-d_G}$.

Corollary 8.4.7. *If G is connected,*

$$\Lambda(G) \le \sqrt{2|E|-|V|+1} = \sqrt{m+\xi(G)}.$$

Proof. Use the formula for the cyclomatic number (Corollary 6.2.14). □

Lower bounds

Theorem 8.4.8. *Let Δ_G be the largest and $d_G = \frac{1}{n}\sum_{x\in G}\deg(x)$ the average vertex degree in G. Then*

$$d_G \le \Lambda(G) \le \Delta_G.$$

Proof. For $v = (1,\dots,1)$ one has ${}^tA = {}^t(\deg(x_1),\dots,\deg(x_n))$. We calculate that

$$\Lambda(G) \overset{\text{Theorem 8.1.2}}{\ge} \frac{\langle v, Av\rangle}{\langle v, v\rangle} \overset{v=(1,\dots,1)}{=} \frac{\sum_{x\in G}\deg(x)}{n} = d_G.$$

Let $v = (v_1,\dots,v_n)$ be an eigenvector for $\Lambda(G)$, where we may assume that $v_i > 0$ for all $i = 1,\dots,n$, and let $v_p := \max\{v_1,\dots,v_n\}$. Because $Av = \Lambda v$, for the pth component we have that

$$\Lambda v_p = \sum_{j=1}^{n} a_{pj}v_j \le v_p\sum_{j=1}^{n} a_{pj} \le v_p\Delta_G. \qquad \square$$

Theorem 8.4.9. *If G is connected with $|V| = n \geq 2$, then*

$$\Lambda(G) \geq 2\cos\frac{\pi}{n+1}.$$

Equality holds exactly for the path P_{n-1} with n vertices. In particular, the graph G is not connected if $\Lambda(G) < 2\cos\frac{\pi}{n+1}$.

Proof. See L. Collatz and U. Singowitz [14]. □

Exercise 8.4.10. Find a definition of the chromatic number χ and prove that $\chi(G) \leq 1 + \Lambda(G)$.

Exerceorem 8.4.11. For a connected graph G, one has $\chi(G) = 1 + \Lambda(G)$ if and only if $G = K_n$ or $G = C_{2m+1}$ with suitable n, m.

Project 8.4.12. Find other bounds in the literature or from the internet and collect the different bounds in a table; include all the information necessary for each bound.

8.5 Spectral determinability

In this section, we collect some results on graphs which are determined by the spectrum up to isomorphism.

Spectral uniqueness of K_n and $K_{p,q}$

Theorem 8.5.1. *For the graph G with eigenvalues $\lambda = \lambda_1 \leq \cdots \leq \lambda_n = \Lambda$, we have:*
(a) $G \cong K_n \Leftrightarrow \Lambda = n-1$;
(b) $G \cong K_{p,q} \Leftrightarrow \Lambda = -\lambda = \sqrt{pq}$ *and* $\lambda_i = 0$ *for all* $1 < i < n = p+q$.

Proof. (a) The "$\Rightarrow$" statement is just Proposition 2.5.10.
For "$\Leftarrow$,"

$$\begin{aligned}\Lambda = n-1 \quad &\overset{\text{Theorem 8.4.5}}{\Rightarrow} \quad n-1 \leq \sqrt{2m-n+1}\\ &\Leftrightarrow \quad n^2-2n+1 \leq 2m-n+1\\ &\Leftrightarrow \quad \frac{n^2-n}{2} \leq m\\ &\Leftrightarrow \quad \frac{n(n-1)}{2} \leq m.\end{aligned}$$

Then

$$G \cong K_n \quad \text{as} \quad \frac{n(n-1)}{2} = |E(K_n)|.$$

(b) The "⇒" statement is Theorem 2.5.11.

To prove "⇐," note that as $\mathrm{chapo}(G) = (t^2 + a_2)\, t^{p+q-2}$, coefficients with odd index are zero, and so by Corollary 8.2.5 the graph G is bipartite. From $-a_2 = pq$, which is the number of edges, we conclude that G is complete bipartite. From $p + q = n$, we can determine p and q. □

Theorem 8.5.2. *Let $G \neq K_1$ be connected with* $\mathrm{chapo}(G) = \sum_{i=0}^{n} a_i t^{n-i}$ *and $\lambda = \lambda_1 \leq \lambda_2 \cdots \leq \lambda_n = \Lambda$. The following are equivalent:*

(i) *G is bipartite.*

(ii) *$a_{2i-1} = 0$ for all $1 \leq i \leq \lceil \frac{n}{2} \rceil$.*

(iii) *$\lambda_i = -\lambda_{n+1-i}$ for $1 \leq i \leq n$; i. e., $-\lambda_1 = \lambda_n$, $-\lambda_2 = \lambda_{n-1}$, and so on.*

(iv) *$\Lambda = -\lambda$.*

Moreover, $m(\lambda_i) = m(-\lambda_i)$.

Proof. (i) ⇔ (ii) is Corollary 8.2.5.

(i) ⇒ (iii): Let $\{x_1, \ldots, x_s\}$, $\{x_{s+1}, \ldots, x_n\}$ be a bipartition. In the adjacency matrix A of a bipartite graph,

$$A = \begin{pmatrix} 0 & \\ & 0 \end{pmatrix},$$

we have $a_{ij} = 0$ for $i, j \leq s$ and for $i, j \geq s + 1$.

Let λ be an eigenvalue with the eigenvector v, i. e., $Av = \lambda v$. Then

$$Av = \begin{pmatrix} \sum_{i=s+1}^{n} a_{1i} v_i \\ \vdots \\ \sum_{i=s+1}^{n} a_{si} v_i \\ \sum_{i=1}^{s} a_{(s+1)i} v_i \\ \vdots \\ \sum_{i=1}^{s} a_{ni} v_i \end{pmatrix} = \lambda \begin{pmatrix} v_1 \\ \vdots \\ v_s \\ v_{s+1} \\ \vdots \\ v_n \end{pmatrix}.$$

This implies that $\tilde{v} = {}^t(v_1, \ldots, v_s, -v_{s+1}, \ldots, -v_n)$ is an eigenvector corresponding to the eigenvalue $-\lambda$. The ordering of the eigenvalues gives (iii).

The mapping $v \mapsto \tilde{v}$ provides an isomorphism between $\mathrm{Eig}(G, \lambda)$ and $\mathrm{Eig}(G, -\lambda)$. Diagonalizability implies that $\dim \mathrm{Eig}(G, \lambda) = m(\lambda)$ for all eigenvalues λ. Then $m(\lambda) = m(-\lambda)$ for all λ and the last statement is also proved.

(iii) ⇒ (iv) is trivial.

(iv) ⇒ (i): For $Av = \lambda_1 v$, we may assume that $\|v\| = 1$. Then $R(v) = \sum a_{ij} v_i v_j = \lambda_1 = -\lambda_n$ by hypothesis, and the triangle inequality gives

$$\sum a_{ij} |v_i| |v_j| \geq |R(v)| \geq \lambda_n.$$

Using Theorem 8.1.2, we also get the converse relation, i. e., we have equality everywhere. Moreover, $\tilde{v}$ is an eigenvector for λ_n (see Theorem 8.1.3) and all its coordinates are nonzero. For v, we then have $v_1, \ldots, v_s > 0$, say, and $v_{s+1}, \ldots, v_n < 0$, where $s \neq 0, n$ because eigenvectors corresponding to different eigenvalues of a symmetric matrix are orthogonal. On the other hand, $\sum a_{ij}|v_i||v_j| = |\sum a_{ij}v_iv_j|$ is possible only if no two summands on the right have opposite signs.

If all the summands are negative, say, then because $a_{ij} \geq 0$, we get that $a_{ij} = 0$ for all $i \leq s$ and $j > s$, and vice versa. Thus

$$A = \begin{pmatrix} & 0 \\ 0 & \end{pmatrix},$$

and consequently G would not be connected.

If all the summands are nonpositive, it follows that $a_{ij} = 0$ for all $i, j \leq s$ and $i, j > s$. Then we have

$$A = \begin{pmatrix} 0 & \\ & 0 \end{pmatrix},$$

and G is bipartite. □

Exerceorem 8.5.3. The Petersen graph with characteristic polynomial $(t-3)(t-1)^5(t+2)^4$ is uniquely determined by its spectrum. Is the same true for the dodecahedron graph? Its characteristic polynomial is $(t-3)(t^2-5)^3(t-1)^5t^4(t+2)^4$.

Exerceorem 8.5.4. If G is connected with $\Lambda(G) = 2$, then

$G \in \{K_{1,4}, C_p,$ , , , $\}$.

If G is connected with $\Lambda(G) < 2$, then G is a subgraph of one of these graphs.

Exerceorem 8.5.5. A connected graph has exactly one positive eigenvalue if it is ***complete multipartite***, i. e., the vertex set has an r-partition $V_1, \ldots, V_r$ such that there are no edges inside one V_i and any two vertices from different V_i are adjacent.

8.6 Eigenvalues and group actions

First, we use the concept of a group orbit on a graph to gain some more information about the characteristic polynomial of certain graphs. This, in turn, gives information about the automorphism group of a graph in some cases.

Groups, orbits, and eigenvalues

Theorem 8.6.1. *Let $G = (V, E)$ be a (directed) graph, where $V_1, \dots, V_k$ are the* Aut(G) *orbits in G. Take the matrix $T = (t_{ij})$, where t_{ij} is the number of edges between V_i and V_j. Then* chapo(T) *is of degree k and divides the characteristic polynomial* chapo(G).

Proof. (See A. Mowshowitz [65], and also [Godsil/Royle 2001], Theorem 9.3.3 on p. 197.) By Lemma 7.6.5 the Aut(G)-orbits form a partition $V_1, \dots, V_k$, i. e., G is multipartite. Consider the vector ${}^t z = (z_1, \dots, z_1, \dots, z_k, \dots, z_k)$, where z_i appears exactly $|V_i|$ times for $i = 1, \dots, k$. Let $T = (t_{ij})$ be the matrix where t_{ij} is the number of edges from V_i to V_j for $i, j = 1, \dots, k$; cf. Lemma 7.6.6. Then for $A = A(G)$, we get

$$Az = \begin{pmatrix} t_{11}z_1 + \cdots + t_{1k}z_k \\ t_{21}z_1 + \cdots + t_{2k}z_k \\ \vdots \\ t_{k1}z_1 + \cdots + t_{kk}z_k \end{pmatrix} = T\overline{z}$$

where $\overline{z} = {}^t(z_1, \dots, z_k)$. Now choose for $\overline{z} \in \mathbb{C}^k$ an eigenvector corresponding to some eigenvalue λ of T, i. e., $T\overline{z} = \lambda\overline{z}$. Thus $Az = \lambda z$. This means that every eigenvalue of T is an eigenvalue of A, and thus chapo(T) divides chapo(A). □

Corollary 8.6.2. *If the characteristic polynomial* chapo(G) *is irreducible over $\mathbb{Q}$ (but not only in this case), we have* $|\text{Aut}(G)| = 1$.

Proof. See A. Mowshowitz [63], as stated in [Cvetković et al. 1979], page 153, Exercise 5.51.

If chapo(G) is irreducible over $\mathbb{Q}$, and hence over $\mathbb{Z}$, we automatically get chapo(A) = chapo(T). Then $k = n$ and all orbits are one-element orbits. □

Example 8.6.3. The converse is not true in general, i. e., there exist asymmetric graphs with reducible characteristic polynomial.

Consider a tree T with an odd number of vertices and $|\text{Aut}(T)| = 1$, i. e., a path with six vertices where vertex 3 has a pending edge. This gives a bipartite graph with seven vertices, i. e., the constant coefficient a_7 in the characteristic polynomial is zero; cf. Corollary 8.2.5. Then chapo(T) has a factor t and, therefore, is not irreducible.

Another example is the following asymmetric graph:

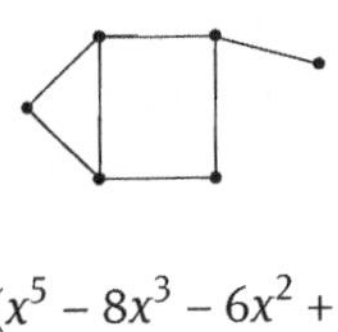

It has characteristic polynomial $x(x^5 - 8x^3 - 6x^2 + 8x + 6)$.

Exerceorem 8.6.4. If G has an automorphism with s odd and t even orbits, then the number of simple eigenvalues of G is no greater than $s + 2t$. There are examples with equality and with strict inequality.

8.7 Transitive graphs and eigenvalues

In Section 2.7, we presented a result connecting automorphisms and eigenvectors, which we used already for line graphs in Section 5.3. This Theorem 2.7.7 will be used again now.

Theorem 8.7.1. *Let G be connected, d-regular, undirected, and* Aut(G)*-vertex transitive. Let λ be a simple eigenvalue. If $|V| = n$ is even, then $\lambda \in \{2\alpha - d \mid \alpha \in \{0, \dots, d\}\}$. If $|V|$ is odd, then $\lambda = d$.*

Proof. See [Cvetković et al. 1979], Theorem. 5.2 on page 136, originally due to M. Petersdorf and H. Sachs [73]. Let P be the matrix of an automorphism $p \in \mathrm{Aut}(G)$. Let v be an eigenvector of λ. From Theorem 2.7.7, we get $Pv = \pm v$. If, say, $p(x_i) = x_j$ for $x_i, x_j \in V(G)$, we get for the components v_i, v_j of v that $v_i = (Pv)_j = \pm v_j$. As G is Aut(G)-vertex transitive, we can find such p for each pair of vertices. Thus $v_i = \pm v_j$ for all components of the above eigenvector corresponding to λ.

Now, if n is odd, Theorem 8.1.6 implies that $u = {}^t(1, \dots, 1)$ is an eigenvector for d. If $\lambda \neq d$ we get $\langle u, v\rangle = 0$, since eigenvectors for different eigenvalues are orthogonal. Moreover, a calculation gives $\sum v_i = 0$, which is not possible for an odd number of summands of the same nonnegative value. Therefore, in this case, $\lambda = d$.

If n is even, we set $\alpha := |\{x_j \in N_G(x_i) : v_j = v_i\}|$ for $x_i \in G$, and thus $d - \alpha = |\{x_j \in N_G(x_i) : v_j = -v_i\}|$. Because $Av = \lambda v$, we get $(Av)_i = \lambda v_i$, where the components of v are added which correspond to the neighbors of x_i. Consequently, $(Av)_i = \alpha v_j - (d - \alpha)v_j = (2\alpha - d)v_j = \lambda v_j$, i. e., $\lambda = 2\alpha - d$. □

In the following result, we relate the investigation of transitivity to eigenvalues.

Theorem 8.7.2. *Let G be a d-regular, undirected and* Aut(G)*-vertex transitive graph with $|V| = 2^q k = n$, where k is odd. Then the following hold:*

(a) *If $q = 0$, then $\lambda = d$ is the only simple eigenvalue of the graph G.*
(b) *If $q = 1$, then G has at most one simple eigenvalue $\lambda \neq d$ and, if so, then $\lambda = 4\beta - d$ where $\beta \in \{0, 1, \dots, \frac{1}{2}(d - 1)\}$.*
(c) *If $q \geq 2$, then G has at most 2^q simple eigenvalues including $\lambda = d$; they are all of the form $\lambda = 2\alpha - d$ for $\alpha \in \{0, 1, \dots, d\}$.*

Proof. See [Cvetković et al. 1979], Theorem 5.3 and footnote on page 137; the result was obtained by H. Sachs and M. Stiebitz. Let $v \in \mathbb{R}^n$ be an eigenvector corresponding to a simple eigenvalue λ, and let P be the matrix of an automorphism $p \in \mathrm{Aut}(G)$. Then Theorem 2.7.7 implies $Pv = \pm v$. If we suppose that $p(x_i) = x_j$, then for the components

v_i, v_j of v we get that $v_i = (Pv)_j = \pm v_j$. As G is vertex transitive, such p exists for every pair of vertices of G. Thus $v_i = \pm v_j$ for components of the eigenvector v of λ.

(a) This is the case where n is odd. Theorem 2.7.5 implies that $u = {}^t(1, \dots, 1)$ is an eigenvector to the eigenvalue d. If $\lambda \neq d$, we get $\langle u, v \rangle = \sum v_i = 0$, as eigenvectors corresponding to different eigenvalues are orthogonal. The second equality is not possible for an odd number of summands with the same nonzero absolute value. So we have $\lambda = d$ in this case.

(c) This is the case where n is even. For $x_i \in G$, set $\alpha := |\{x_j \in N_G(x_i) : v_j = v_i\}|$. Then $d - \alpha = |\{x_j \in N_G(x_i) : v_j = -v_i\}|$. Now $Av = \lambda v$ implies $(Av)_i = \lambda v_i$. Here, those components of v are added which correspond to the neighbors of x_i. So we get $(Av)_i = \alpha v_i - (d - \alpha)v_i = (2\alpha - d)v_i = \lambda v_i$, i. e., $2\alpha - d = \lambda$.

We omit the proof of the special case (b). □

Exercise 8.7.3. Find examples of each case. Cases (b) and (c) need two examples each, because of the "at most."

Theorem 8.7.4. *Let G be undirected, d-regular and* Aut(G)*-vertex transitive, and let λ be a simple eigenvalue. Then $\lambda = \pm d$.*

Proof. Take $x_j, x_\ell \in N(x_i)$. By hypothesis, there exists an automorphism p with $p(x_i) = x_i$ and $p(x_j) = x_\ell$. This implies that for the permutation matrix P of p, we have $Pv = v$. Therefore, $x_j = x_\ell$. Consequently, $\alpha = 0$ or d and $\lambda = \pm d$, with the notation of Theorem 8.7.1. □

Corollary 8.7.5. *Under the conditions of Theorem* 8.7.4*, d and $-d$ are the only possible simple eigenvalues and $-d$ arises exactly when G is bipartite.*

Proof. The first statement is clear; the second follows from Theorem 8.5.2. □

Derogatory graphs

The following definition for matrices originates from linear algebra. It raises some questions for graphs that are "natural" in the mathematical sense. Can we describe (some) derogatory and nonderogatory graphs?

Definition 8.7.6. A graph G is said to be ***derogatory*** if its minimal and characteristic polynomials do not coincide, i. e., if mipo(G) $\neq$ chapo(G).

It is clear that graphs whose eigenvalues are all simple are not derogatory. The next theorem characterizes such graphs if they are assumed to be Aut(G)-vertex transitive.

Theorem 8.7.7. *Let G be directed and* Aut(G)*-vertex transitive. Then G is not derogatory if and only if all eigenvalues of G are simple. All undirected d-regular graphs other than K_2 are derogatory.*

Proof. Suppose that G is undirected and all eigenvalues of G are simple. Then $A(G)$ is diagonalizable. From Theorem 8.7.2(a), we get that G has only one simple eigenvalue if $|V(G)|$ is odd. So $|V(G)|$ must be even. Then Theorem 8.7.2(c) gives $d+1$ simple eigenvalues with d-regularity. But then $d+1=n$. This implies $G=K_n$, and thus $n=2$ since all eigenvalues have to be simple.

The converse is trivial.

For the case of directed graphs, see Theorem 15 on page 87 of H. Sachs and M. Stiebitz [82]. □

Exercise 8.7.8. Compute the minimal polynomials for some derogatory graphs such as K_3, as well as for directed Aut-vertex transitive graphs with nonsimple eigenvalues.

Graphs with Abelian groups

Here, we impose commutativity as an algebraic restriction on the group of a graph G with n vertices.

Theorem 8.7.9. *Let G be undirected and* Aut(G)*-vertex transitive, and suppose that* Aut(G) *is Abelian. Then* Aut(G) *acts strictly fixed point-free on G and consists entirely of involutions, i. e., $g^2 = 1_G$ for all $g \in G$. These groups are the so-called elementary Abelian 2-groups.*

Proof. See W. Imrich [40], and also [Imrich/Klavžar 2000]. □

Exerceorem 8.7.10. All groups $\mathbb{Z}_2^s$ can be obtained in this way, except when $s=2,3,4$; cf. Remark 7.7.7.

Theorem 8.7.11. *Let G be* Aut(G)*-vertex transitive with n vertices (either directed or undirected). If G has more than $\frac{n}{2}$ simple eigenvalues, then* Aut(G) *is Abelian.*

Proof. See Theorem 15 on page 87 of H. Sachs, M. Stiebitz [82]. □

Exercise 8.7.12. Find a negative and a positive example.

Corollary 8.7.13. *Let G be undirected and* Aut(G)*-vertex transitive with n vertices. If G has more than $\frac{n}{2}$ simple eigenvalues, then* Aut(G) *acts strictly fixed point-free on G and consists entirely of involutions.*

Proof. Use the two previous theorems. □

Remark 8.7.14. Note that in the cases where G is undirected and Aut(G)-vertex transitive and Aut(G) is Abelian, the action is regular, and thus G is a Cayley graph; cf. Theorem 7.7.13.

Question. It seems clear that a graph G has an Abelian strong monoid if and only if $\mathrm{SEnd}(G) = \mathrm{Aut}(G)$ and $\mathrm{Aut}(G)$ is Abelian. If $\mathrm{SEnd}(G) \neq \mathrm{Aut}(G)$, there exist two idempotent strong endomorphisms which do not commute. Which graphs have an Abelian endomorphism monoid? Apparently such graphs must fulfill $\mathrm{SEnd}(G) = \mathrm{Aut}(G)$.

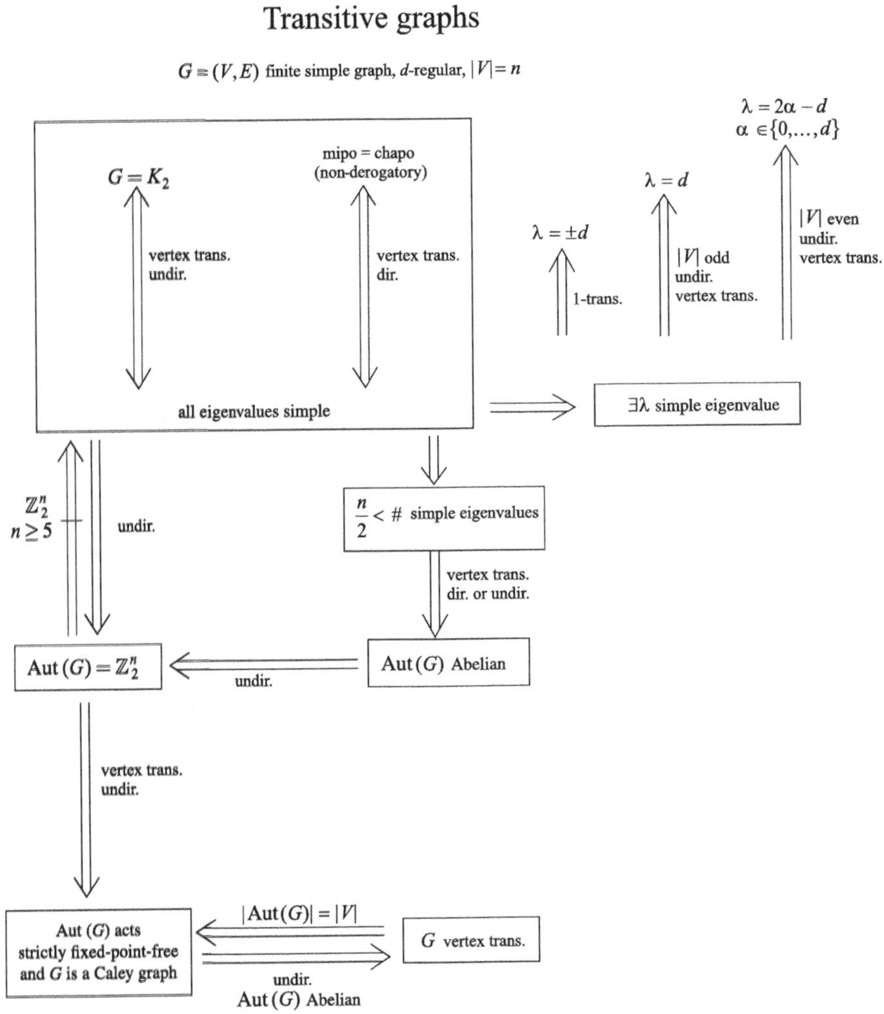

8.8 Comments

As in the previous chapter, one might want to consider generalizations from Aut to SEnd on one side and replace groups by semigroups which are close to groups on the other side.

Note that the question of spectral determinability has parallels with the determinability by the endospectrum (Section 1.7) and also with the determinability of certain semigroups by their Cayley graphs, as studied in later chapters.

We collected some of the results about transitive graphs in the diagram on page 180. This is a possible way to obtain an overview of different results, and is recommended also in other situations where there are many results on a topic which at first glance may seem confusing. It would be a good exercise to identify in the text the results corresponding to the implications shown in the diagram.

9 Graphs and semigroups

There are various connections between graphs and semigroups and, in particular, groups. Some of these have been discussed in previous chapters. In this chapter, after a brief review of semigroup theory we study von Neumann regularity of the endomorphisms of bipartite graphs and related properties, locally strong semigroups of paths, and strong semigroups in general. The latter concept is closely related to lexicographic products of graphs and wreath products of semigroups over acts. All three topics show the close links between algebraic properties and geometric/combinatorial properties of the graphs.

9.1 Semigroups

The content of this first section is of purely algebraic nature, the ideas will be applied to graphs later. We give some notation, definitions and results, all of which can be found, for example, in [Petrich/Reilly 1999]. The reader may skip this section initially, and consult it later when needed.

An element m of a semigroup M is said to be ***(von Neumann) regular*** if there exists $n \in M$ with $mnm = m$. In this case, for $p = nmn$ one has $mpm = m$ and $pmp = p$. An element p with this property is called a ***pseudo-inverse*** of m. Note that sometimes the word *inverse* is used, even for n. If all elements of M are regular, M is called a ***regular semigroup***.

An element m of a semigroup M is said to be ***completely regular*** if it has a commuting pseudo-inverse, i. e., there exists $p \in M$ with $pm = mp$. A semigroup M is said to be ***completely regular*** if all of its elements are completely regular.

We denote by $\mathrm{Idpt}(M)$ the idempotent elements of M and by $C(M)$ the elements of the center of M, i. e., elements which commute with all other elements of M.

Definition 9.1.1. A regular semigroup M is said to be:
- ***orthodox*** if $\mathrm{Idpt}(M)$ is a semigroup;
- ***left inverse*** if $ee'e = e'e$ for all $e, e' \in \mathrm{Idpt}(M)$;
- ***right inverse*** if $ee'e = ee'$ for all $e, e' \in \mathrm{Idpt}(M)$;
- ***inverse*** if $\mathrm{Idpt}(M)$ is commutative;
- a ***Clifford semigroup*** if $\mathrm{Idpt}(M) \subseteq C(M)$, i. e., the elements of $\mathrm{Idpt}(M)$ commute with all elements of M.

Note that only regular and completely regular are concepts which apply also to individual elements of a semigroup. Taking into account the following theorem, we could also call elements of a semigroup "inverse," but this is unusual, because of possible confusion.

https://doi.org/10.1515/9783110617368-009

Theorem 9.1.2. *The following implications hold:*
- *group ⇒ Clifford semigroup ⇔ completely regular and inverse;*
- *completely regular ⇔ union of maximal subgroups;*
- *inverse ⇔ regular and every element has a unique pseudo-inverse ⇔ left inverse and right inverse ⇒ orthodox.*

A semigroup S is called a ***right zero semigroup*** if $xy = y$ for all $x, y \in S$; it is called a ***left zero semigroup*** if $xy = x$ for all $x, y \in S$. We denote by R_n (resp., L_n) the n-element right (resp., left) zero semigroup, for $n \in \mathbb{N}$.

A semigroup S is called a ***right*** (resp., *left*) *group*, if it is uniquely right (resp., left) solvable, i. e., for all $r, t \in S$ there exists a unique $s \in S$ such that $rs = t$ (resp., $sr = t$). It turns out that right groups are always of the form $A \times R_n$ and left groups of the form $A \times L_n$ where A is a group.

As usual, multiplication on $S = A \times R_n$ is defined componentwise by

$$(g, r)(g', r') = (gg', r') \quad \text{for } g, g' \in A \text{ and } r, r' \in R_n.$$

This is why we call the semigroup $S = A \times R_n$ also a *right zero union of groups* (RZUG) over A and $S = L_n \times A$ a *left zero union of groups* (LZUG) over A: the multiplication has the same structure as in right or left zero semigroups, i. e., the right or left factor is dominant and determines the group in which we play.

Exercise 9.1.3. Prove that right groups are always of the form $A \times R_n$ where A is a group.

Prove that a multiplication of the form $g_1 g_2 \in A_2$ for $g_i \in A_i$, $i = 1, 2$, leads to a semigroup, i. e., it is associative only if $A_1 \cong A_2$.

A ***band*** is a semigroup that consists entirely of commuting idempotents.

A semigroup S is said to be ***right*** (resp., ***left***) ***cancellative*** if for all $x, y, z \in S$ we have that $xy = xz$ (resp., $yx = zx$) implies $y = z$.

A nonempty subset I of S is called a ***right*** (resp., ***left***) ***ideal*** of S if $s \in S$ and $a \in I$ imply that $as \in I$ (resp., $sa \in I$); I is a **(*two-sided*) *ideal*** of S if it is both a left and a right ideal of S. A (right or left) ideal I of S is proper if $I \neq S$.

Let $s \in S$. The right, left and two-sided ideals sS, Ss and sSs of S are called the ***principal*** right, left and two-sided ideals of S generated by s. A semigroup S is said to be ***right simple*** if it has no proper right ideals and ***left simple*** if it has no proper left ideals and ***simple*** if it has no ideals.

A completely regular semigroup S is ***completely simple*** if it is simple.

We have the following implications:

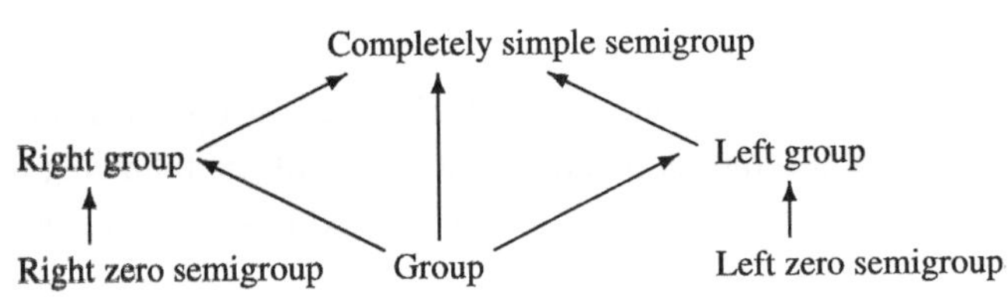

Remark 9.1.4. We note that completely simple semigroups are exactly the so-called Rees matrix semigroups. They are defined as follows.

Suppose that A is a group, I and Λ are nonempty sets, and P is a $\Lambda \times I$ matrix over A. The *Rees matrix semigroup* $\mathcal{M}(A, I, \Lambda, P)$ with *sandwich matrix* P consists of all triples (g, i, λ) where $i \in I$, $\lambda \in \Lambda$ and $g \in A$, with multiplication defined by

$$(g_1, i_1, \lambda_1)(g_2, i_2, \lambda_2) = (g_1 p_{\lambda_1 i_2} g_2, i_1, \lambda_2),$$

with $p_{\lambda_1 i_2} \in P$. If there exists an element $1 \in I \bigcap \Lambda$ such that for all $i \in I$ and $\lambda \in \Lambda$ we have $p_{\lambda 1} = p_{1i} = 1_A$, the identity of A, we say that P is *normalized*.

It is a well-known result that a semigroup S is completely simple if and only if S is isomorphic to a Rees matrix semigroup with a normalized sandwich matrix. Moreover, S is a right (resp., left) group if and only if $|I| = 1$ (resp., $|\Lambda| = 1$).

Exercise 9.1.5. Check these statements, possibly referring to the literature.

Let X be a partially ordered set and let $Y \subseteq X$. An element b of X is called a *lower bound* for Y if $b \leq y$ for every y in Y. A lower bound c of Y is called a *greatest lower bound (meet)* for Y if $b \leq c$ for every lower bound b of Y. An *upper bound* and a *least upper bound (join)* are defined analogously. A partially ordered set X is called a *meet* (resp., *join*) *semilattice* if every two-element subset $\{a, b\}$ of X has a meet, denoted by $a \wedge b$, (resp., join denoted by $a \vee b$) in X. If $a \not\geq b$ and $b \not\geq a$, then a and b are called *incomparable*, written $a \| b$. A partially ordered set X is called a *semilattice* if it is a meet semilattice or a join semilattice. Here, all semilattices will be meet semilattices.

A semigroup S is said to be a ***semilattice of (disjoint) semigroups*** $(S_\alpha, \circ_\alpha)$, $\alpha \in Y$, if:

(1) Y is a semilattice;
(2) $S = \bigcup_{\alpha \in Y} S_\alpha$;
(3) $S_\alpha S_\beta \subseteq S_{\alpha \wedge \beta}$.

It is a ***strong semilattice of semigroups*** if, in addition, for all $\beta \geq \alpha$ in Y there exists a semigroup homomorphism $f_{\beta,\alpha} : S_\beta \to S_\alpha$, called the ***defining homomorphism*** or ***structure homomorphism***, such that:

(4) $f_{\alpha,\alpha} = \mathrm{id}_{S_\alpha}$, the identity mapping, for all $\alpha \in Y$;
(5) $f_{\beta,\alpha} \circ f_{\gamma,\beta} = f_{\gamma,\alpha}$ for all $\alpha, \beta, \gamma \in Y$ with $\alpha \leq \beta \leq \gamma$, where the multiplication of $x \in S_\alpha$ and $y \in S_\beta$ in $S = \bigcup_{\alpha \in Y} S_\alpha$ is defined by

$$x * y = f_{\alpha,\alpha\wedge\beta}(x) f_{\beta,\alpha\wedge\beta}(y).$$

It might seem more natural, in these definitions, to write the defining homomorphisms as mappings on the right of the argument; but we do not do this, since it would be the only place in the book where it comes up.

Theorem 9.1.6. *For a semigroup S, the following are equivalent:*

(i) *S is a Clifford semigroup.*

(ii) *S is a semilattice of groups.*
(iii) *S is a strong semilattice of groups,* $[Y; A_\alpha, f_{\beta,\alpha}]$, *where Y is a semilattice, the* A_α *are groups and the* $f_{\beta,\alpha}$ *are defining homomorphisms, for* $\alpha, \beta \in Y$.

We observe that a semilattice of semigroups may not be strong if the semigroups are not groups. An example of this situation will appear in Theorem 9.3.10.

Besides the above, we will need the following standard definitions and notation. If C is a nonempty subset of the semigroup S, then $\langle C \rangle$ or even $\langle C \rangle_S$ denotes the *subsemigroup of S generated by C*. The subsemigroup $\langle C \rangle$ consists of all elements of S that can be expressed as finite products of elements of C.

An element s of a semigroup S is said to be ***periodic*** if there exist positive integers m, n such that $s^{m+n} = s^m$. A subset A of S is ***periodic*** if every element of A is periodic. In particular, if all principal left ideals of a semigroup are finite, or even more obvious if the semigroup is finite, then the semigroup is periodic.

As usual, a constant mapping $c_y : X \to Y$ for $y \in Y$ is defined by $c_y(x) = y$ for all $x \in X$. The identity mapping on X is denoted by id_X.

In what follows, we will use the term strong semilattice of groups as well as Clifford semigroup. We will also use the properties of and formulate results for the special case of strong chains of semigroups.

To get a better feeling for semigroups, the reader may want to look at some tables that show the number of nonisomorphic semigroups with a given (small) number of elements. These tables can be found in P. Grillet [27, 28].

Theorem 9.1.7. *The number of nonisomorphic and non-antiisomorphic n-element semigroups having certain properties are given in the following table. Among them are all* 17 *groups with less than* 11 *elements, namely* $\mathbb{Z}_2, \mathbb{Z}_3, \mathbb{Z}_4, \mathbb{Z}_2 \times \mathbb{Z}_2 \cong D_2, \mathbb{Z}_5, \mathbb{Z}_6 \cong \mathbb{Z}_2 \times \mathbb{Z}_3, D_3 \cong S_3, \mathbb{Z}_7, \mathbb{Z}_8, \mathbb{Z}_2^3, \mathbb{Z}_4 \times \mathbb{Z}_2, \mathbb{Z}_9, \mathbb{Z}_3 \times \mathbb{Z}_3, \mathbb{Z}_{10} \cong \mathbb{Z}_5 \times \mathbb{Z}_2, D_5$, *and the noncommutative quaternions. Besides them only* D_3, D_4, D_5 *are noncommutative.*

n	2	3	4	5	6	7	8
All	4	18	126	1160	15973	836021	1843120128
Commutative	3	12	58	325	2143	17291	221805

n	9	10
Commutative	11545843	3518930337
Commutative Clifford	25284	161698

From https://arxiv.org/pdf/1301.6023 we take the total number of semigroups with nine elements which are not isomorphic or antiisomorphic (i. e., nonequivalent) to be 52989400714478. The result is due to A. Distler, T. Kelsey, and J. Mitchell. There are 12418001077381302684 nonequivalent semigroups of order 10. This in "The Semigroups of Order 10" by Andreas Distler, Chris Jefferson, Tom Kelsey, Lars Kotthoff, Lecture Notes in Computer Science (LNCS, volume 7514).

Project 9.1.8. How many "small" noncommutative Clifford semigroups/monoids (which are not groups) exist? Which of them are monoids? For this, you have to use at least one of the noncommutative groups, the smallest is S_3. For the defining homomorphisms, it will be helpful to see that, for instance, there is no nontrivial homomorphism from S_3 onto $\mathbb{Z}_3$, as $\mathbb{Z}_2$ is not normal in S_3.

You can answer the same question for all commutative Clifford semigroups with two, three, four elements and so on.

9.2 End-regular bipartite graphs

In this section, we present some results on bipartite graphs with regular endomorphism semigroups. Some early results in this direction were published in E. Wilkeit [91].

As a tool we will use factorizations of endomorphisms according to the Homomorphism Theorem (Theorem 1.6.10), the so-called epi-mono factorizations (see Remark 1.6.11) and retract-coretract factorizations (cf. Definition 1.5.8).

Regular endomorphisms and retracts

Theorem 9.2.1. *The endomorphism $f \in \mathrm{End}(G)$ of a graph G is regular if and only if every epi-mono factorization of f is a retract-coretract factorization.*

Proof. To prove necessity, for $f \in \mathrm{End}(G)$ we first get an epi-mono factorization $f = \overline{f}\pi_{\rho_f}$ by the homomorphism theorem (Theorem 1.6.10). From the defining formula for the regularity of f, we then get that f and thus also π_{ρ_f} is a retraction with coretraction g or $g\overline{f}$.

The sufficiency is clear. □

Corollary 9.2.2. *The endomorphism monoid of any graph G is regular if and only if for every graph congruence ρ on G the canonical epimorphism $G \to G/\rho$ is a retraction and every monomorphism $G/\rho \to G$ is a coretraction.*

Proposition 9.2.3. *The following are equivalent for every graph G and any integer $n \geq 1$:*
(i) *The graph G is bipartite and has diameter greater than or equal to k.*
(ii) *The path P_k of length k is a retract of G.*
(iii) *The path P_k of length k is a factor graph of G.*

Proof. (i) ⇒ (ii): Let ℓ be the diameter of G and choose a vertex $u \in G$ with eccentricity ℓ, i. e., u is a starting point of a shortest path of length ℓ. For $0 \leq i \leq \ell$, set

$$N_i(u) := \{v \in G \mid d(u,v) = i\},$$

$$R_i(u) := N_i(u) \quad \text{for } 0 \leq i \leq k-2,$$
$$R_{k-1}(u) := \bigcup\{N_{k+2j-1}(u) \mid 0 \leq j \leq (\ell - k + 1)/2\},$$
$$R_k(u) := \bigcup\{N_{k+2j}(u) \mid 0 \leq j \leq (\ell - k)/2\}.$$

Since G is bipartite, there are no adjacent vertices in $N_i(u)$ or in $R_i(u)$. Therefore,

$$x \rho y \Leftrightarrow \exists i \in \mathbb{N},\ 0 \leq i \leq k : x, y \in R_i(u)$$

defines a congruence ρ on G. Obviously, $G/\rho \cong P_k$ and the canonical surjection $G \to G/\rho$ is a retraction. Thus, any path of length k beginning in u is a possible image under a corresponding coretraction.

(ii) ⇒ (iii) is trivial.

(iii) ⇒ (i): By contraposition, we see that if G contains a circuit of odd length or if $\operatorname{diam}(G) < k$, then every factor graph of G has the respective property. □

End-regular and End-orthodox connected bipartite graphs

Corollary 9.2.4. *Any bipartite graph G with regular endomorphism monoid has diameter less than 5.*

Proof. By Proposition 9.2.3, every bipartite graph with diameter 5 or greater has P_5 as a retract and P_3 as a subgraph. The surjection of $P_5 = \{0, 1, 2, 3, 4, 5\}$ onto P_3 which identifies 1 and 3 as well as 2 and 4 is obviously not a retraction. Then, by Theorem 9.2.1, the monoid of G is not regular. □

The following observation is not difficult to understand. A proof can be found in R. Nowakowski and I. Rival [66].

Lemma 9.2.5. *Every circuit of minimal length in a bipartite graph G is a retract of G.*

Theorem 9.2.6. *The connected bipartite graphs with a regular endomorphism monoid are exactly the following:*
(a) *$K_{m,n}$, including K_1, K_2, C_4 and the trees of diameter 2, i. e., the stars;*
(b) *the trees of diameter 3, which are the double stars, and C_6;*
(c) *C_8 and the path P_4 of length 4.*

Proof. By Corollary 9.2.4, a bipartite graph with regular endomorphism monoid has diameter at most 4.

In this case, we infer that G does not contain a subgraph $K_{1,3}$ if G has a factor graph P_4 or C_6, and G does not contain a subgraph C_4 if it has a factor graph P_3.

(a) If G has diameter less than or equal to 2, then G is complete bipartite. If G is a tree, then it is a star, i. e., $G = K_{1,n}$ with $n > 1$.

(b) Suppose G has diameter 3. If G is a tree, we get the double stars, namely $P_3[\overline{K}_n, K_1, K_1, \overline{K}_m]$ with $m, n \geq 1$; see Theorem 1.7.5. Since G has a retract P_3, by Proposition 9.2.3 we get that C_4 is a forbidden subgraph of G. So if G is not a tree, it must contain C_6 as a circuit of minimal length, which is a retract of G as stated in Lemma 9.2.5. We infer that G does not contain any subgraph $K_{1,3}$, and hence contains no vertex of degree ≥ 3. Therefore, it is C_6.

(c) Suppose now that G has diameter 4. Then P_4 is a retract of G and, in analogy to (b), we get that G does not contain a vertex of degree 3 or greater. Therefore, G is C_8 or P_4.

Using Theorem 9.2.1, it is routine to show that the given graphs have regular endomorphism monoids. □

Theorem 9.2.7. *The connected bipartite graphs with an orthodox endomorphism monoid are exactly the following:*
(a) *K_1 and K_2, with $|\mathrm{End}(K_1)| = 1$ and $\mathrm{End}(K_2) = \mathbb{Z}_2$;*
(b) *C_4 and the path P_2 of length 2, with End = SEnd in both cases;*
(c) *the path P_3 of length 3.*

The endomorphism monoids are not inverse except for the trivial cases of K_1 and K_2.

Proof. We have to examine only the graphs from Theorem 9.2.6. This inspection shows that only in the given cases do the idempotents of the respective endomorphism monoid form themselves a monoid. As an example, consider the following two idempotent endomorphisms of $C_6 = \{0, 1, 2, 3, 4, 5\}$: (1) map 0 to 2 and 5 to 3, while the rest remains fixed; (2) map 3 and 5 to 1, map 4 to 0, while the rest is fixed. Application of the second after the first is not an idempotent.

For the last statement, it is clear that idempotents do not commute. □

Question. Which of the above endomorphisms are locally strong, quasi-strong, or strong? How do these properties relate to algebraic properties?

Question. Investigate bipartite graphs with an idempotent closed endomorphism monoid which is not necessarily regular, i. e., not orthodox.

9.3 Locally strong endomorphisms of paths

In Theorem 1.7.5, it was proved that all endomorphisms of paths (as special trees) which are not automorphisms are locally strong or half-strong, i. e., paths are of endotype 6, End = HEnd ≠ LEnd ≠ QEnd = SEnd = Aut.

Recall that an endomorphism of a graph is locally strong if it reflects edges "locally." This means that if the vertices in $X = \{x_1, \dots, x_n\}$ are mapped onto x and the vertices in $X' = \{x'_1, \dots, x'_n\}$ are mapped onto x', where x and x' are adjacent, then each x_i is adjacent to at least one x'_j and vice versa. Half-strong means that there exists at least one edge between X and X'. Strong endomorphisms, as used in later sections of the chapter, reflect all edges, i. e., all vertices of X are adjacent to all vertices of X'.

In S. Arworn [2], there is an algorithm for determining the cardinalities of the endomorphism monoids of finite undirected paths. All endomorphisms of undirected paths can also be counted by first counting the congruence classes, compare U. Knauer and M. Michels [54].

We now present an algorithm to determine the cardinalities of the set of locally strong endomorphisms of finite undirected and directed paths. We show, moreover, that the set of locally strong endomorphisms on an undirected path will form a monoid if and only if the length of the path is a prime number or equal to 4. For directed paths, the condition turns into "length prime, 4 or 8." Theorems 9.3.10 and 9.3.12 give algebraic descriptions of these monoids. This section is based on S. Arworn, U. Knauer and S. Leeratanavalee [4].

Undirected paths

Let $P_n = \{0, \ldots, n\}$ denote the undirected path of length n with $n+1$ vertices.

Let $f : P_n \to P_n$ be an endomorphism. The length of the image path of f is called the ***length*** of f. We denote the set of endomorphisms of length l by $\mathrm{End}_l(P_n)$, or $\mathrm{LEnd}_l(P_n)$ if the endomorphisms are locally strong.

An endomorphism $f : P_n \to P_n$ is called a ***complete folding*** if the congruence relation $\ker f = \{(x,y) \in P_n \times P_n \mid f(x) = f(y)\}$ partitions P_n into $l+1$ classes where $l|n$ and the equivalence classes are of the form

$$\begin{aligned}
[0] &= \{2ml \in P_n \mid m = 0, 1, \ldots\}, \\
[l] &= \{(2m+1)l \in P_n \mid m = 0, 1, 2, \ldots\}, \\
[r] &= \{2ml + r \in P_n \mid m = 0, 1, \ldots\} \bigcup \{2ml - r \in P_n \mid m = 1, 2, \ldots\} \\
&\qquad \text{for } r \text{ such that } 0 < r < l.
\end{aligned}$$

Clearly, in this case a complete folding has length l.

In the following picture, we have a complete folding with $l = 5$ of P_{20}:

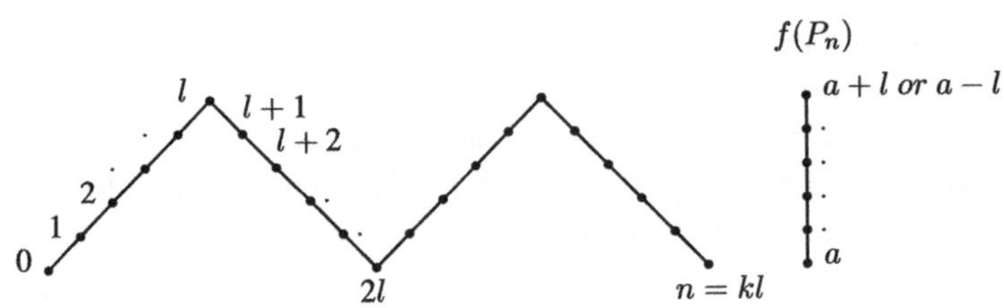

Remark 9.3.1. An undirected path has exactly two automorphisms.

It is clear that every complete folding of an undirected path is locally strong.

Moreover, if f is a locally strong endomorphism and $f(P_n) = \{a, a+1, \ldots, a+l\} \subseteq P_n$, then $f(0) = a$ or $a + l$.

Lemma 9.3.2. *Every locally strong endomorphism on P_n is a complete folding.*

Proof. Let $f : P_n \to P_n$ be a locally strong endomorphism on P_n, and let $f(P_n) = \{a, a + 1, \ldots, a + l\}$. By Remark 9.3.1, we get $f(0) = a$ or $a + l$. Suppose that $f(0) = a$; then $f(1) = a+1$. Next, we show that $f(r) = a+r$ for all r with $0 \le r \le l$. Suppose there exists t, $0 < t < l$, such that $f(r) = a + r$ for all r with $0 \le r \le t$ but $f(t + 1) = a + t - 1$.

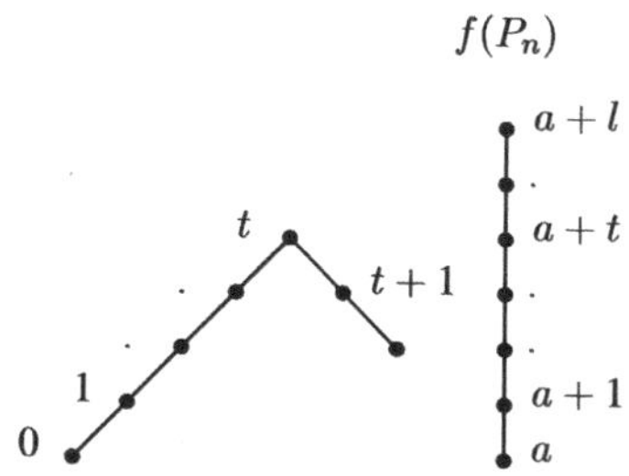

Since $\{a + t, a + t + 1\} \in E$, $t \in f^{-1}(a + t)$ and $t - 1, t + 1 \in f^{-1}(a + t - 1)$, there is no $x \in f^{-1}(a + t + 1)$ such that $\{t, x\} \in E$. So f is not a locally strong endomorphism. Thus $f(r) = a + r$ for all $r = 0, 1, \ldots, l$.

Suppose now that $f(l+r) = a+l-r$ for all $r = 0, 1, \ldots, t'$ but $f(l+t'+1) = a+l-t'+1$ for some t' with $0 < t' < l$.

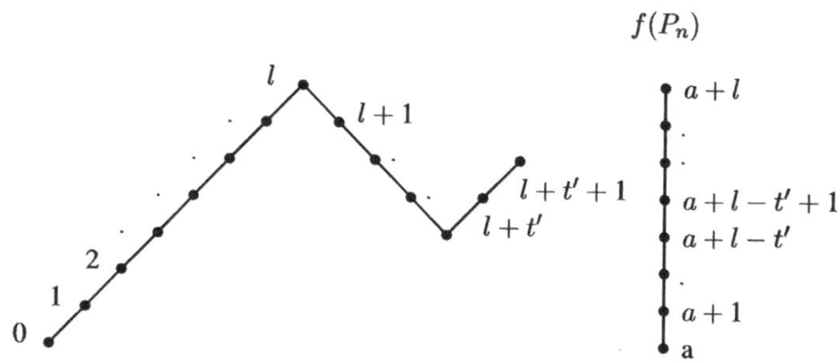

Then $f(l + t' + 1) = f(l + t' - 1) = a + l - t' + 1$. Hence there is no $x \in f^{-1}(a + l - t' - 1)$ such that $\{x, l + t'\} \in E$. So f is not a locally strong endomorphism.

If l does not divide n, then $n \in [r]$ for some r with $0 < r < l$. Hence $f(n) = a + r$ and $f(n - 1) = a + r - 1$ (or $a + r + 1$). Then $\{a + r,\ a + r + 1\} \in E$ (or $a + r - 1,\ a + r \in E$) but there is no $x \in f^{-1}(a + r + 1)$ (or $x \in f^{-1}(a + r - 1)$) such that $\{n, x\} \in E$. This contradicts the assumption of f being locally strong. Thus $l | n$. □

From Remark 9.3.1 and Lemma 9.3.2, we then get the following result.

Theorem 9.3.3. *An endomorphism of an undirected path is locally strong if and only if it is a complete folding.*

We will denote a locally strong endomorphism $f : P_n \to P_n$ of length l which maps 0 to a and 1 to $a+1$ (resp., $a-1$) by f_{l,a^+} (resp., f_{l,a^-}).

For example,

$$f_{3,2^+} : P_9 \to P_9 \quad \text{is}$$

$$f_{3,2^+} = \begin{pmatrix} 0 & 1 & 2 & 3 & 4 & 5 & 6 & 7 & 8 & 9 \\ 2 & 3 & 4 & 5 & 4 & 3 & 2 & 3 & 4 & 5 \end{pmatrix};$$

$$f_{3,6^-} : P_9 \to P_9 \quad \text{is}$$

$$f_{3,6^-} = \begin{pmatrix} 0 & 1 & 2 & 3 & 4 & 5 & 6 & 7 & 8 & 9 \\ 6 & 5 & 4 & 3 & 4 & 5 & 6 & 5 & 4 & 3 \end{pmatrix}.$$

Theorem 9.3.4. *Denote by* $\mathrm{LEnd}_l(P_n)$ *all locally strong endomorphisms of length* l *of the undirected path* P_n. *Then* $|\mathrm{LEnd}_l(P_n)| = 2(n-l+1)$ *and* $|\mathrm{LEnd}(P_n)| = 2\sum_{l|n}(n-l+1)$.

Proof. This is quite clear, since every divisor of n determines a congruence on P_n, which in turn is determined by a locally strong endomorphism, and the respective factor graph can be embedded in P_n exactly $2(n-l+1)$ times. This argument is of course based on the homomorphism theorem (Theorem 1.6.10). □

Directed paths

We consider "up-up" directed paths $\overrightarrow{P}_n$ of length n as follows:

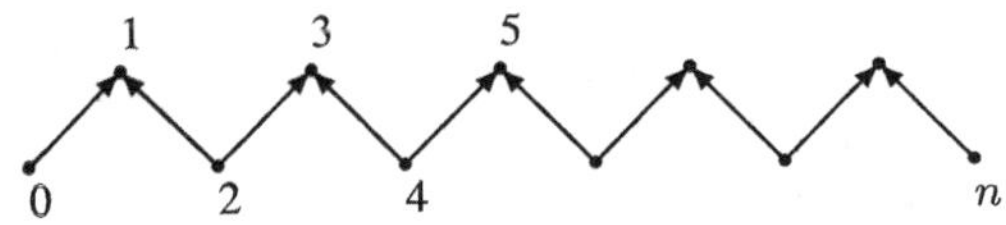

if n is even, and

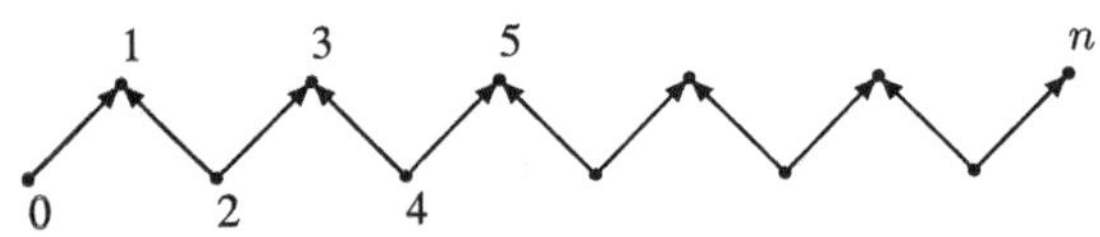

if n is odd.

This corresponds to directed bipartite graphs and is a way of defining directed paths such that there exist nontrivial endomorphisms.

Remark 9.3.5. If $f : \overrightarrow{P}_n \to \overrightarrow{P}_n$ is an endomorphism of the directed path $\overrightarrow{P}_n$, then $f(x)$ is odd if and only if x is odd. And $|\mathrm{Aut}(\overrightarrow{P}_n)| = 1$ if n is odd, and $= 2$ if n is even.

In the same manner as for undirected paths case, we obtain the following result.

Theorem 9.3.6. *An endomorphism on the directed path is locally strong if and only if it is a complete folding.*

Now the formula for the number of locally strong endomorphisms becomes a little more complicated.

Theorem 9.3.7. *Denote by* $\mathrm{LEnd}_l(\overrightarrow{P}_n)$ *the set of all locally strong endomorphisms of length l of the directed path* $\overrightarrow{P}_n$*, where l divides n. Then*

$$|\mathrm{LEnd}_l(\overrightarrow{P}_n)| = \begin{cases} n-l+1 & \text{if } l \text{ is odd,} \\ n-l+2 & \text{if } l \text{ is even.} \end{cases}$$

Also,

$$|\mathrm{LEnd}(\overrightarrow{P}_n)| = \begin{cases} \sum_{l|n}(n-l+1) & \text{if } n \text{ is odd,} \\ \sum_{l|n,\mathrm{odd}}(n-l+1) + \sum_{l|n,\mathrm{even}}(n-l+2) & \text{if } n \text{ is even.} \end{cases}$$

Proof. Case 1. Suppose that n is odd and $l|n$.

In the picture, we have $n = 15$ and $l = 5$:

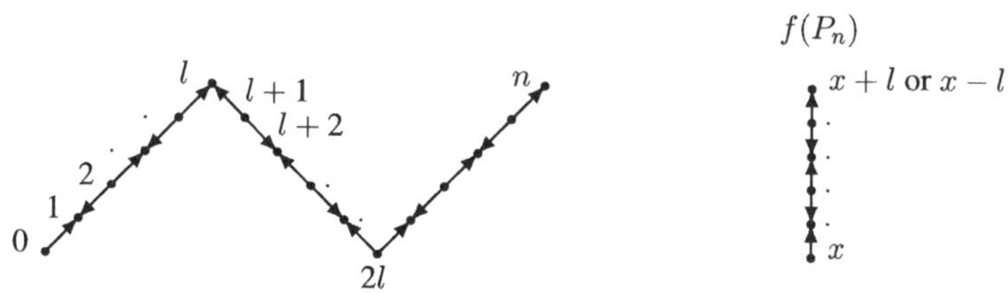

Then

$$\begin{aligned} |\mathrm{LEnd}_l(P_n)| &= |\{f_{l,x^+} : P_n \to P_n \mid x = 0,2,4,\ldots,n-l\}| \\ &\quad + |\{f_{l,x^-} : P_n \to P_n \mid x = n-1, n-3, n-5, \ldots, l+1\}| \\ &= \left|\left\{f_{l,x^+} : P_n \to P_n \;\middle|\; x = 0,2,4,\ldots,2\left(\frac{n-l}{2}\right)\right\}\right| \\ &\quad + \left|\left\{f_{l,x^-} : P_n \to P_n \;\middle|\; x = n-1, n-3, n-5, \ldots, n-\left(2\left(\frac{n-l}{2}\right)-1\right)\right\}\right| \\ &= \left(\frac{n-l}{2}+1\right) + \left(\frac{n-l}{2}\right) \\ &= n-l+1. \end{aligned}$$

Case 2. Suppose that n is even and $l|n$.

Case 2(a). Here, l is odd; in the picture we have $n = 20$ and $l = 5$:

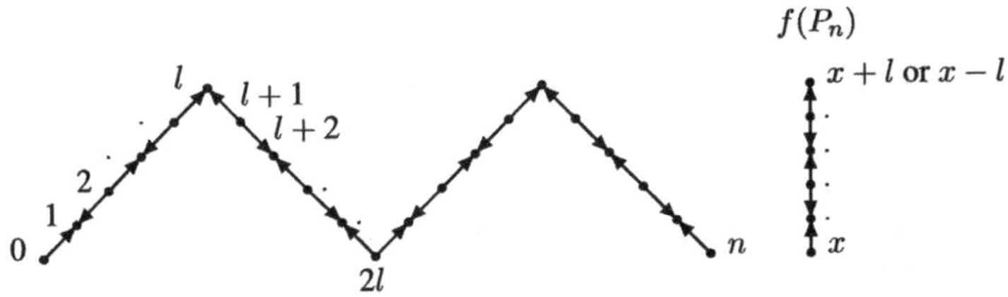

Then

$$\begin{aligned}|\mathrm{LEnd}_l(P_n)| &= |\{f_{l,x^+} : P_n \to P_n \mid x = 0, 2, 4, \dots, n-l-1\}| \\ &\quad + |\{f_{l,x^-} : P_n \to P_n \mid x = n, n-2, n-4, \dots, l+1\}| \\ &= \left|\left\{f_{l,x^+} : P_n \to P_n \mid x = 0, 2, 4, \dots, 2\left(\frac{n-l-1}{2}\right)\right\}\right| \\ &\quad + \left|\left\{f_{l,x^-} : P_n \to P_n \mid x = n, n-2, n-4, \dots, \right.\right. \\ &\qquad\qquad \left.\left. n-2\left(\frac{n-l-1}{2}\right)\right\}\right| \\ &= \left(\frac{n-l-1}{2}+1\right)+\left(\frac{n-l-1}{2}+1\right) \\ &= n-l-1+2 \\ &= n-l+1.\end{aligned}$$

Case 2(b). Now suppose that l is even.

In the picture below, we have $n = 16$ and $l = 4$. (Note that if n/l is odd it would end with n on the top as in Case 1.)

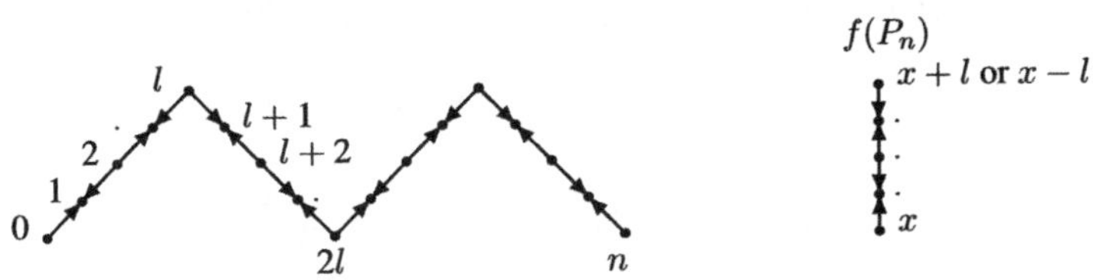

Then

$$\begin{aligned}|\mathrm{LEnd}_l(P_n)| &= |\{f_{l,x^+} : P_n \to P_n \mid x = 0, 2, 4, \dots, n-l\}| \\ &\quad + |\{f_{l,x^-} : P_n \to P_n \mid x = n, n-2, n-4, \cdots, l\}| \\ &= \left|\left\{f_{l,x^+} : P_n \to P_n \mid x = 0, 2, 4, \dots, 2\left(\frac{n-l}{2}\right)\right\}\right| \\ &\quad + \left|\left\{f_{l,x^-} : P_n \to P_n \mid x = n, n-2, n-4, \dots, n-2\left(\frac{n-l}{2}\right)\right\}\right|\end{aligned}$$

$$= \left(\frac{n-l}{2}+1\right)+\left(\frac{n-l}{2}+1\right)$$
$$= n-l+2.$$

Therefore, we get

$$|\mathrm{LEnd}(P_n)| = \begin{cases} \sum_{l|n}(n-l+1) & \text{if } n \text{ is odd,} \\ \sum_{\substack{l|n \\ l \text{ odd}}}(n-l+1)+\sum_{\substack{l|n \\ l \text{ even}}}(n-l+2) & \text{if } n \text{ is even.} \end{cases}$$ □

Algebraic properties of LEnd

The following two observations are clear.

Lemma 9.3.8. *Every endomorphism $f : P_n \to P_n$ of length 1 of a path P_n is a locally strong endomorphism. Moreover, in this case $f \circ g$ and $g \circ f$ are of length 1 for any $g : P_n \to P_n$.*

Remark 9.3.9. The above lemma implies that the set of endomorphisms of length 1 is always a left group. This left group forms the infimum in the not necessarily strong semilattice of subsets of $\mathrm{LEnd}(P_n)$ which are not necessarily groups or semigroups.

Recall that in unions of groups, i. e., in completely regular semigroups, the multiplication of elements from different groups cannot be described easily. Here, we are in a more comfortable situation if n is prime.

Theorem 9.3.10. *The set $\mathrm{LEnd}(P_n)$ forms a monoid if and only if n is a prime number or 4. If n is prime, then $\mathrm{LEnd}(P_n)$ is a left group consisting of copies of $\mathbb{Z}_2$ together with the automorphism group $\mathbb{Z}_2$. The monoid $\mathrm{LEnd}(P_4)$ is a union of groups if we delete the two elements $f_{2,1^+}$ and $f_{2,3^-}$, which are not even regular in $\mathrm{LEnd}(P_4)$. This union of groups is a (nonstrong) semilattice of left groups with infimum $\mathrm{LEnd}_1(P_4)$, the left group of endomorphisms of length 1.*

Proof. If $p > 2$ is a prime which divides n, consider

$$f_{p,0^+} \circ f_{p,2^+} = \begin{pmatrix} 0 & 1 & 2 & 3 & \cdots & p-1 & p & p+1 & \cdots \\ 2 & 3 & 4 & 5 & \cdots & p-1 & p-2 & p-1 & \cdots \end{pmatrix}.$$

This is not a complete folding, thus $f_{p,0^+} \circ f_{p,2^+}$ is not a locally strong endomorphism.

If $n = 2^k$ with $k \geq 3$, consider

$$f_{2,0^+} \circ f_{4,1^+} = \begin{pmatrix} 0 & 1 & 2 & 3 & 4 & 5 & 6 & 7 & 8 & \cdots \\ 1 & 2 & 1 & 0 & 1 & 0 & 1 & 2 & 1 & \cdots \end{pmatrix},$$

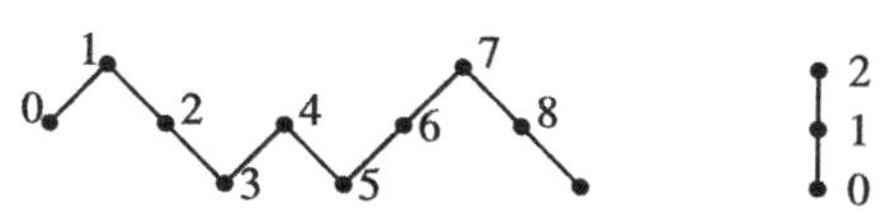

Again, $f_{2,0^+} \circ f_{4,1^+}$ is not a locally strong endomorphism. This proves the necessity.

To prove sufficiency, observe that if n is prime we get the statement from Lemma 9.3.8 and Theorem 9.3.6. The algebraic structure of the monoid is obviously of the described form. On P_4 there are eight locally strong endomorphisms of length 1, namely $f_{1,0^+}, f_{1,1^-}, f_{1,1^+}, f_{1,2^-}, f_{1,2^+}, f_{1,3^-}, f_{1,3^+}$ and $f_{1,4^-}$; there are six locally strong endomorphisms of length 2, namely $f_{2,0^+}, f_{2,1^+}, f_{2,2^-}, f_{2,2^+}, f_{2,3^-}$ and $f_{2,4^-}$; and there are only two locally strong endomorphisms of length 4, namely $f_{4,0^+}$ and $f_{4,4^-}$, which are the automorphisms.

Upon deleting the elements $f_{2,1^+}$ and $f_{2,3^-}$ we get a union of groups, more precisely a chain of left groups, with $L_2 \times \mathbb{Z}_2 \xrightarrow{f} L_4 \times \mathbb{Z}_2$. The two automorphisms form another group $\mathbb{Z}_2$ which would be the supremum to the foregoing. □

Remark 9.3.11. Note that the two locally strong endomorphisms $f_{2,1^+}$ and $f_{2,3^-}$, which are not regular in $\mathrm{LEnd}(P_4)$, are regular in $\mathrm{End}(P_4)$. So $f_{2,1^+}$ has the two inverses

$$g_1 = \begin{pmatrix} 0 & 1 & 2 & 3 & 4 \\ 1 & 2 & 1 & 0 & 1 \end{pmatrix} \quad \text{and} \quad g_2 = \begin{pmatrix} 0 & 1 & 2 & 3 & 4 \\ 3 & 2 & 3 & 4 & 3 \end{pmatrix} \quad \text{in } \mathrm{End}(P_4),$$

and similarly for $f_{2,3^-}$.

For directed paths, we have the following theorem.

Theorem 9.3.12. *The set* $\mathrm{LEnd}(\vec{P}_n)$ *of a directed path* $\vec{P}_n$ *forms a monoid if and only if* n *is a prime number or* 4 *or* 8.

Proof. In the case where the length n of the directed path has a prime divisor greater than 2, we use the same proof as for undirected paths. Locally strong endomorphisms of length 2 satisfy the conditions of Lemma 9.3.8, stated there for locally strong endomorphisms of length 1 for undirected paths. To see this, we interpret two successive directed arcs, such as $(0,1)$ and $(1,2)$ as a single undirected arc.

With this argument, we can use the first part of the proof of Theorem 9.3.10 to see that $\mathrm{LEnd}(\vec{P}_{2^k})$ is not closed starting with $\vec{P}_{16}$.

Consequently, for $\mathrm{LEnd}(\vec{P}_8)$ we get the same multiplication table as for $\mathrm{LEnd}(P_4)$ in Theorem 9.3.10; we merely have to add the eight endomorphisms of $\vec{P}_8$ of length 1, which again are locally strong.

For $\vec{P}_4$, consider the multiplication table of $\mathrm{LEnd}(\vec{P}_4)$, after deleting the two automorphisms, which is displayed below on page 197. In the table, we write l_{x^+} for f_{l,x^+} and l_{x^-} for f_{l,x^-}, for all $l, x \in P_4$. So, e. g., we write 1_{0^+} for $f_{1,0^+}$.

We see that this is a union of groups, i. e., a completely regular semigroup: four one-element groups and two copies of $\mathbb{Z}_2$. More precisely we have a strong chain of left groups $L_2 \times \mathbb{Z}_2 \xrightarrow{f} L_4$, where the structure homomorphism f operates as follows $f(2_{0^+}) = 1_{0^+}, f(2_{2^-}) = 1_{2^-}, f(2_{4^-}) = 1_{4^-}, f(2_{2^+}) = 1_{2^+}$. Again L_n denotes the left zero semigroup with n elements. The two automorphisms form a group $\mathbb{Z}_2$ which would be the supremum to the foregoing.

$\circ$	1_{0^+}	1_{2^-}	1_{4^-}	1_{2^+}	2_{0^+}	2_{2^-}	2_{4^-}	2_{2^+}
1_{0^+}	1_{0^+}	1_{0^+}	1_{0^+}	1_{0^+}	1_{0^+}	1_{0^+}	1_{0^+}	1_{0^+}
1_{2^-}	1_{2^-}	1_{2^-}	1_{2^-}	1_{2^-}	1_{2^-}	1_{2^-}	1_{2^-}	1_{2^-}
1_{4^-}	1_{4^-}	1_{4^-}	1_{4^-}	1_{4^-}	1_{4^-}	1_{4^-}	1_{4^-}	1_{4^-}
1_{2^+}	1_{2^+}	1_{2^+}	1_{2^+}	1_{2^+}	1_{2^+}	1_{2^+}	1_{2^+}	1_{2^+}
2_{0^+}	1_{0^+}	1_{2^-}	1_{0^+}	1_{2^-}	2_{0^+}	2_{2^-}	2_{0^+}	2_{2^-}
2_{2^-}	1_{2^-}	1_{0^+}	1_{2^-}	1_{0^+}	2_{2^-}	2_{0^+}	2_{2^-}	2_{0^+}
2_{4^-}	1_{4^-}	1_{2^+}	1_{4^-}	1_{2^+}	2_{4^-}	2_{2^+}	2_{4^-}	2_{2^+}
2_{2^+}	1_{2^+}	1_{4^-}	1_{2^+}	1_{4^-}	2_{2^+}	2_{4^-}	2_{2^+}	2_{4^-}

Multiplication table $\mathrm{LEnd}(\vec{P}_4)$, without the 2 automorphisms. □

9.4 Wreath product of monoids over an act

This section again focuses on algebraic aspects, which will later be applied to graphs. Recall that for a monoid S and a nonempty set A, the ***set of all mappings*** S^A from A to S with the multiplication $(fg)(a) = f(a)g(a)$ for $f, g \in S^A$ and all $a \in A$ forms a monoid. Again, for $s \in S$ we denote by $c_s \in S^A$ the constant mapping which maps all elements of A onto s.

Let R be a monoid (or semigroup), and let A be a set. Recall the definition of a left (or right) R-act from Definition 7.6.1. We write ${}_RA$ if R operates on A from the left by $(rr')a = r(r'a) \in A$ (and $1_R\, a = a$ for $1_R \in R$) for all $r, r' \in R$ and $a \in A$; operations from the right are defined analogously.

Most of the following concepts can be found, e. g., in [Kilp et al. 2000].

Construction 9.4.1. Let R and S be monoids and let ${}_RA$ be a left R-act. On the set $R \times S^A$, consider the multiplication defined by

$$(r,f)(p,g) = (rp, f_p g)$$

for $r, p \in R$ and $f, g \in S^A$, where for $a \in A$ we set

$$(f_p g)(a) := f(pa)g(a).$$

Lemma 9.4.2. *With the above multiplication, $R \times S^A$ becomes a monoid with identity $1_{R\times S^A} = (1_R, c_1)$.*

Proof. Let $a \in A$, $p, q, r \in R$ and $f, g, h \in S^A$. Then

$$\begin{aligned}((f_p g)_q h)(a) &= (f_p g)(qa)h(a)\\ &= f(pqa)g(qa)h(a) = (f_{pq} g_q h)(a)\end{aligned}$$

and, therefore,

$$\begin{aligned}((r,f)(p,g))(q,h) &= (rp,f_pg)(q,h) = (rpq,(f_pg)_qh)\\ &= (rpq,f_{pq}g_qh) = (r,f)(pq,g_qh)\\ &= (r,f)((p,g)(q,h)),\end{aligned}$$

i. e., multiplication in $(R \times S^A)$ is associative.

Since

$$(r,f)(1,c_1) = (r,f_1c_1) = (r,f) = (r,c_1f) = (1,c_1)(r,f)$$

for all $r \in R$ and $f \in S^A$, we have that $1_{R\times S^A} = (1_R, c_1)$ is the identity element of the semigroup $R \times S^A$, so $R \times S^A$ is a monoid. □

Definition 9.4.3. We denote the above monoid by $(R \wr S|_R A)$ and call it the ***wreath product*** of R by S through $_RA$.

Example 9.4.4. For monoids R, S, and an R-act $_RA$, it is clear that

$$(R \wr S|_R A) \cong \begin{cases} R \times S & \text{if } |_RA| = 1,\\ S^A & \text{if } |R| = 1,\\ R & \text{if } |S| = 1.\end{cases}$$

Therefore, the smallest nontrivial example needs R, S, $_RA$ with two elements at least, and has eight elements; the next larger one will have 12 elements. For a concrete example, take the complete graph K_2. A computation shows that $(\operatorname{Aut}(K_2) \wr \operatorname{Aut}(K_2)|_{\operatorname{Aut} K_2} K_2)$ is isomorphic to the eight-element dihedral group D_4.

Lemma 9.4.5. *The canonical mapping*

$$\begin{aligned}(R \wr S|_R A) &\to R\\ (r,f) &\mapsto r,\end{aligned}$$

which is surjective, and the canonical mappings

$$\begin{aligned}R &\to (R \wr S|_R A)\\ r &\mapsto (r,c_1),\end{aligned}$$

$$\begin{aligned}S &\to (R \wr S|_R A)\\ s &\mapsto (1,c_s),\end{aligned}$$

which are injective, are monoid homomorphisms.

Moreover, the canonical mapping

$$R \prod S \to (R \wr S|_R A)$$

$$(r,s) \mapsto (r, c_s)$$

is a monoid homomorphism.

Proof. Note that

$$(r, c_s)(r', c_{s'}) = (rr', (c_s)_{r'} c_{s'}) = (rr', c_{ss'}).$$

The rest is clear. □

Lemma 9.4.6. *If $\delta : S \to S'$ is a monoid homomorphism, then the mapping*

$$(\mathrm{id}_R \wr \delta | \mathrm{id}_A) : (R \wr S|_R A) \longrightarrow (R \wr S'|_R A)$$

such that

$$(\mathrm{id}_R \wr \delta | \mathrm{id}_A)((r,f)) = (r, \delta f) \quad \text{for } r \in R,\ f \in S^A$$

is a monoid homomorphism.

Moreover, $(\mathrm{id}_R \wr \delta|\mathrm{id}_A)$ is injective (resp., surjective) if and only if δ is injective (resp., surjective).

Proof. First, note that $\delta f \in S'^A$ with the usual composition of mappings. Take $(r,f), (p,g) \in (R \wr S|_R A)$. For every $a \in {}_R A$, we have

$$\delta(f(pa)g(a)) = \delta(f(pa))\delta(g(a)) = ((\delta f)(pa))((\delta g)(a)),$$

so we get that $\delta(f_p g) = (\delta f)_p (\delta g)$. Then

$$\begin{aligned}(\mathrm{id}_R \wr \delta|\mathrm{id}_A)((r,f)(p,g)) &= (\mathrm{id}_R \wr \delta|\mathrm{id}_A)((rp, f_p g)) = (rp, \delta(f_p g))\\ &= (rp, (\delta f)_p(\delta g)) = (r, \delta f)(p, \delta g)\\ &= ((\mathrm{id}_R \wr \delta|\mathrm{id}_A)((r,f)))((\mathrm{id}_R \wr \delta|\mathrm{id}_A)((p,g))).\end{aligned}$$

Moreover,

$$(\mathrm{id}_R \wr \delta|\mathrm{id}_A)((1_R, c_1)) = (1_R, \delta c_1) = (1_R, c_1) \in (R \wr S'|_R A).$$

Thus we see that $(\mathrm{id}_R \wr \delta|\mathrm{id}_A)$ is a monoid homomorphism.

Finally, note that the mapping $S^A \to S'^A$ with $f \mapsto \delta f$ is injective (surjective) if and only if $\delta : S \to S'$ is injective (surjective). Thus we have that $(\mathrm{id}_R \wr \delta|\mathrm{id}_A)$ is injective (surjective) if and only if δ is injective (surjective). □

Lemma 9.4.7. *If $\alpha : {}_R A \to {}_R A'$ is a homomorphism of left R-acts, then the mapping*

$$(\mathrm{id}_R \wr \mathrm{id}_S | \alpha) : (R \wr S|_R A') \longrightarrow (R \wr S|_R A)$$

such that

$$(\mathrm{id}_R \wr \mathrm{id}_S\, |\alpha)((r,f')) = (r,f'\alpha) \quad \textit{for } r \in R,\ f' \in S^{A'}$$

is a monoid homomorphism.

Moreover, if $|S| \geq 2$, *then* $(\mathrm{id}_R \wr \mathrm{id}_S\, |\alpha)$ *is injective (resp., surjective) if and only if* α *is surjective (resp., injective).*

Proof. First, note that $f'\alpha \in S^A$ with the usual composition of mappings. Since $\alpha : {}_RA \to {}_RA'$ is a homomorphism of left R-acts, for every $a \in {}_RA$, $p \in R$ and $f', g' \in S^{A'}$ we have that

$$\begin{aligned}((f'_p g')\alpha)(a) &= (f'_p g')(\alpha(a)) = f'(p\alpha(a))g'(\alpha(a)) \\ &= f'(\alpha(pa))g'(\alpha(a)) = (f'\alpha)(pa)(g'\alpha)(a) = ((f'\alpha)_p(g'\alpha))(a),\end{aligned}$$

i. e., $(f'_p g')\alpha = (f'\alpha)_p(g'\alpha)$. Then

$$\begin{aligned}(\mathrm{id}_R \wr \mathrm{id}_S\, |\alpha)((r,f')(p,g')) &= (\mathrm{id}_R \wr \mathrm{id}_S\, |\alpha)((rp, f'_p g')) \\ &= (rp, (f'\alpha)_p(g'\alpha)) = (r,f'\alpha)(p,g'\alpha) \\ &= ((\mathrm{id}_R \wr \mathrm{id}_S\, |\alpha)((r,f')))((\mathrm{id}_R \wr \mathrm{id}_S\, |\alpha)((p,g')))\end{aligned}$$

and

$$(\mathrm{id}_R \wr \mathrm{id}_S\, |\alpha)((1_R, c_1)) = (1_R, c_1\alpha) = (1_R, c_1) \in (R \wr S|_R A).$$

Therefore $(\mathrm{id}_R \wr \mathrm{id}_S\, |\alpha)$ is a monoid homomorphism.

Finally, note that if $|S| \geq 2$, then the mapping $S^{A'} \to S^A$ with $f' \mapsto f'\alpha$ is surjective if and only if α is injective and it is injective if and only if α is surjective. □

9.5 Structure of the strong monoid

We know that every monoid is isomorphic to the endomorphism monoid of a graph; see Theorem 7.4.4. In contrast, observe that not every monoid is isomorphic to the strong monoid of a graph, since a strong monoid has at least two idempotents not equal to 1, if it is not a group (recall Corollary 1.5.7).

Here, we consider only graphs without loops and, therefore, all congruences are loop-free congruences; see Definition 1.6.4.

The canonical strong decomposition of G

Definition 9.5.1. Take $G = (V,E)$, finite or infinite. Define the relation $\nu \in V \times V$ by $x\nu x' \Leftrightarrow N_G(x) = N_G(x')$; it is called the ***canonical strong congruence***. We will write ν_G

if necessary, for instance when several graphs are involved. The factor graph $G/\nu :=$ $(G_\nu, E_\nu) := (\{x_\nu \mid x \in G\}, \{\{x_\nu, y_\nu\} \mid \{x, y\} \in E\})$ is called the ***canonical strong factor graph*** of G.

As a consequence of this definition, we get that for the edge $\{x_\nu, y_\nu\} \in E_\nu$, all preimages of x_ν have all preimages of y_ν as neighbors and vice versa. It follows from the definition of the relation $x\nu x'$ that x and x' are not neighbors, since otherwise x would have to be a neighbor of itself.

Lemma 9.5.2. *The canonical surjection $\pi_\nu : G \to G/\nu$ is a strong graph homomorphism.*

Theorem 9.5.3. *The canonical strong factor graph G/ν is S-A unretractive, i. e.,* $\operatorname{SEnd}(G/\nu) = \operatorname{Aut}(G/\nu)$, *if G/ν is finite.*

Proof. If G/ν would have a nonbijective strong endomorphism, there would exist two vertices in G/ν with the same neighborhood; cf. Proposition 1.5.5. This is not possible since their preimages would then also have the same neighborhood in G, and thus the congruence ν would identify them. □

Example 9.5.4. We show that $\operatorname{SEnd}(G/\nu) \neq \operatorname{Aut}(G/\nu)$ is possible if G/ν is not finite. Take $|\mathbb{N}|$ copies of the path P_3 of length 3. This is already a canonical strong factor graph since ν is trivial; see Theorem 1.7.5. Moving the whole graph one step to the right is a strong endomorphism which is clearly not surjective, and thus not an automorphism.

It is clear from the definition of the canonical strong factor graph G_ν of G that G has the following decomposition in a generalized lexicographic product, compare Definition 4.4.3.

Theorem 9.5.5. *For every graph G, we have a decomposition in a generalized lexicographic product $G = U[(Y_u)_{u\in U}]$ where $U = G/\nu$ is the canonical strong factor graph and $V(Y_u) = \{x \in G \mid \pi_\nu(x) = u \in U\}$, $E(Y_u) = \emptyset$, $u \in U$.*

Proof. Assume that $\{x, y\} \in E(Y_u)$. Then $\{x, y\} \in E(G)$ and $x\nu y$. This is not possible, as x is not a neighbor of itself. □

The theorem can also be considered as a construction which enables us to construct all graphs with a given canonical strong factor graph. It works as follows.

Start with an S-A unretractive graph U, i. e., $\operatorname{SEnd}(U) = \operatorname{Aut}(U)$. Examples of S-A unretractive graphs are K_n and also all paths P_n of length $n > 2$; compare Theorem 1.7.5.

Insert in place of each vertex u of U a set V_u.

Connect all points in V_u with all points in $V_{u'}$ by edges if (u, u') is an edge in U.

Example 9.5.6. We consider the graph $G \cong K_3[(K_1, \overline{K}_3, \overline{K}_2)]$, compare Definition 4.4.3 and Construction 4.4.5. Here, K_3 is the canonical strong factorgraph of G. Start with K_3, blow up its three points to bubbles, insert $\overline{K}_3$, $\overline{K}_2$, and K_1 each in one bubble. Now connect all points of $\overline{K}_3$ with all points of $\overline{K}_2$ and similarly, and both sets of points with K_1. The result is the complete 3-partite graph $K_{1,2,3}$.

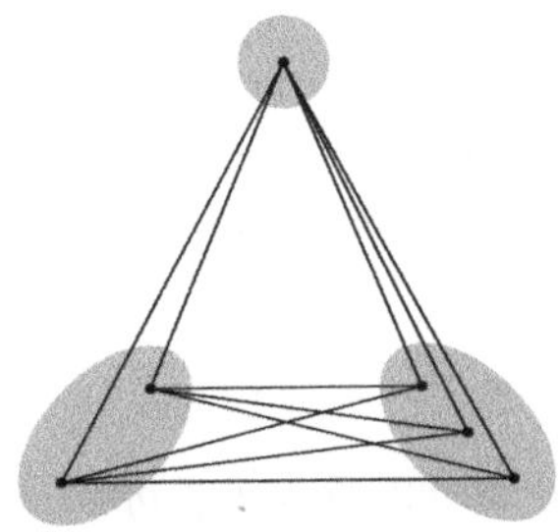

Theorem 9.5.5 implies the following corollary.

Corollary 9.5.7. *Let G be finite. Then* $|\mathrm{SEnd}(G)| = 1$ *if and only if* $|\mathrm{Aut}(G)| = 1$.

Proof. The existence of a nonbijective strong endomorphism of G implies that at least one Y_u has more than one element. But then the permutation of these two vertices gives a nontrivial automorphism of G; see Proposition 1.7.3.

The converse is Theorem 9.5.3. □

Decomposition of SEnd

The canonical strong decomposition of a graph G gives a decomposition of the monoid $\mathrm{SEnd}(G)$ which makes it possible to analyze algebraic properties of $\mathrm{SEnd}(G)$ in a very convenient way.

The multiplication of these decomposed strong endomorphisms of G can be interpreted algebraically as the composition in a so-called generalized wreath product with a small category, as will be shown in Application 9.5.14. However, except in Exercise 9.5.17, we will not make use of this interpretation in what follows.

We use Definition 9.5.1.

Lemma 9.5.8. *For* $f \in \mathrm{SEnd}(G)$ *consider the equivalence relation* $\nu_{f(G)}$ *on* $f(G)$ *defined by* $f(x)\nu_{f(G)}f(x') \Leftrightarrow N_{f(G)}(f(x)) = N_{f(G)}(f(x'))$ *for* $x, x' \in G$. *Then* $|f(G)/\nu_{f(G)}| = |G/\nu|$.

Proof. We use the homomorphism theorem (Theorem 1.6.10), factorizing the composition $\pi_{\nu_{f(G)}} f$ as $f'\pi_\nu$ where now $f' : G/\nu \to f(G)/\nu_{f(G)}$, $[x]_\nu \mapsto [f(x)]_{\nu_{f(G)}}$. Here f' is well-defined if $\nu \subseteq \mathrm{Ker}(\pi_{\nu_{f(G)}} f)$ and injective if we have the equality $\nu = \mathrm{Ker}(\pi_{\nu_{f(G)}} f)$; it is then even bijective since $\pi_{\nu_{f(G)}} f$ is of course surjective.

For the first statement, we show that $N_{f(G)}(f(x)) = N_{f(G)}(f(x'))$ if $x\nu_{f(G)}x'$. Take $f(y) \in N_{f(G)}(f(x))$; then $y \in N_G(f(x)) = N_G(f(x'))$ as f is strong, and $f(y) \in N_{f(G)}(f(x'))$ as f is a homomorphism. Thus $N_{f(G)}(f(x)) \subseteq N_{f(G)}(f(x'))$, and similarly for the converse implication.

For the second statement, we prove that $N_G(x) = N_G(x')$, i. e., xvx' if $f(x)v_{f(G)}f(x')$ or, equivalently, if $N_{f(G)}(f(x)) = N_{f(G)}(f(x'))$. Assume that $[x]_v \neq [x']_v$. Then there exists $y \in G$ with $\{x,y\} \in E, \{x',y\} \notin E$, and hence $\{f(x),f(y)\} \in E, \{f(x'),f(y)\} \notin E$ as f is strong, contradicting the assumption. So together we have that f' is a bijective mapping, and thus $|f(G)/v_{f(G)}| = |G/v|$. □

Lemma 9.5.9. *Let G/v be finite and $f \in \mathrm{SEnd}(G)$. Then, for $x, x' \in G$,*

$$N_{f(G)}(f(x)) = N_{f(G)}(f(x')) \quad \textit{implies} \quad N_G(f(x)) = N_G(f(x')).$$

Proof. We know that $N_G(f(x)) \supseteq N_{f(G)}(f(x)) = N_{f(G)}(f(x')) \subseteq N_G(f(x'))$. This means that possibly $|f(G)/v_{f(G)}| \leq |f(G)/v| \leq |G/v|$, but then Lemma 9.5.8 implies the equality $N_G(f(x)) = N_G(f(x'))$, using finiteness. □

Exercise 9.5.10. Prove that the result is different if $|G/v|$ is infinite. Take the union of one $P_2 = \{1_0, 2_0, 3_0\}$ and infinitely many $(P_3)_i = \{0_i, 1_i, 2_i, 3_i\}$ and f such that every path is mapped one step to the right while preserving the numbers (i. e., $n_i \mapsto n_{i+1}$). Then $N_G(f(1_0)) = \{0_1, 2_1\} \neq \{2_1\} = N_G(f(3_0))$. But G/v just identifies 1_0 and 3_0, and thus $N_{f(G)}(f(1_0)) = \{f(2_0)\} = N_{f(G)}(f(3_0))$.

Theorem 9.5.11. *Take a graph G with the canonical strong decomposition $G = U[(Y_u)_{u\in U}]$ where U is finite. Then for every $f \in \mathrm{SEnd}(G)$ and $(u, y_u) \in U[(Y_u)_{u\in U}]$ we have*

$$f((u, y_u)) = (s(u), f_u(y_u)).$$

This way every $f \in \mathrm{SEnd}(G)$ is a pair $(s, (f_u)_{u\in U})$ where $s \in \mathrm{Aut}(U)$ and $f_u : Y_u \to Y_{s(u)}$ is a mapping for all $u \in U$. Conversely, all such pairs are strong endomorphisms of G. With this notation, we have the following multiplication in $\mathrm{SEnd}(U[(Y_u)_{u\in U}])$:

$$(s, (f_u)_{u\in U})(t, (g_u)_{u\in U}) = (st, (f_{tu}g_u)_{u\in U}),$$

that is,

$$\begin{array}{ccccc} (u, (y_u)) & \xmapsto{(t,(g_u))} & (tu, (g_u(y_u))) & \xmapsto{(s,(f_{tu}))} & (stu, (f_{tu}(g_u(y_u)))) \\ \in Y_u & & \in Y_{tu} & & \in Y_{stu} \end{array}$$

Note that associativity of this multiplication is established once we prove the theorem, since it is based on a composition of two mappings, namely the multiplication in $\mathrm{Aut}(U)$ and the action of $\mathrm{Aut}(U)$ on U. Note moreover the similarity to the multiplication in the wreath product (Construction 9.4.1). There we had in the second component one mapping $f \in S^A$, now we have a family of mappings.

Proof. It is clear that every pair $(s, (f_u)_{u\in U})$ is a strong endomorphism of $U[(Y_u)_{u\in U}]$.

Take $f \in \mathrm{SEnd}(U[(Y_u)_{u\in U}])$ with $f((u, y_u)) = (v, y_v) \in U[(Y_u)_{u\in U}]$. Define $s : U \to U$ by $su := v = p_1(f(u, y_u))$ for an arbitrary $y_u \in Y_u$. We show that this is a correct definition. To do this, suppose that $f((u, y_u)) = (v, y_v)$ and $f((u, y'_u)) = (v', y'_{v'})$, where according to the decomposition of G we have that $(u, y_u)\, v\, (u, y'_u)$, i. e., $N_G(u, y_u) = N_G(u, y'_u)$, which implies that $f(N_G(u,y_u)) = f(N_G(u,y'_u))$. By the definition of $N_{f(G)}$, this gives the equality $N_{f(G)}(f(u,y_u)) = N_{f(G)}(f(u,y'_u))$. Consequently, we get from Lemma 9.5.9 that $N_G(f(x)) = N_G(f(x'))$. This implies $p_1(f(u,y_u)) = p_1(f(u,y'_u))$, which proves the correctness of the definition of s.

Now define $f_u : Y_u \to Y_{su}$ by $y_u \mapsto p_2(f(u,y_u))$, which clearly is a correct definition.

Since $\mathrm{Aut}(U) = \mathrm{SEnd}(U)$, we have to show that s is strong, i. e., that $\{u,v\} \in E(U)$ if and only if $\{su,sv\} \in E(U)$. Now, $\{u,v\} \in E(U)$ means that

$$\{(u,y_u),(v,y_v)\} \in E(G) \quad \text{for all } y_u \in Y_u,\ y_v \in Y_v.$$

As f is strong, this is equivalent to

$$\{f(u,y_u),f(v,y_v)\} = \{(su,f_u(y_u)),(sv,f_v(y_v))\} \in E(G),$$

which is the case if and only if $\{su,sv\} \in E(U)$. □

Exercise 9.5.12. Find an example which shows that quasi-strong endomorphisms in general do not preserve ν-classes.

A generalized wreath product with a small category

The semigroup side of this decomposition procedure in Theorem 9.5.11 can be described in a more abstract way as a generalized wreath product. This, however, is rather complicated and technical, and may appeal only to specialists; if you choose to skip it, nothing serious will be lost. Application 9.5.14 is just a reformulation of parts of Theorem 9.5.11.

Construction 9.5.13 (The generalized wreath product $W = R \wr \boldsymbol{K}$). Let $\boldsymbol{K}$ be a small category and R a monoid such that $X := \mathrm{Ob}\,\boldsymbol{K} \in R\text{-}\boldsymbol{Act}$. Write $M := \mathrm{Morph}\,\boldsymbol{K} := \bigcup_{x,y\in X} \boldsymbol{K}(x,y)$ and consider

$$W := \{(r,f) \mid r \in R, f \in M^X, f(x) \in \boldsymbol{K}(x,rx) \text{ for } x \in X\}.$$

Then, for $(r,f),(p,g) \in W$ define

$$(r,f)(p,g) := (rp, f_p g),$$

where $(f_p g)(x) := f(px)g(x)$ for any $x \in X$ and $f(px)g(x)$ is the composition of morphisms in $\boldsymbol{K}$.

Application 9.5.14. Take a simple undirected graph G, and let $U := G/\nu$ be the canonical strong factor graph of G. Then $G = U[(Y_u)_{u\in U}]$ is the canonical strong decomposition of G and Y_u denotes the equivalence class of $u \in U$ with respect to ν. Define the small category $K = K_{G/v}$ by $\operatorname{Ob} K := U$ and $K(u,v) := \textbf{\textit{Set}}(Y_u, Y_v)$ with composition of morphisms as in the category ***Set*** for $u, v \in U$, and take $R = \operatorname{Aut}(U)$. Then

$$\begin{aligned} \alpha : \operatorname{SEnd}(G) &\to \operatorname{Aut}(U) \wr K \\ f &\mapsto (p, (f_u)) \end{aligned}$$

defines an isomorphism of monoids, where p is the permutation of U induced by f and $f_u := f|_{Y_u} : Y_u \to Y_{pu}$ is the corresponding mapping induced by f. (See [Kilp et al. 2000] pp. 175–178.)

The chances to characterize monoids which are strong endomorphism monoids of a graph seem to be not very good, nevertheless.

Example 9.5.15 (The (strong) endomorphism monoid of P_2). Consider $P_2 = U[(Y_u)_{u\in U}] = K_2[(K_1, \overline{K}_2)]$, i. e., $U = K_2 = \{a, b\}$ is the canonical strong factor graph of P_2, $Y_a = K_1 = \{a\}$ and $Y_b = \overline{K}_2 = \{b_1, b_2\}$. Thus $\operatorname{Ob} K := \{a, b\}$. The morphism sets (sets of mappings) in K are as follows:

$K(a,a) := \textbf{\textit{Set}}(Y_a, Y_a)$ is 1-element,
$K(a,b) := \textbf{\textit{Set}}(Y_a, Y_b)$ is 2-element,
$K(b,a) := \textbf{\textit{Set}}(Y_b, Y_a)$ is 1-element, and
$K(b,b) := \textbf{\textit{Set}}(Y_b, Y_b)$ is 4-element, the transformations of a 2-element set.

We get the following 6 strong endomorphisms – there are no other endomorphisms, i. e., P_2 has Endotype 16, compare Theorem 1.7.5. We give the multiplication table. Observe that $\operatorname{SEnd}(P_2)$ is not a union of groups, i. e., it is not completely regular.

$$id, p = \begin{pmatrix} a & b_1 & b_2 \\ a & b_2 & b_1 \end{pmatrix},\quad f_1 = \begin{pmatrix} a & b_1 & b_2 \\ a & b_1 & b_1 \end{pmatrix},\quad f_2 = \begin{pmatrix} a & b_1 & b_2 \\ a & b_2 & b_2 \end{pmatrix},\quad f_3 = \begin{pmatrix} a & b_1 & b_2 \\ b_1 & a & a \end{pmatrix},\quad f_4 = \begin{pmatrix} a & b_1 & b_2 \\ b_2 & a & a \end{pmatrix}.$$

$\circ$	id	p	f_1	f_2	f_3	f_4
id	id	p	f_1	f_2	f_3	f_4
p	p	id	f_1	f_2	f_3	f_4
f_1	f_1	f_2	f_1	f_2	f_3	f_4
f_2	f_2	f_1	f_1	f_2	f_3	f_4
f_3	f_3	f_4	f_3	f_4	f_1	f_2
f_4	f_4	f_3	f_3	f_4	f_1	f_2

Cardinality of SEnd(*G*)

The analysis of Application 9.5.14 and a simple counting argument gives the following theorem.

Theorem 9.5.16. *Let G be finite, with $G = U[(Y_u)_{u\in U}]$. Then*

$$|\operatorname{SEnd}(G)| = \sum_{s\in\operatorname{Aut}(U)} \prod_{u\in U} |Y_{su}|^{|Y_u|}.$$

Exercise 9.5.17. There exists no graph G such that $\mathrm{SEnd}(G) \neq \mathrm{Aut}(G)$, i. e., with endotype greater than 15 and $|\mathrm{SEnd}(G)| \in \{2,3,5,\dots,25,29\}$ upon analyzing the possible right-hand sides of the formula.

Example 9.5.18. Take $G = K_3[(K_1, \overline{K}_3, \overline{K}_2)]$ from Example 9.5.6. Then $Aut(K_3) \cong S_3$. Set $K_3 = \{1,3,2\}$. Then for the sets of respective mappings we have $|\boldsymbol{Set}(1,3)| = 3$, $|\boldsymbol{Set}(1,2)| = 2$, $|\boldsymbol{Set}(1,1)| = |\boldsymbol{Set}(3,1)| = |\boldsymbol{Set}(2,1)| = 1$ and so on. With id_{S_3} we have $1 \cdot 2^2 \cdot 3^3 = 108$ strong endomorphisms, 12 of which are bijective.

Regularity and more for T_A

First, we collect some easy facts about the transformation monoid of a set A.

Theorem 9.5.19. *Let A be a set and $T_A = A^A$ the full transformation monoid of A (i. e., all mappings from A to A). Then T_A is always regular and, moreover, the following implications hold:*
(a) *completely regular ⇔ orthodox ⇔ left inverse ⇔ $|A| \leq 2$;*
(b) *right inverse ⇔ inverse ⇔ Clifford ⇔ group ⇔ commutative ⇔ idempotent ⇔ $|A| = 1$.*

Proof. Regularity is well known and easy to prove.

Sufficiency is obvious in all cases.

Necessity in each case is proved by exhibiting a counterexample. Take $A = \{1,2,3\}$. For "completely regular" consider $f(1) = 2, f(2) = f(3) = 3$. Then any pseudo-inverse g must satisfy $g(2) = 1$, and then $gf(1) = 1$ but $1 \notin \mathrm{Im}\, fg$.

For "orthodox" and "left inverse," consider the two idempotents $h(1) = 1$, $h(2) = h(3) = 3$ and $g(1) = g(2) = 2$, $g(3) = 3$. Then gh is not idempotent and $hgh \neq gh$.

The other cases are treated similarly, but using $A = \{1,2\}$. □

Corollary 9.5.20. *For T_A, the implications in Theorem 9.1.2 reduce to:*

$$\left.\begin{array}{l} \textit{group} \\ \Leftrightarrow \textit{inverse} \\ \Leftrightarrow \textit{right inverse} \\ \Leftrightarrow \textit{Clifford monoid} \end{array}\right\} \Rightarrow \left\{\begin{array}{l} \textit{completely regular} \\ \Leftrightarrow \textit{orthodox} \\ \Leftrightarrow \textit{left inverse} \end{array}\right\} \Rightarrow \textit{regular.}$$

Regularity and more for SEnd(G)

Theorem 9.5.21. *Take $G = U[(Y_u)_{u \in U}]$ with $|U|$ finite. Then* SEnd(G) *is a regular monoid, i. e., for every element $(s,(f_u)) \in \mathrm{SEnd}(U[(Y_u)_{u\in U}])$ one has*

$$(s,(f_u))(s^{-1},(f'_u))(s,(f_u)) = (s,(f_u)),$$

where we choose

$$f'_u(y_u) \in \begin{cases} (f_{s^{-1}u})^{-1}(y_u) & \text{if } y_u \in \operatorname{Im} f_{s^{-1}u}, \\ Y_{s^{-1}u} & \text{if } y_u \in Y_u \setminus \operatorname{Im} f_{s^{-1}u}. \end{cases}$$

Proof. We have to prove that the proposed f' satisfies the equality $ff'f = f$. Note, that $s \in \operatorname{Aut}(U)$ by Theorem 9.5.11. Indeed, for $(u, y_u) \in U[(Y_u)_{u\in U}]$ we get

$$(u, y_u) \overset{f}{\mapsto} (su, f_u(y_u)) \overset{f'}{\mapsto} (s^{-1}su, f'(f_u(y_u))) \overset{f}{\mapsto} (su, f_u(y_u)). \qquad \square$$

Example 9.5.22. If $U = \{0, 1, 2, \ldots\}$ is an infinite chain, then $\operatorname{SEnd}(U) \cong (\mathbb{N}, +)$, which obviously is not regular.

For convenience, we formulate the following lemma, which is clear from the structure of $U[(Y_u)_{u\in U}]$.

Lemma 9.5.23. *An element $(s, (f_u)) \in \operatorname{SEnd}(U[(Y_u)_{u\in U}])$ is idempotent if and only if $s = \mathrm{id}_U$ and f_u is idempotent for all $u \in U$.*

Theorem 9.5.24. *The monoid $\operatorname{SEnd}(U[(Y_u)_{u\in U}])$ is regular and:*
(a) *completely regular $\Leftrightarrow$ $|Y_u| \leq 2$ for all $u \in U$, and $|Y_u| = 2$ implies $|Y_{su}| = 1$ for all $s \in \operatorname{Aut}(U)$ with $su \neq u$;*
(b) *orthodox $\Leftrightarrow$ left inverse $\Leftrightarrow$ $|Y_u| \leq 2$ for all $u \in U$;*
(c) *right inverse $\Leftrightarrow$ inverse $\Leftrightarrow$ Clifford $\Leftrightarrow$ group $\Leftrightarrow$ $|Y_u| = 1$ for all $u \in U$, i.e., $\operatorname{SEnd}(U[(Y_u)_{u\in U}]) \cong \operatorname{Aut}(U)$ is a group;*
(d) *commutative $\Leftrightarrow$ $|Y_u| = 1$ for all $u \in U$, i.e., $\operatorname{SEnd}(U[(Y_u)_{u\in U}]) \cong \operatorname{Aut}(U)$ which is commutative;*
(e) *idempotent $\Leftrightarrow$ $|Y_u| = 1$ for all $u \in U$, $|Y_u| = 1$, and $|\operatorname{Aut}(U)| = 1$, i.e., $|\operatorname{SEnd}(U[(Y_u)_{u\in U}])| = 1$.*

Proof. (a) Sufficiency is obvious; any pseudo-inverse constructed in Theorem 9.5.21 will do in this case. To prove necessity, note that the first part of the condition follows from the corresponding part of Theorem 9.5.19. Now assume that $|Y_u| = 2$ and take $\mathrm{id}_U \neq s \in \operatorname{Aut}(U)$ such that $|Y_{su}| = 2$. Consider $(s, (f_u))$ where $f_u : Y_u \to Y_{su}$ is surjective and $f_{s^{-1}u} : Y_{s^{-1}u} \to Y_u$ is not surjective. Any pseudo-inverse of $(s, (f_u))$ is of the form $(s^{-1}, (g_u))$, and because of the complete regularity we have $(s, (f_u))(s^{-1}, (g_u)) = (\mathrm{id}_U, (f_{s^{-1}u}g_u)) = (\mathrm{id}_U, (g_{su}f_u)) = (s^{-1}, (g_u))(s, (f_u))$ for some pseudo-inverse. Now bijectivity of f_u implies that $g_{su} : Y_{su} \to Y_u$ actually satisfies $g_{su} = f_u^{-1}$, i.e., $g_{su}f_u$ is surjective on Y_u. On the other hand, as $f_{s^{-1}u}$ is not surjective onto Y_u we get that $g_{su}f_u \neq f_{s^{-1}u}g_u$, which is a contradiction.

For parts (b) to (e), sufficiency is obvious; use Lemma 9.5.23 for (b). Necessity is also obvious in all cases, owing to Theorem 9.5.19; one uses $(\mathrm{id}_U, (f_u))$ where the f_u are from the respective counterexamples in the proof of Theorem 9.5.19. $\square$

Corollary 9.5.25. *The implication structure of Corollary* 9.5.20 *is slightly different for* $\mathrm{SEnd}(U[(Y_u)_{u\in U}])$ *and becomes the following:*

$$\left.\begin{array}{l}\textit{group}\\ \Leftrightarrow \textit{Clifford monoid}\\ \Leftrightarrow \textit{inverse}\\ \Leftrightarrow \textit{right inverse}\end{array}\right\} \Rightarrow \begin{array}{l}\textit{completely}\\ \textit{regular}\end{array} \Rightarrow \left\{\begin{array}{l}\textit{orthodox}\\ \Leftrightarrow\\ \textit{left inverse}\end{array}\right\} \Rightarrow \textit{regular.}$$

9.6 Comments

The decomposition of SEnd is very useful for the algebraic investigation of the strong monoid of a graph, as we have seen. Using the strong decomposition of a graph G, it should easier to describe all automorphisms of G.

One important special case deals with the so-called ***split graphs***; cf. e. g., Ulrich Knauer and Apirat Wanichsombat [55]. These are graphs which have a complete graph K_n as a core and, in addition, a set I of mutually independent vertices which are adjacent only to vertices of K_n. Here, regular, idempotent closed, orthodox, and completely regular endomorphism monoids are investigated. It would be interesting to replace the complete graph by an asymmetric graph or even a rigid graph and ask the same questions about the endomorphisms.

These constructions also point toward possibilities of building graphs whose endomorphism monoids are Clifford monoids, in which case the structure semilattice under the Clifford semigroup is a lattice and the identity element of the top group figures as the identity element of the endomorphism monoid.

For an opposite view, namely to describe monoids which are strong endomorphism monoids of a graph, we can use Theorem 9.5.24. We see, that such monoids must be regular, and they are already groups or even 1-element in the Cases (c), (d), and (e) of Theorem 9.5.24.

10 Compositions, unretractivities, and monoids

We consider various compositions of graphs or operations with graphs and investigate various unretractivities of these compositions. Moreover, we ask under what conditions a composition of graphs leads to suitable composition of monoids. The idea behind this question, which is quite familiar in mathematics, is a sort of distributivity of End or Aut over the operations.

Unfortunately, it is not convenient to consider End or Aut as functors, since they are functors in two variables and, for instance, $\mathrm{End}(G) = \mathrm{Hom}(G, G)$ is contravariant in the first variable and covariant in the second; similarly for Aut.

In this chapter, all graphs are without loops.

10.1 Lexicographic products

First, we look at some simple properties associated with wreath products of monoids, lexicographic products of graphs. Recall that we do not have categorical descriptions for lexicographic and wreath products. Let G and H be graphs.

Lemma 10.1.1. *Take $f^2 = f \in \mathrm{End}(G[H])$ and $\{x, x'\} \in E(G)$. For $H_x := \{(x, y) \in G[H] \mid y \in H\}$, the H-**layer at** x, one has $f(H_{x'}) \bigcap H_x = \emptyset$. In other words, for $f^2 = f \in \mathrm{End}(G[H])$, the equality $f(x', y') = f(x, y) = (x, y)$ implies $\{x, x'\} \notin E(G)$.*

Proof. Assume that $f(x', y') = (x, y) \in f(H_{x'}) \bigcap H_x$ for some $y' \in V(H)$ and $\{x, x'\} \in E(G)$. Then $\{(x', y'), f(x', y')\} = \{(x', y'), (x, y)\} \in E(G[H])$, and applying f again gives a loop, which is impossible. □

Lemma 10.1.2. *Take $G \in \{K_n, C_{2n+1}\}$. For $f^2 = f \in \mathrm{End}(G[H])$ we have that $f(x, y) = (x, y')$ for all $(x, y) \in G[H]$, with $y' \in H$, i. e., $p_1 f = p_1$.*

Proof. With Lemma 10.1.1, this is clear for K_n. Suppose that there exist $(x, y), (x', y') \in V(C_{2n+1}[H])$, with $x \neq x'$, such that $f(x, y) = (x', y') = f(x', y')$. Let P_1 and P_2 be the two different paths in C_{2n+1} connecting x and x'. We construct two paths in $C_{2n+1}[H]$ connecting (x, y) and (x', y') using the first components from P_1 and P_2, possibly with some of them used more than once. Then $p_1 f : P_1 \to C_{2n+1}$ and $p_1 f : P_2 \to C_{2n+1}$ are graph homomorphisms, which when combined give an endomorphism $p_1 f : C_{2n+1} \to C_{2n+1}$. This is not bijective with $p_1 f(x) = p_1 f(x')$, which is impossible as $\mathrm{End}(C_{2n+1}) = \mathrm{Aut}(C_{2n+1})$. □

Lemma 10.1.3. *For $\xi^2 = \xi \in \mathrm{SEnd}(G)$ and $x \in G$, one has $N(x) = N(\xi(x))$.*

Proof. Take $x' \in N(x)$. Then $\{\xi(x), \xi(x')\} = \{\xi^2(x), \xi(x')\} \in E(G)$, and hence $\{\xi(x), x'\} \in E(G)$ as ξ is strong. Consequently, $N(x) \subseteq N(\xi(x))$. Conversely, take $x' \in N(\xi(x))$. Then

https://doi.org/10.1515/9783110617368-010

$\{\xi^2(x), \xi(x')\} = \{\xi(x), \xi(x')\} \in E(G)$, and thus $\{x, x'\} \in E(G)$ as ξ is strong. Consequently, $N(x) \supseteq N(\xi(x))$. □

Remark 10.1.4. For $M \in \{\mathrm{End}, \mathrm{SEnd}\}$, we have that $(M(G) \wr M(H)|G)$ is a group if and only if $M(G)$ and $M(H)$ are groups.

Exerceorem 10.1.5. Let G and H be (arbitrary) graphs and recall Definition 9.4.3. Then:
(1) $(\mathrm{End}(G) \wr \mathrm{End}(H)|G) \subseteq \mathrm{End}(G[H])$;
(2) $(\mathrm{Aut}(G) \wr \mathrm{Aut}(H)|G) \subseteq \mathrm{Aut}(G[H])$.

Note that a corresponding result is not true for strong endomorphisms; neither are the converse implications, as the following example shows.

Example 10.1.6. As usual, T_2 denotes the full transformation monoid on two elements, S_2 the permutation group on two elements and D_4 the dihedral group on four elements.

Now $(\mathrm{SEnd}(\overline{K}_2) \wr \mathrm{SEnd}(K_2)|\overline{K}_2) \cong (T_2 \wr S_2|\overline{K}_2)$, which is not a group, and thus not contained in $\mathrm{SEnd}(\overline{K}_2[K_2]) = \mathrm{Aut}(\overline{K}_2[K_2]) = \mathrm{Aut}(C_4) \cong D_4$. So $(\mathrm{SEnd}(\overline{K}_2) \wr \mathrm{SEnd}(K_2)|\overline{K}_2) \nsubseteq \mathrm{SEnd}(\overline{K}_2[K_2])$.

Observe that $\mathrm{Aut}(\overline{K}_2 \wr \mathrm{Aut}(K_2)|\overline{K}_2) \cong \mathrm{Aut}(\overline{K}_2[K_2])$.

For the converse implication, observe that

$$(\mathrm{SEnd}(K_2) \wr \mathrm{SEnd}(K_2)|K_2) = (\mathrm{Aut}(K_2) \wr \mathrm{Aut}(K_2)|K_2)$$

has eight elements and, therefore, does not contain $\mathrm{SEnd}(K_2[K_2]) = \mathrm{SEnd}(K_4) = \mathrm{Aut}(K_4) \cong S_4$, which has 24 elements. So

$$(\mathrm{SEnd}(K_2) \wr \mathrm{SEnd}(K_2)|K_2) \nsupseteq \mathrm{SEnd}(K_2[K_2]).$$

This also shows that the converse implication of (2) in Exerceorem 10.1.5 is not true in general.

Equality in (1) of Exerceorem 10.1.5 will turn out to be sufficient for one implication with SEnd (see Theorem 10.3.1), which, in turn, is sufficient for equality in (2) of Exerceorem 10.1.5 (see Theorem 10.3.2). This equality in (2) is characterized in Theorem 10.3.5. A similar characterization of the corresponding equality for SEnd is given in Theorem 10.3.10.

Next, we consider six types of strong endomorphisms of lexicographic products $G[H]$ of graphs G and H which can be constructed from strong endomorphisms of the components (1, 2a, 2b) and vice versa (3, 4a, 4b). The straightforward proofs use the preceding two lemmas. We leave them as exercises.

Construction 10.1.7. Take $(x, y) \in G[H]$.
(1) For $\eta \in \mathrm{SEnd}(H)$ set $f((x, y)) := (x, \eta(y))$. Then $f \in \mathrm{SEnd}(G[H])$. Moreover, η is injective if and only if f is injective.

(2a) Take $\xi^2 = \xi \in \mathrm{SEnd}(G)$ and set $f((x,y)) := (\xi(x), y)$. Then $f \in \mathrm{SEnd}(G[H])$. Moreover, f is injective if and only if ξ is injective.

(2b) Take $y_0 \in H$, an isolated vertex, and $\xi^2 = \xi \in \mathrm{SEnd}(G)$.

Set $f((x,y)) := \begin{cases} (\xi(x), y_0) & \text{for } y = y_0, \\ (x,y) & \text{otherwise.} \end{cases}$

Then $f \in \mathrm{SEnd}(G[H])$.

Moreover, ξ is injective if and only if f is injective.

(3) Take $f^2 = f \in \mathrm{SEnd}(G[H])$. For $x \in G$ set $\xi(x) := p_1 f(x, y_0)$ for some isolated vertex $y_0 \in H$. Then $\xi \in \mathrm{SEnd}(G)$.

(4a) Take $f^2 = f \in \mathrm{SEnd}(G[H])$. For $x \in G$ set $\eta_x(y) := p_2 f((x,y))$ for $y \in H$. Then $\eta_x \in \mathrm{SEnd}(H)$.

(4b) Take $f^2 = f \in \mathrm{SEnd}(G[H])$. Suppose that $H = H_1 \bigcup H_2$ where H_1 is connected. For $x \in G$ set $\eta_x(y) := p_2 f((x,y))$ if $y \in H_1$, and $\eta_x(y) := y$ if $y \in H_2$. Then $\eta_x \in \mathrm{SEnd}(H)$.

Question. Can these constructions be extended to End, HEnd, LEnd, and QEnd of lexicographic products?

10.2 Unretractivities and lexicographic products

In this section, we present results about the E-S unretractivity, E-A unretractivity, and S-A unretractivity of the lexicographic product (of certain finite graphs); i. e., we consider lexicographic products such that all endomorphisms are strong or automorphisms or such that all strong endomorphisms are bijective, i. e., automorphisms. Recall from Definition 1.7.1 that a graph has endotype 0 if it is E-A unretractive, endotype 16 if it is E-S unretractive, and endotype less than 16 if it is S-A unretractive.

The first results on this topic can be found in U. Knauer [50, 51].

Theorem 10.2.1. *Take* $G \in \{K_n, C_{2n+1}\}$*. Then* $\mathrm{End}(G[H]) = \mathrm{Aut}(G[H])$ *if and only if* $\mathrm{End}(H) = \mathrm{Aut}(H)$.

Proof. To prove necessity, note that by Exerceorem 10.1.5 we have

$$(\mathrm{End}(G) \wr \mathrm{End}(H)|G) \subseteq \mathrm{End}(G[H]) = \mathrm{Aut}(G[H]).$$

Then $\mathrm{End}(G)$ and $\mathrm{End}(H)$ are groups, and thus G and H are unretractive.

To prove sufficiency, take $G \in \{K_n, C_{2n+1}\}$ and suppose that $\mathrm{End}(H) = \mathrm{Aut}(H)$. By finiteness, we may assume for $f \in \mathrm{End}(G[H])$ that $f^2 = f$.

Case 1. There exist $(x,y), (x,y') \in V(G[H])$ such that $f(x,y') = (x,y)$ with $y \neq y'$. Then $f(x,y) = (x,y)$ since f is idempotent. Now take $\eta_x \in \mathrm{End}(H)$ defined as in (4b) of Construction 10.1.7. Then η_x is not injective as $y \neq y'$. This contradicts $\mathrm{End}(H) = \mathrm{Aut}(H)$.

Now, if $G = K_n$, then $f^2 = f \in \mathrm{End}(K_n[H])$ implies $p_1 f(x,y) = x$ by Lemma 10.1.2, so Case 1 is done.

Case 2. There exist $(x,y),(x',y') \in V(C_{2n+1}[H])$, with $x \neq x'$, such that $f(x,y) = (x',y') = f(x',y')$. Then Lemma 10.1.2 implies that $p_1 f = p_1$, which is a contradiction. □

We give two definitions next, one of which is known from Definition 9.5.1. Both will be used again later, for example in Theorem 10.3.5, where they originated.

Definition 10.2.2. The relation $\nu_G \subseteq G \times G$ is defined by

$$x \,\nu_G\, x' \Leftrightarrow N_G(x) = N_G(x').$$

The relation $\sigma_G \subseteq G \times G$ is defined by

$$x \,\sigma_G\, x' \Leftrightarrow N_G(x) \bigcup \{x\} = N_G(x') \bigcup \{x'\}.$$

Now $x\nu_G x'$ means that x and x' are not adjacent and have the same neighbors, and $x\sigma_G x'$ means that x and x' are adjacent and have the same neighbors. So $\nu_G = \Delta$ or $\sigma_G = \Delta$ mean that different nonadjacent or adjacent vertices do not have the same neighbors in G. The smallest examples with nontrivial relations are the path $P_2 = \{0,1,2\}$ of length 2, where $1\nu_{P_2}3$, and the complete graph $K_3 = \{1,2,3\}$, where $1\sigma_{K_3}3$ and the same for any other pair of points in K_3.

The notation comes from Sabidussi's original paper *The composition of graphs* [77]. Later, the relation ν_G was mostly called R_G and σ_G was mostly called S_G.

We have the following results under certain conditions; see R. Kaschek [43].

Remark 10.2.3. $\mathrm{End}(G[H]) = \mathrm{Aut}(G[H])$ if and only if:

(a) $\mathrm{End}(G[H]) = (\mathrm{Aut}(G) \wr \mathrm{Aut}(H)|G)$, under the condition that $\overline{H}$ is connected and $\sigma_G = \Delta$, where as usual Δ denotes the diagonal of $G \times G$;

(b) $\mathrm{End}(G) = \mathrm{Aut}(G)$ and $\mathrm{End}(H) = \mathrm{Aut}(H)$, under the condition that G has no triangles and no isolated vertices.

Note that Theorem 10.2.1 is a special case of (b).

Proposition 10.2.4. *If* $\mathrm{End}(G[H]) = \mathrm{SEnd}(G[H])$, *then* $\mathrm{End}(G) = \mathrm{SEnd}(G)$ *and* $\mathrm{End}(H) = \mathrm{SEnd}(H)$.

Proof. Take $\xi \in \mathrm{End}(G)$ and suppose that $\{\xi(x), \xi(x')\} \in E(G)$ for $x, x' \in G$. Define $f := (\xi, \mathrm{id}_Y) \in \mathrm{End}(G[H]) = \mathrm{SEnd}(G[H])$, i. e., $f(x,y) = (\xi(x), y)$ for $(x,y) \in G[H]$. Since $\{f(x,y), f(x',y)\} = \{(\xi(x),y),(\xi(x'),y)\} \in E(G[H])$, we get $\{(x,y),(x',y)\} \in E(G[H])$, and thus $\{x,x'\} \in E(G)$. This proves that ξ is strong.

Take $\eta \in \mathrm{End}(H)$ and suppose that $\{\eta(y), \eta(y')\} \in E(H)$ for $y, y' \in H$. Define $f := (\mathrm{id}_X, \eta) \in \mathrm{End}(G[H]) = \mathrm{SEnd}(G[H])$, i. e., $f(x,y) = (x, \eta(y))$ for $(x,y) \in G[H]$. Since $\{f(x,y), f(x,y')\} = \{(x,\eta(y)),(x,\eta(y'))\} \in E(G[H])$, we get $\{(x,y),(x',y)\} \in E(G[H])$, and thus $\{y,y'\} \in E(H)$. This proves that η is strong. □

Corollary 10.2.5. *If* $|H| \geq 2$ *and* $\mathrm{End}(G[H]) = \mathrm{SEnd}(G[H])$, *then* $\mathrm{End}(G) = \mathrm{Aut}(G)$ *or* $E(H) = \emptyset$.

Proof. If there exists $\xi \in \mathrm{SEnd}(G)$ with $\xi(x) = \xi(x')$ for $x \neq x'$, then (ξ, id_Y) is not strong if $\{y, y'\} \in E(H)$. □

Lemma 10.2.6. *If* $\mathrm{End}(H) = \mathrm{SEnd}(H)$, *then* $\mathrm{End}(C_{2n+1}[H]) = \mathrm{SEnd}(C_{2n+1}[H])$.

Proof. Take $f \in \mathrm{End}(C_{2n+1}[H]) \setminus \mathrm{SEnd}(C_{2n+1}[H])$. Then there exists

$$\begin{aligned}\{(x_1, y_1), (x_2, y_2)\} &\notin E(C_{2n+1}[H])\\ \text{such that } \{(x_1', y_1'), (x_2', y_2')\} &= \{f(x_1, y_1), f(x_2, y_2)\} \in E(C_{2n+1}[H])\end{aligned}$$

This is true for any power of f and so we suppose that f is idempotent. Then by Lemma 10.1.2 we obtain that $x_1 = x_1'$ and $x_2 = x_2'$. Now $\{(x_1, y_1), (x_2, y_2)\} \notin E(C_{2n+1}[H])$ and $\{(x_1, y_1'), (x_2, y_2')\} \in E(C_{2n+1}[H])$ imply that $x_1 = x_2$, $\{y_1, y_2\} \notin E(H)$, $\{y_1', y_2'\} \in E(H)$. Defining $\eta(y) := p_2 f(x_1, y)$ gives an endomorphism of H which is not strong, contradicting the hypothesis. □

Theorem 10.2.7 (E-S unretractive). *Take* $G \in \{K_n, C_{2n+1}\}$. *Then* $\mathrm{End}(G[H]) = \mathrm{SEnd}(G[H])$ *if and only if* $\mathrm{End}(H) = \mathrm{SEnd}(H)$.

Proof. This follows from Lemmas 10.2.6 and 10.2.4. □

Here, too, we have some further results under certain conditions; compare again R. Kaschek [43].

Remark 10.2.8. $\mathrm{End}(G[H]) = \mathrm{SEnd}(G[H])$ if and only if:

(a) $\mathrm{End}(G) = \mathrm{SEnd}(G)$, under the condition that $H = \overline{K}_n$;

(b) $\mathrm{End}(G) = \mathrm{Aut}(G)$ and $\mathrm{End}(H) = \mathrm{SEnd}(H)$ and $\mathrm{Idpt}(G) \subseteq (\mathrm{End}(G) \wr \mathrm{End}(H)|G)$, under the condition that $H \neq \overline{K}_n$;

(c) $\mathrm{End}(G) = \mathrm{Aut}(G)$ and $\mathrm{End}(H) = \mathrm{SEnd}(H)$, under the condition that G has no triangles and no isolated vertices.

Note that Theorem 10.2.7 is a special case of (a) and possibly of (b).

Theorem 10.2.9 (S-A unretractive). $\mathrm{SEnd}(G[H]) = \mathrm{Aut}(G[H])$ *if and only if:*

(a) $\mathrm{SEnd}(G) = \mathrm{Aut}(G)$ *and* $\mathrm{SEnd}(H) = \mathrm{Aut}(H)$; *or*

(b) $\mathrm{SEnd}(H) = \mathrm{Aut}(H)$ *and* H *has no isolated vertex.*

Proof. For the necessity, we show first that H is S-A unretractive. To do this, we use (1) in Construction 10.1.7, and we take any $\eta \in \mathrm{SEnd}(H)$. Then the constructed f must be injective, and so η is injective, and thus in $\mathrm{Aut}(H)$.

If now H has an isolated vertex y_0, then G is S-A unretractive. This is obtained by using the statement from (2b) in Construction 10.1.7, since for any idempotent $\xi \in \mathrm{SEnd}(G)$ the constructed f is injective, and thus ξ is injective and, therefore, in $\mathrm{Aut}(G)$.

To prove sufficiency, take $f^2 = f \in \mathrm{SEnd}(G[H])$, i. e., suppose there exists $(x, y) \neq (x', y') \in V(G[H])$ with $f(x', y') = (x, y) = f(x, y)$.

(a) Let y_0 be an isolated vertex of H. If $x = x'$, take $\eta_x \in \mathrm{SEnd}(H) = \mathrm{Aut}(H)$ as in (4) of Construction 10.1.7, which is not injective as $y \neq y'$. Therefore, $x \neq x'$.

If $y = y_0 = y'$, take $\xi \in \mathrm{SEnd}(G) = \mathrm{Aut}(G)$ as in (3) of Construction 10.1.7, which again is not injective.

So let $y' \neq y_0$. Then there exists $y_1 \in V(H)$ with $\{y', y_1\} \in E(H)$, since an S-A unretractive graph cannot have more than one isolated vertex. Then we have that $\{(x', y'), (x', y_1)\} \in E(G[H])$. Let $f(x', y_1) = (x_2, y_2)$; then $\{(x, y)(x_2, y_2)\} \in E(G[H])$ but $\{x_2, x'\} \notin E(G)$ by Lemma 10.1.1. Thus $x_2 = x'$ and $\{(x, y), (x', y_2)\} \in E(G[H])$, but again $\{x, x'\} \notin E(G)$ by Lemma 10.1.1. So $x = x'$ follows, contradicting the assumption that $x \neq x'$.

Now let $y \neq y_0$. Then there exists $y_1 \in V(H)$ with $\{y, y_1\} \in E(H)$, and then $\{(x, y), (x, y_1)\} \in E(G[H])$. Consequently, we have that $\{f(x, y), f(x, y_1)\} = \{f(x', y'), f(x, y_1)\} \in E(G[H])$, and thus $\{(x', y'), (x, y_1) \in E(G[H])\}$. This is impossible, as by assumption $x \neq x'$ and $\{x, x'\} \notin E(G)$, again by Lemma 10.1.1. This completes the proof of (a).

(b) Now H has no isolated vertex, so there exists $y'' \in V(H)$ with $\{y', y''\} \in E(H)$. If $x \neq x'$, then $\{x, x'\} \notin E(G)$ by Lemma 10.1.1. Consequently, we have $\{(x', y'), (x', y'')\} \in E(G[H])$, and thus $\{f(x', y'), f(x', y'')\} = \{f(x, y), f(x', y'')\} \in E(G[H])$. Then $\{(x, y), (x', y'')\} \in E(G[H])$ as f is strong. Since $x \neq x'$, this implies $\{x, x'\} \in E(G)$, which is a contradiction.

If $x = x'$ but $y \neq y'$, then take η_x from (4a) in Construction 10.1.7, which is not injective in this case; this contradicts the S-A unretractivity of H. □

10.3 Monoids and lexicographic products

Here, we consider the question of how End operates on the lexicographic product End$(G[H])$ of two graphs G and H. It turns out that the appropriate composition of monoids here is the wreath product; see Definition 9.4.3.

We present the results of U. Nummert [68].

Theorem 10.3.1. $\mathrm{End}(G[H]) = (\mathrm{End}(G) \wr \mathrm{End}(H)|G)$ *implies* $\mathrm{SEnd}(G[H]) \subseteq (\mathrm{SEnd}(G) \wr \mathrm{SEnd}(H)|G)$ *(where G and H are without loops).*

Proof. Take $\varphi \in \mathrm{SEnd}(G[H]) \subseteq \mathrm{End}(G[H]) = (\mathrm{End}(G) \wr \mathrm{End}(H)|G)$ with $\varphi = (r, f)$, presented as an element of the wreath product, with the notation of Construction 9.4.1. Consider $\{r(x), r(x')\} \in E(G)$. Then $\{\varphi(x, y), \varphi(x', y)\} = \{(r(x), f(x)(y)), (r(x'), f(x')(y))\} \in E(G[H])$, which implies that $\{(x, y), (x', y)\} \in E(G[H])$ as φ is strong. This means that $\{x, x'\} \in E(G)$.

Suppose now that $\{f(x)(y), f(x)(y')\} \in E(H)$. Then $\{(r(x), f(x)(y)), (r(x), f(x)(y'))\} \in E(G[H])$, which implies that $\{(x, y), (x, y')\} \in E(G[H])$ as φ is strong. Thus $\{y, y'\} \in E(H)$. □

Theorem 10.3.2. *If G and H are finite, then* $\mathrm{SEnd}(G[H]) \subseteq (\mathrm{SEnd}(G) \wr \mathrm{SEnd}(H)|G)$ *implies* $\mathrm{Aut}(G[H]) = (\mathrm{Aut}(G) \wr \mathrm{Aut}(H)|G)$.

Proof. Take $\varphi \in \mathrm{Aut}(G[H]) \subseteq (\mathrm{SEnd}(G) \wr \mathrm{SEnd}(H)|G)$ with $\varphi = (r,f)$, which is bijective. We show that r and $f(x)$ are bijective for all $x \in G$. This implies that $(r,f) \in (\mathrm{Aut}(G) \wr \mathrm{Aut}(H)|G)$. The converse is true by Exerceorem 10.1.5. Then for all $(x',y') \in G[H]$ there exists $(x,y) \in G[H]$ with $(r,f)(x,y) = (r(x),f(x)(y)) = (x',y')$. Thus r is surjective and, therefore, bijective if G is finite.

Suppose now that $(r,f)(x,y) = (r,f)(x,y')$; then $f(x)(y) = f(x)(y')$. Now injectivity of (r,f) implies $y = y'$. Therefore, $f(x)$ is injective for all $x \in G$, and thus bijective if H is finite. □

Remark 10.3.3. It can be seen that the previous theorem is also true if only one of G or H is finite.

Theorem 10.3.4. *Take arbitrary (i. e., not necessarily finite) graphs G and H, where G is without loops. Then* $(\mathrm{SEnd}(G) \wr \mathrm{SEnd}(H)|G) \subseteq \mathrm{SEnd}(G[H])$ *if and only if* $H = \overline{K}_{|H|}$ *or* $\nu_G = \Delta$, *i. e.,* $\{x,x'\} \notin E(G)$ *and* $N_G(x) = N_G(x')$ *implies* $x = x'$ *for* $x,x' \in G$.

Proof. To prove necessity, suppose there exists $r \in \mathrm{SEnd}(G) \setminus \mathrm{Aut}(G)$, i. e., there exist $x \neq x' \in G$ with $r(x) = r(x')$; then $\{x,x'\} \notin E(G)$ since G has no loops. Consider $(r,\mathrm{id}) \in (\mathrm{SEnd}(G) \wr \mathrm{SEnd}(H)|G)$, where $\mathrm{id}(x) = \mathrm{id}_H$ for all $x \in G$. Then $\{(r,\mathrm{id})(x,y),(r,\mathrm{id})(x',y')\} = \{(r(x),y),(r(x'),y')\} \notin E(G[H])$, and thus $\{y,y'\} \notin E(H)$ for any $y,y' \in H$. Consequently, $H = \overline{K}_{|H|}$.

To prove sufficiency, consider $(r,f) \in (\mathrm{SEnd}(G) \wr \mathrm{SEnd}(H)|G) \subseteq \mathrm{End}(G[H])$, by Exerceorem 10.1.5. Suppose that $\{(r(x),f(x)(y)),(r(x'),f(x')(y'))\} \in E(G[H])$. If $\{r(x), r(x')\} \in E(G)$, then $\{x,x'\} \in E(G)$ since r is strong, and thus we have $\{(x,y),(x',y')\} \in E(G[H])$. If $r(x) = r(x')$ and $\{f(x)(y),f(x')(y')\} \in E(H)$, i. e., $H \neq \overline{K}_{|H|}$, we have that $\nu_G = \Delta$ implies $x = x'$. Moreover, we get that $\{y,y'\} \in E(H)$, and thus $\{(x,y),(x,y')\} \in E(G[H])$, using the fact that $f(x) = f(x')$ is strong. □

The following result is due to G. Sabidussi [77]. It uses the relations ν_G and σ_G from Definition 10.2.2. As usual, Δ denotes the diagonal of $G \times G$. A nice proof can be found in [Imrich/Klavžar 2000].

Theorem 10.3.5. $(\mathrm{Aut}(G) \wr \mathrm{Aut}(H)|G) \cong \mathrm{Aut}(G[H])$ *if and only if* $\nu_G \neq \Delta$ *implies that H is connected and* $\sigma_G \neq \Delta$ *implies that $\overline{H}$ is connected.*

In words, this theorem says that H must be connected if G has two nonadjacent vertices with the same neighborhood, i. e., G is not S-A unretractive, and $\overline{H}$ must be connected if G has two adjacent vertices with the same neighborhood.

We illustrate the necessity of the conditions with examples.

Example 10.3.6. Consider $P_2[\overline{K}_2] = \{0a,0b,1a,1b,2a,2b\}$ with $P_2 = \{0,1,2\}$ and $\overline{K}_2 = \{a,b\}$. Then the permutation of $0a$ and $2a$, e. g., is an automorphism which does not belong to $(\mathrm{Aut}(P_2) \wr \mathrm{Aut}(\overline{K}_2)|P_2)$ since it does not preserve layers (see Lemma 10.3.8), and, indeed, $\nu_{P_2} \neq \Delta$ and $\overline{K}_2$ is not connected.

Now consider $K_3[K_2] = \{1a, 1b, 2a, 2b, 3a, 3b\} \cong K_6$ with $K_3 = \{1, 2, 3\}$ and $K_2 = \{a, b\}$. Then the permutation of $1a$ and $3a$, e. g., is an automorphism which does not belong to $(\mathrm{Aut}(K_3) \wr \mathrm{Aut}(K_2)|K_3)$ since it does not preserve layers (see Lemma 10.3.8), and, indeed, $\sigma_{K_3} \neq \Delta$ and $\overline{K}_2$ is not connected.

Corollary 10.3.7. $\mathrm{Aut}(G[H]) = \{1\}$ *if and only if* $\mathrm{Aut}(G) = \mathrm{Aut}(H) = \{1\}$*; that is,* $G[H]$ *is asymmetric if and only if* G *and* H *are asymmetric.*

Proof. First, $\mathrm{Aut}(G[H]) = \{1\}$ implies that $\nu_G = \sigma_G = \Delta$. Then $\mathrm{Aut}(G) = \mathrm{Aut}(H) = \{1\}$ by Theorem 10.3.5, and vice versa. □

Lemma 10.3.8. $\mathrm{Aut}(G[H]) \cong (\mathrm{Aut}(G) \wr \mathrm{Aut}(H)|G)$ *if and only if for every* $x \in G$ *and* $\varphi \in \mathrm{Aut}(G[H])$ *there exists* $x' \in G$ *such that* $\varphi(H_x) \subseteq H_{x'}$; $\mathrm{SEnd}(G[H]) \cong (\mathrm{SEnd}(G) \wr \mathrm{SEnd}(H)|G)$ *if and only if for every* $x \in G$ *and* $\varphi \in \mathrm{SEnd}(G[H])$ *there exists* $x' \in G$ *such that* $\varphi(H_x) \subseteq H_{x'}$*. So in both cases* φ *preserves* H*-layers.*

Proof. Necessity is obvious.

To prove sufficiency, take $\varphi \in \mathrm{SEnd}(G[H])$. Then, by the hypothesis, $\varphi = (r,f)$ and it is easy to see that $f(x) : H \to H$ is a strong endomorphism for every $x \in G$. We show that $r(x) \neq r(x')$ if $\{x, x'\} \in E(G)$. The strong subgraph with vertex set $H_x \bigcup H_{x'}$ is $K_2[H]$. In the case of $r(x) = r(x')$, we would get $\varphi(K_2[H]) \subseteq H_{r(x)}$, which is impossible. Consequently, r is a strong endomorphism of G.

For $\varphi \in \mathrm{Aut}(G[H])$, the bijectivities of r and f follow from finiteness and the bijectivity of φ. □

Corollary 10.3.9. $\mathrm{SEnd}(G[H]) \subseteq (\mathrm{SEnd}(G) \wr \mathrm{SEnd}(H)|G)$ *if and only if* $\mathrm{Aut}(G[H]) \cong (\mathrm{Aut}(G) \wr \mathrm{Aut}(H)|G)$.

Proof. The necessity comes from part (b) of Theorem 10.3.1.

For sufficiency, note that from Theorem 10.3.5 we get two conditions on H which are inherited by the canonical strong factor graph H/ν_H. This implies that $(\mathrm{Aut}(G) \wr \mathrm{Aut}(H/\nu_H)|G) \cong \mathrm{Aut}(G[H/\nu_H])$ by Theorem 10.3.5. Moreover, the structure of the lexicographic product implies that $G[H/\nu_H] = (G[H])/\nu_{G[H]}$. Thus $\mathrm{SEnd}(G[H/\nu_H]) = \mathrm{Aut}(G[H/\nu_H])$.

By Lemma 10.3.8, for every $x \in G$ and $\varphi \in \mathrm{Aut}(G[H/\nu_H])$ there exists $x' \in G$ with $\varphi((H/\nu_H)_x) \subseteq (H/\nu_H)_{x'}$ for the respective H/ν_H-layers and, consequently, $\varphi(H_x) \subseteq H_{x'}$ for the respective H-layers.

Now, by Lemma 10.3.8, this is equivalent to $\mathrm{SEnd}(G[H]) \subseteq (\mathrm{SEnd}(G) \wr \mathrm{SEnd}(H)|G)$. □

Theorem 10.3.10. *We have* $(\mathrm{SEnd}(G) \wr \mathrm{SEnd}(H)|G) \cong \mathrm{SEnd}(G[H])$ *if and only if* $\nu_G = \Delta$ *and* $\sigma_G \neq \Delta$ *implies that* $\overline{H}$ *is connected.*

Proof. Necessity: $(\mathrm{SEnd}(G) \wr \mathrm{SEnd}(H)|G) \cong \mathrm{SEnd}(G[H])$ implies first that $\nu_G = \Delta$ or $H = \overline{K}_{|H|}$ by Theorem 10.3.4, and second that $(\mathrm{Aut}(G) \wr \mathrm{Aut}(H)|G) \cong \mathrm{Aut}(G[H])$ by Theorem 10.3.1. So we can apply Theorem 10.3.5 and get that $\nu_G = \Delta$. Applying Theorem 10.3.5 again gives the rest of the statement.

Sufficiency: by Theorem 10.3.5 we get $(\mathrm{Aut}(G) \wr \mathrm{Aut}(H)|G) \cong \mathrm{Aut}(G[H])$, and thus $(\mathrm{SEnd}(G) \wr \mathrm{SEnd}(H)|G) \supseteq \mathrm{SEnd}(G[H])$ by Corollary 10.3.9. Moreover, $\nu = \Delta$ implies the converse implication by Theorem 10.3.4. □

Again, we illustrate the necessity of the conditions as in Example 10.3.6.

Example 10.3.11. Consider $P_2[H]$ for an arbitrary graph H with at least one edge and with $P_2 = \{0, 1, 2\}$. Then, mapping the layer H_1 identically onto the layer H_3 and fixing the rest is an element in $(\mathrm{SEnd}(G) \wr \mathrm{SEnd}(H)|G)$ which is not strong, i. e., does not belong to $\mathrm{SEnd}(G[H])$.

For the second condition, we can use the same graphs as in Example 10.3.6.

Remark 10.3.12. End-regularity of graphs has been investigated by a number of researchers. We mention Suohai Fan. In [18], e. g., he characterizes End-orthodox lexicographic products of graphs.

Under the assumptions that G is connected, has odd girth, i. e., its shortest cycle has odd length, and does not contain triangles, $\mathrm{End}(G[H])$ is regular, orthodox, left inverse, right inverse, inverse, or completely regular if $\mathrm{End}(H)$ has the same property.

Moreover, under the assumptions that both G and H do not have triangles and either one of them has odd girth, we have $\mathrm{End}(G[H]) \cong (\mathrm{End}(G) \wr \mathrm{End}(H)|G)$ – which is along the same lines as the results of Sabidussi (Theorem 10.3.5) and Nummert (Theorem 10.3.10).

In this case, $\mathrm{End}(G[H])$ is regular if and only if $\mathrm{End}(G)$ is a group and $\mathrm{End}(H)$ is regular or vice versa.

Most of this was proved by Suohai Fan in his dissertation in 1993.

10.4 The union and the join

Most of the results in this section come from Apirat Wanichsombat [89]. We suggest that the reader considers the proofs of the results and the open questions as exercises, which are at various levels of difficulty.

The sum of monoids

For unions and also for joins of graphs, we introduce another composition of monoids, which looks like the Cartesian product but differs from it when we consider their actions on sets (such as vertices of graphs); cf. Definition 7.6.1.

Definition 10.4.1. Take monoids M and N and left acts (M,X) and (N,Y). The ***sum*** of the monoids $M+N=\{m+n \mid m\in M, n\in N\}$ has multiplication defined by $(m+n)(m'+n') := mm' + nn'$ and the identity element $1+1$.

Remark 10.4.2. The sum $M+N$ operates on $X \bigcup Y$ by $(m+n)x := mx$ and $(m+n)y := ny$ for $x \in X$, $y \in Y$, $m \in M$ and $n \in N$. In this way, we get the left $(M+N)$-act $X \bigcup Y$. This construction is slightly different from the product $M\times N$ of the monoids M and N, which is used for the operation on the product $X \times Y$. So although the monoids $M+N$ and $M\times N$ are isomorphic, the left $(M+N)$-act $X\bigcup Y$ and the left $(M\times N)$-act $X\times Y$ are not semilinearly isomorphic.

Lemma 10.4.3. *An element $h = h_X + h_Y \in M(X) + M(Y)$ is idempotent if and only if h_X and h_Y are idempotent.*

The sum of endomorphism monoids

Lemma 10.4.4. *If $f^2 = f \in \mathrm{End}(G+H)$, then $f \in \mathrm{End}(G) + \mathrm{End}(H)$, where G has no loops.*

Theorem 10.4.5. *Let G and H be graphs and consider $M \in$ {End, HEnd, LEnd, QEnd, SEnd, Aut}. Then $M(G) + M(H) \subseteq M(G+H)$ and $M(G) + M(H) \subseteq M(G\bigcup H)$, but not conversely. Moreover, the right-hand sides, i. e., the sets $M(G+H)$ and $M(G\bigcup H)$ may be incomparable.*

Project 10.4.6. Construct examples for all possible M, showing incomparability and that converses are not true.

To start, we have examples for some of the M: $\mathrm{End}(K_2+K_3) = \mathrm{End}(K_5) = \mathrm{Aut}(K_5) \cong S_5$, which is not a subset of $\mathrm{End}(K_2) + \mathrm{End}(K_3) = \mathrm{Aut}(K_2) + \mathrm{Aut}(K_3) \cong S_2 \times S_3$; nor is $\mathrm{End}(K_2 \bigcup K_3) = \mathrm{HEnd}(K_2\bigcup K_3)$, which is not a group. Note that $\mathrm{LEnd}(K_2\bigcup K_3) = \mathrm{Aut}(K_2\bigcup K_3) = \mathrm{Aut}(K_2) + \mathrm{Aut}(K_3) \cong S_2 \times S_3$.

For "moreover," we see that $\mathrm{End}(K_2+K_2) = \mathrm{Aut}(K_2+K_2) \cong S_4$ but $\mathrm{End}(K_2\bigcup K_2)$ is not a group.

All of these examples will be positive and negative examples to Theorem 10.4.9, so they can help to understand and possibly improve the results.

Corollary 10.4.7. *If $M(G)$ is not closed as a monoid, then $M(G+H)$ and $M(G\bigcup H)$ are not closed for $M \in$* {HEnd, LEnd, QEnd}.

Corollary 10.4.8. *If $M(G) \neq M'(G)$, then $M(G\bigcup H) \neq M'(G+H)$ for $M, M' \in$* {HEnd, LEnd, QEnd}.

The earliest and famous results of [Harary 1969] are hidden in (6) of the following theorem. Some of the other results were also in M. Frenzel [23].

Theorem 10.4.9. *Let the graphs be connected, finite, and without loops.*

(1) $\mathrm{End}(G)\bigcup \mathrm{End}(H) \cong \mathrm{End}(G+H)$ *if and only if* $\mathrm{Hom}(G,H) = \emptyset$ *and* $\mathrm{Hom}(H,G) = \emptyset$.

(2) $\mathrm{HEnd}(G)\bigcup\mathrm{HEnd}(H) \cong \mathrm{HEnd}(G+H)$ *if and only if* $\mathrm{HHom}(G,H) = \emptyset$ *and* $\mathrm{HHom}(H,G) = \emptyset$.
(3) $\mathrm{LEnd}(G)\bigcup\mathrm{LEnd}(H) \cong \mathrm{LEnd}(G{+}H)$ *if and only if* $\mathrm{LHom}(G,H) = \emptyset$ *and* $\mathrm{LHom}(H,G) = \emptyset$ *and* $h(H)\bigcap N_G g(G) \neq \emptyset$ *and* $h(H) \neq g(G)$ *for* $h \in \mathrm{LHom}(H,G)$ *and* $g \in \mathrm{LEnd}(G)$, *or vice versa.*
(4) $\mathrm{QEnd}(G)\bigcup\mathrm{QEnd}(H) \cong \mathrm{QEnd}(G+H)$ *if and only if* $\mathrm{QHom}(G,H) = \emptyset$ *and* $h(H)\bigcap N_G g(G) \neq \emptyset$ *for* $h \in \mathrm{QHom}(H,G)$ *and* $g \in \mathrm{QEnd}(G)$, *or vice versa.*
(5) $\mathrm{SEnd}(G)\bigcup\mathrm{SEnd}(H) \cong \mathrm{SEnd}(G{+}H)$ *if and only if for all components* $\mathrm{SHom}(G,H) = \emptyset$ *or* $\mathrm{SHom}(HG,G) = \emptyset$
(6) $\mathrm{Aut}(G)\bigcup\mathrm{Aut}(H) \cong \mathrm{Aut}(G+H)$ *if and only if* $G \not\cong H$.

The situation for the join is much easier, as one might expect. The following should be rather easy to prove.

Exerceorem 10.4.10. Let the graphs be finite without loops, and take $M \in \{\mathrm{End}, \mathrm{HEnd}, \mathrm{LEnd}, \mathrm{QEnd}, \mathrm{SEnd}, \mathrm{Aut}\}$. Then $M(G) + M(H) \cong M(G+H)$ if and only if $f(G) \subseteq G$ and $f(H) \subseteq H$ for all $f \in M(G+H)$.

Unretractivities

We repeat some known facts first.

Lemma 10.4.11.
(1) *Idempotent endomorphisms of G are in* $\mathrm{HEnd}(G)$, *i. e.,* $\mathrm{Idpt}(G) \subseteq \mathrm{HEnd}(G)$.
(2) *If G is finite with* $\mathrm{End}(G) \neq \mathrm{HEnd}(G)$, *then* $\mathrm{HEnd}(G) \neq \mathrm{SEnd}(G)$.

Proof. (1) follows from direct calculation; cf. Remark 1.5.10.
(2) follows from the fact that endotypes 1 and 17 do not exist; cf. Proposition 1.7.2. □

We consider E-A unretractivities, E-S unretractivities, and S-A unretractivities, i. e., graphs of endotypes 0, 16 and less than 16. Some of the results in the following theorem can be found in U. Knauer [50] (parts (2a) and (3a)), and U. Knauer [51] (part (1a)). Some more were also in M. Stamer [87].

Theorem 10.4.12. *Let G, H be finite graphs without loops, not both* K_1. *Then*
(1a) $\mathrm{End}(G\bigcup H) = \mathrm{Aut}(G\bigcup H)$ *if and only if* $\mathrm{End}(G) = \mathrm{Aut}(G)$ *and* $\mathrm{End}(H) = \mathrm{Aut}(H)$ *and* $\mathrm{Hom}(G,H) = \mathrm{Hom}(H,G) = \emptyset$.
(1b) $\mathrm{End}(G\bigcup H) = \mathrm{SEnd}(G\bigcup H)$ *if and only if* $\mathrm{End}(G) = \mathrm{SEnd}(G)$ *and* $\mathrm{End}(H) = \mathrm{SEnd}(H)$ *and* $\mathrm{Hom}(G,H) = \mathrm{Hom}(H,G) = \emptyset$.
(2a) $\mathrm{HEnd}(G\bigcup H) = \mathrm{Aut}(G\bigcup H)$ *if and only if* $\mathrm{HEnd}(G) = \mathrm{Aut}(G)$ *and* $\mathrm{HEnd}(H) = \mathrm{Aut}(H)$ *and* $\mathrm{HHom}(G,H) = \mathrm{HHom}(H,G) = \emptyset$.

(2b) $\text{End}(G \bigcup H) = \text{SEnd}(G \bigcup H)$ *if and only if* $\text{End}(G) = \text{SEnd}(G)$ *and* $\text{End}(H) = \text{SEnd}(H)$ *and* $\text{Hom}(G, H) = \text{Hom}(H, G) = \emptyset$.

(3a) *(Hypothesis)* $\text{LEnd}(G \bigcup H) = \text{Aut}(G \bigcup H)$ *if and only if* $\text{LEnd}(G) = \text{Aut}(G)$ *and* $\text{LEnd}(H) = \text{Aut}(H)$ *and* $\text{LHom}(G, H) = \text{LHom}(H, G) = \emptyset$.

(4a) $\text{QEnd}(G \bigcup H) = \text{Aut}(G \bigcup H)$ *if and only if* $\text{QEnd}(G) = \text{Aut}(G)$ *and* $\text{QEnd}(H) = \text{Aut}(H)$.

(4b) *(Hypothesis)* $\text{QEnd}(G \bigcup H) = \text{SEnd}(G \bigcup H)$ *if and only if* $\text{QEnd}(G) = \text{SEnd}(G)$ *and* $\text{QEnd}(H) = \text{SEnd}(H)$.

(5) $\text{SEnd}(G \bigcup H) = \text{Aut}(G \bigcup H)$ *if and only if* $\text{SEnd}(G) = \text{Aut}(G)$ *and* $\text{SEnd}(H) = \text{Aut}(H)$.

Question. Can you find a statement (3b)?

Can you simplify the conditions in (2a) and (2b) by dropping all occurrences of "H" after the "if and only if," in view of the fact that endotypes 1 and 17 do not exist?

If you consider C_9 and C_3, it becomes clear that in (3a), emptiness of only one of the two LEnd sets is not sufficient.

Theorem 10.4.13. *Let G and H be finite graphs without loops, not both* K_1.

(1a) $\text{End}(G + H) = \text{Aut}(G + H)$ *if and only if* $\text{End}(G) = \text{Aut}(G)$ *and* $\text{End}(H) = \text{Aut}(H)$.

(1b) $\text{End}(G + H) = \text{SEnd}(G + H)$ *if and only if* $\text{End}(G) = \text{SEnd}(G)$ *and* $\text{End}(H) = \text{SEnd}(H)$.

(2a) $\text{HEnd}(G + H) = \text{Aut}(G + H)$ *if and only if* $\text{HEnd}(G) = \text{Aut}(G)$ *and* $\text{HEnd}(H) = \text{Aut}(H)$.

(2b) $\text{HEnd}(G + H) = \text{SEnd}(G + H)$ *if and only if* $\text{HEnd}(G) = \text{SEnd}(G)$ *and* $\text{HEnd}(H) = \text{SEnd}(H)$.

(3a) $\text{LEnd}(G + H) = \text{Aut}(G + H)$ *if and only if* $\text{LEnd}(G) = \text{Aut}(G)$ *and* $\text{LEnd}(H) = \text{Aut}(H)$.

(4a) $\text{QEnd}(G + H) = \text{Aut}(G + H)$ *if and only if* $\text{QEnd}(G) = \text{Aut}(G)$ *and* $\text{QEnd}(H) = \text{Aut}(H)$.

(5) $\text{SEnd}(G + H) = \text{Aut}(G + H)$ *if and only if* $\text{SEnd}(G) = \text{Aut}(G)$ *and* $\text{SEnd}(H) = \text{Aut}(H)$.

Question. Can you find statements (3b) and (4b)?

Can you simplify the conditions in (2a) and (2b) by dropping all occurrences of "H" after the "if and only if," in view of the fact that endotypes 1 and 17 do not exist?

Project 10.4.14. From the inner logic, the cases (1c), (2c), (3c), (1d), (2d), and (1e) are missing from Theorems 10.4.12 and 10.4.13. Can you formulate them and try to state and prove theorems?

To begin with, you may find it helpful to represent the general situation of this section by a "fish bone" diagram similar to the one at the beginning of the next section. Using the diagram, associate the results obtained with the appropriate arrows.

10.5 The box product and the cross product

Most of the results in this section have, as far as we know, not been proved in easily accessible publications. Again, we propose that they be considered as exercises at various different levels.

The results and questions in this section are concerned with the following diagram of inclusions. Those which are not always valid are gray. Note that in each of the three columns, the upper vertical arrows can be subdivided twice with LEnd and HEnd. This subdivision will produce the corresponding slanted arrows to both sides.

The "multiplicativity" of forming endomorphism sets is symbolized by equalities along the slanted lines; unretractivities are equalities in the vertical directions.

Equalities or implications in the horizontal direction (e. g., whether $\mathrm{Aut}(G \,\square\, H)$ contains or is contained in $\mathrm{Aut}(G \times H)$) are, as far as we know, open questions.

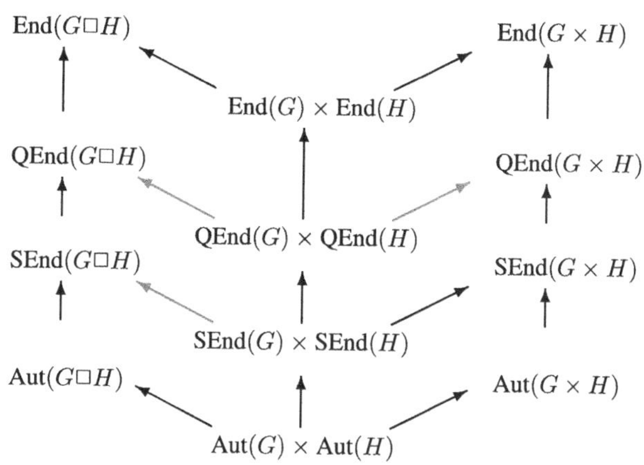

Unretractivities

Again, we consider E-A unretractivities, E-S unretractivities, and S-A unretractivities, i. e., graphs of endotypes 0, 16 and less than 16. These correspond to the vertical lines in the above diagram.

For one of the E-S retractivities, we have the following result.

Proposition 10.5.1. *Let χ denote the chromatic number. Suppose that for the graphs G and H one of the following conditions holds:*

(a) *$K_n = G$ and $\chi(H) \leq n$, and not both are K_2;*

(b) *$G = C_{2m+1}$ and $H = C_{2n+1}$;*

(c) *G is r-cyclically connected (i. e., all points from a cycle and all points with distance r on this cycle are connected by an additional edge) and $\chi(H) \leq \chi(G)$.*

Then $\mathrm{End}(G \,\square\, H) \neq \mathrm{SEnd}(G \,\square\, H)$.

For E-A unretractivity, we have a set of sufficient conditions that are weaker than those in Proposition 10.5.1.

Proposition 10.5.2. *Let $\chi(H)$ denote the chromatic number of H. Suppose that one of the following conditions holds:*
(a) *K_n is a subgraph of G and $\chi(H) \leq n$;*
(b) *$G = C_{2m+1}$ and C_{2n+1} is a strong subgraph of H;*
(c) *H is a strong subgraph of G.*

Then $\mathrm{End}(G \times H) \neq \mathrm{Aut}(G \times H)$.

Theorem 10.5.3. *Assume that the graphs are connected, finite, and without loops.*
(1a) $\mathrm{SEnd}(G \,\square\, H) = \mathrm{Aut}(G \,\square\, H)$ *if and only if* $G \neq K_2$ *or* $H \neq K_2$.
(1b) $\mathrm{SEnd}(G \times H) = \mathrm{Aut}(G \times H)$ *if and only if* $\mathrm{SEnd}(G) = \mathrm{Aut}(G)$ *and* $\mathrm{SEnd}(H) = \mathrm{Aut}(H)$.
(2a) $\mathrm{End}(G \,\square\, H) = \mathrm{SEnd}(G \,\square\, H)$ *implies* $\mathrm{End}(G) = \mathrm{Aut}(G)$ *and* $\mathrm{End}(H) = \mathrm{Aut}(H)$. *One has* $\mathrm{End}(G \,\square\, H) \neq \mathrm{SEnd}(G \,\square\, H)$ *under the conditions of Proposition* 10.5.1.
(2b) $\mathrm{End}(G \times H) = \mathrm{SEnd}(G \times H)$ *implies* $\mathrm{End}(G) = \mathrm{SEnd}(G)$ *and* $\mathrm{End}(H) = \mathrm{SEnd}(M)$.
(3) $\mathrm{End}(G \times H) = \mathrm{Aut}(G \times H)$ *implies* $\mathrm{End}(G) = \mathrm{Aut}(G)$ *and* $\mathrm{End}(H) = \mathrm{Aut}(M)$ *and* $|G \times H| \geq 42$. *One has* $\mathrm{End}(G \times H) \neq \mathrm{Aut}(G \times H)$ *under the conditions of Proposition* 10.5.2.

Question. Can you deduce conditions for nonequality in (1b) and (2b) from the respective conditions in (3)?

Find conditions for the other unretractivities.

Can you find similar results for the boxcross product, the disjunction, and the complete product?

The product of endomorphism monoids

For the monoids of box products as well as of cross products of graphs, the suitable composition of monoids is the cross product of monoids, i. e., categorically speaking the product of monoids. The first result in this direction is hidden in (3a) of Theorem 10.5.5. All of these results concern the slanted arrows in the "fish bone" diagram at the beginning of this section.

Graphs G and H are said to be ***relatively box prime*** if G and H do not admit decompositions as box products with isomorphic box factors not equal to K_1.

Theorem 10.5.4. *We have the following inclusions of products of monoids in the monoids of graph products:*
(1a) $\mathrm{End}(G) \times \mathrm{End}(H) \subseteq \mathrm{End}(G \,\square\, H)$.
(1b) $\mathrm{End}(G) \times \mathrm{End}(H) \subseteq \mathrm{End}(G \times H)$.
(2a) $\mathrm{QEnd}(G) \times \mathrm{QEnd}(H) \subseteq \mathrm{QEnd}(G \,\square\, H)$ *if and only if* $\mathrm{QEnd}(G) = \mathrm{Aut}(G)$ *and* $\mathrm{QEnd}(H) = \mathrm{Aut}(H)$.
(3a) $\mathrm{SEnd}(G) \times \mathrm{SEnd}(H) \subseteq \mathrm{SEnd}(G \,\square\, H)$ *if and only if* $\mathrm{SEnd}(G) = \mathrm{Aut}(G)$ *and* $\mathrm{SEnd}(H) = \mathrm{Aut}(H)$.
(3b) $\mathrm{SEnd}(G) \times \mathrm{SEnd}(H) \subseteq \mathrm{SEnd}(G \times H)$.

(4a) $\mathrm{Aut}(G) \times \mathrm{Aut}(H) \subseteq \mathrm{Aut}(G \,\square\, H)$.
(4b) $\mathrm{Aut}(G) \times \mathrm{Aut}(H) \subseteq \mathrm{Aut}(G \times H)$.

Question. What can be said about (2b) and the missing sets of endomorphisms HEnd and LEnd?

Theorem 10.5.5. *Here, we sharpen the inclusions in the previous theorem.*
(1a) *Under Condition* (a), (b), *or* (c) *of Proposition* 10.5.1 *or if G and H have vertices of degree* 1, *one has* $\mathrm{End}(G) \times \mathrm{End}(H) \subsetneq \mathrm{End}(G \,\square\, H)$.
(1b) *Condition* (a), (b) *or* (c) *of Proposition* 10.5.2 *implies that* $\mathrm{End}(G) \times \mathrm{End}(H) \subsetneq \mathrm{End}(G \times H)$.
(2a) $\mathrm{SEnd}(G) \times \mathrm{SEnd}(M) \cong \mathrm{SEnd}(G \,\square\, H)$ *if and only if G and H are relatively box prime and* $\mathrm{SEnd}(G) = \mathrm{Aut}(G)$, $\mathrm{SEnd}(H) = \mathrm{Aut}(H)$.
(2b) $\mathrm{SEnd}(G) \times \mathrm{SEnd}(H) = \mathrm{SEnd}(G \times H)$ *implies* $G \not\cong H$ *and* $\mathrm{SEnd}(G) = \mathrm{Aut}(G)$ *and* $\mathrm{SEnd}(H) = \mathrm{Aut}(H)$ *and is implied by* $G = K_m$ *and* $H = K_n$ *for* $m, n \geq 1$, $m \neq n$.
(3a) $\mathrm{Aut}(G) \times \mathrm{Aut}(H) \cong \mathrm{Aut}(G \,\square\, H)$ *if and only if G and H are relatively box prime.*
(3b) $\mathrm{Aut}(G) \times \mathrm{Aut}(H) = \mathrm{Aut}(G \times H)$ *implies* $G \not\cong H$ *and* $\mathrm{SEnd}(G) = \mathrm{Aut}(G)$ *and* $\mathrm{SEnd}(H) = \mathrm{Aut}(H)$ *and is implied by* $G = K_m$ *and* $H = K_n$ *for* $m, n \geq 1$, $m \neq n$.

Statements (1a), (1b), (2b), and (3b) are also in: P. Heidemann [34]; (2a) was also in: M. Frenzel [23].

For (3a), see G. Sabidussi [75].

Questions. Find characterizations for the situations in (1a), (1b), (2b), and (3b). What can be said about the "missing" statements between (1) and (2) concerning HEnd, LEnd, and QEnd (which in general are only sets and not monoids)?

What can be said if on the right-hand sides we take the boxcross product, the disjunction or the complete product?

10.6 Comments

In this chapter, and especially in the last section, there are many open questions worthy of investigation. Constructing proofs of the stated results might also be a worthwhile exercise. It may be possible to improve some of the results as well. As usual, one could attempt to find results where End, SEnd, or Aut is replaced by HEnd, LEnd, or QEnd.

We have two types of questions:
1. In which cases do End, etc. "preserve" or "reflect" compositions of graphs or of monoids, respectively? There are many compositions of graphs but only a few of monoids, which makes things complicated.
2. How do unretractivities of composed graphs depend on unretractivities of the factors?

Furthermore, it will be interesting to study the structure of X-Y unretractive graphs for $X, Y \in \{\text{End}, \text{HEnd}, \text{LEnd}, \text{QEnd}, \text{SEnd}, \text{Aut}\}$. This is related to the concept of endotypes of graphs; see Section 1.7. For general graphs, this does not seem very promising, but the situation may be better for special types of graphs such as paths, trees (cf. Theorem 1.7.5), bipartite graphs (cf. Theorem 1.5.4), split graphs, and so on.

11 Cayley graphs of semigroups

Arthur Cayley (1821–1895) introduced graphs of groups in 1878. One of the first investigations of these—later so-called—Cayley graphs of algebraic structures can be found in Maschke's work from 1896 about groups of genus zero, i. e., groups which possess a generating system such that the Cayley graph is planar; see the reference in Theorem 13.1.5.

Cayley graphs of groups have been extensively studied and many interesting results have been obtained—a very fruitful interconnection between algebra and graph theory. Cayley graphs of semigroups have also been considered by many authors. A selection of results on Cayley graphs of semigroups will be the subject of the rest of the book.

The first two sections of the present chapter consist mostly of applied category theory and use the language of category theory, as introduced in Chapter 3, and the categorical definitions of various graph products as given in Chapter 4.

First, we will touch on the categorical question of how to interpret the Cayley construction as a functor and which properties this functor enjoys. Preservation of products under the Cay functor, presented in Section 11.2, has important applications for the construction of Cayley graphs, especially Cayley graphs of certain completely regular semigroups; see Remark 11.2.4. Most of this section is taken from Ulrich Knauer, Yamning Wang and Xia Zhang [56]. Both sections are mainly exercises in category theory.

In Section 11.3, we discuss graph theoretical characterizations of Cayley graphs of semigroups. As an application; we construct Cayley graphs of right and left groups.

After this, we investigate strong semilattices of semigroups and specialize the results to strong semilattices of groups, i. e., Clifford semigroups, and to strong semilattices of right or left groups.

We then focus on Cayley graphs of the above classes of semigroups with generating connection set, which will be of importance for Chapter 13. We close the chapter with several examples.

Much more work has been done and, e. g., J. Meksawang, Sayan Panma, and Ulrich Knauer [60].

Suohai Fan and Y. Zeng [19].

Behnam Khosravi and Bahman Khosravi [46].

Bahman Khosravi [45].

11.1 The Cay functor

We present some elementary results which describe the construction of Cayley graphs starting from semigroups with given connection sets. As usual, we will use set notation also for proper classes.

https://doi.org/10.1515/9783110617368-011

Define a category $\mathbf{SgC}$ of semigroups with connection sets, where $\operatorname{Ob}\mathbf{SgC} = \{(S,C) \mid S \text{ a semigroup}, C \subseteq S\}$. For $(S,C),(T,D) \in \mathbf{SgC}$, we consider the morphism set $\mathbf{SgC}((S,C),(T,D)) = \{f \mid f : S \to T \text{ a semigroup homomorphism with } f|_C : C \to D\}$. Then $\operatorname{Ob}\mathbf{SgC}$ together with $\operatorname{Morph}\mathbf{SgC}$ is a category, where $\operatorname{Morph}\mathbf{SgC}$ denotes the class of all morphism sets in $\mathbf{SgC}$.

Let $\mathbf{D}$ be the category of digraphs, which may have loops and multiple edges, with graph homomorphisms.

As usual, we define the (uncolored) Cayley graph of a semigroup S with connection set $C \subseteq S$, using right action, as $\operatorname{Cay}(S,C) = (S,E)$, where (s,sc) are the arcs, i. e., the elements of $E = E(\operatorname{Cay}(S,C))$ for all $s \in S$ and $c \in C$.

Theorem 11.1.1. *Let S and T be semigroups, with subsets $C \subseteq S$ and $D \subseteq T$. Then $\operatorname{Cay} : \mathbf{SgC} \to \mathbf{D}$ given by*

$$\begin{array}{ccccc}
(S,C) & \longmapsto & \operatorname{Cay}(S,C) & & s \in S \\
\downarrow f & \longmapsto & \downarrow \operatorname{Cay}(f) & & \downarrow \\
(T,D) & \longmapsto & \operatorname{Cay}(T,D) & & f(s) \in T
\end{array}$$

for any $f \in \mathbf{SgC}((S,C),(T,D))$ and $s \in S$ is a covariant functor.

Proof. We show first that Cay produces homomorphisms in $\mathbf{D}$. Suppose (s,sc) is an arc in $\operatorname{Cay}(S,C)$, where $s \in S$, $c \in C$. Then $(f(s),f(sc)) = (f(s),f(s)f(c))$ is an arc in $\operatorname{Cay}(T,D)$ for each $f \in \mathbf{SgC}((S,C),(T,D))$. It follows that $\operatorname{Cay}(f)$ is a homomorphism from $\operatorname{Cay}(S,C)$ to $\operatorname{Cay}(T,D)$.

Now we verify (1) and (2) of Definition 3.3.1.

(1) We have

$$\operatorname{Cay}(\mathrm{id}_{(S,C)}) = \mathrm{id}_{\operatorname{Cay}(S,C)},$$

since $\operatorname{Cay}(\mathrm{id}_S)(s) = \mathrm{id}(s) = s = \mathrm{id}_{\operatorname{Cay}(S,C)}(s)$.

(2) For $f \in \mathbf{SgC}((S,C),(T,D))$ and $g \in \mathbf{SgC}((T,D),(U,E))$, we have

$$\operatorname{Cay}(gf)(s) = gf(s) = g(f(s)) = \operatorname{Cay}(g)\operatorname{Cay}(f)(s),$$

for any $s \in S$. So $\operatorname{Cay}(gf) = \operatorname{Cay}(g)\operatorname{Cay}(f)$. □

The following statement is straightforward; the second part is proved by the subsequent example.

Corollary 11.1.2. *The functor $\operatorname{Cay} : \mathbf{SgC} \to \mathbf{D}$ is faithful. It is full if we consider only right zero semigroups S, but not in general.*

Proof. Note that for right zero semigroups S and T, the functor Cay is full. The reason is that in this case every mapping f from S to T is a semigroup homomorphism.

Moreover, every element in a connection set produces a loop in the respective Cayley graph, and these are the only loops, which we can easily deduce from the results on right or left groups in Chapter 12. Since graph homomorphisms map loops onto loops, the condition $f(C) \subseteq D$ is automatically satisfied in ***SgC***.

In general, however, a morphism in ***D*** between two Cayley graphs which come from semigroups is not a semigroup homomorphism; see the following example. A similar situation shows up in Example 11.1.7. □

Example 11.1.3. Take the semigroup $\mathbb{Z}_3 = \{0, 1, 2\}$ with addition, and let $C = \{2\}$. Define a mapping $f : \mathbb{Z}_3 \to \mathbb{Z}_3$ with $f(0) = 2$, $f(1) = 0$ and $f(2) = 1$. Then f is a morphism in ***D*** from the directed triangle $\mathrm{Cay}(\mathbb{Z}_3, \{2\})$ to $\mathrm{Cay}(\mathbb{Z}_3, \{2\})$, but obviously f is not a semigroup homomorphism. Moreover, the example also shows that the condition $f \in \boldsymbol{D}(\mathrm{Cay}(S, C), \mathrm{Cay}(T, D))$ does not imply that $f|_C$ is a mapping from C to D. For a picture of $\mathrm{Cay}(\mathbb{Z}_3, \{2\})$, see Example 7.3.3.

Reflection and preservation of morphisms

From the definition of the Cay functor and the fact that Cay is covariant and faithful, we get, as usual (see, e. g., [Kilp et al. 2000]), the following result.

Corollary 11.1.4. *The functor* Cay *preserves and reflects injective mappings and surjective mappings. It preserves retractions and coretractions.*

Note that in the category ***SgC*** of semigroups with connection sets, monomorphisms are injective, and as always, surjective mappings are epimorphisms. The converse of the latter is not true for infinite semigroups. Then there exist nonsurjective epimorphisms in the category of semigroups, but they will not turn into epimorphisms in the category of digraphs, since they are not surjective. So by an infinite example we can show that the functor Cay does not preserve epimorphisms.

Example 11.1.5. Recall that $i : (\mathbb{N}, +) \hookrightarrow (\mathbb{Z}, +)$ is not surjective but it is an epimorphism in ***Sgr*** (cf. Example 3.1.14), and also in ***SgC***. Then $\mathrm{Cay}(i) : \mathrm{Cay}(\mathbb{N}, \{1\}) \to \mathrm{Cay}(\mathbb{Z}, \{1, -1\})$ is not surjective. Therefore, it is not an epimorphism, as both concepts coincide in graph categories; cf. Exercise 3.1.8.

Corollary 11.1.6. *The functor* Cay *preserves epimorphisms only in the category of finite semigroups with connection sets.*

The following examples show that the functor Cay does not reflect retractions and coretractions. In the first case, we also use an infinite semigroup.

Example 11.1.7 (Retractions). Let $\pi : (\mathbb{N}_0, \cdot) \to (\mathbb{Z}_6, \cdot) = (\{\overline{0}, \overline{1}, \ldots, \overline{5}\}, \cdot)$ be the canonical surjection (mod 6). Take $C = \{0\} \subseteq \mathbb{N}_0$ and $\overline{C} = \{\overline{0}\} \subseteq \mathbb{Z}_6$. Then π is a not a retraction in ***SgC*** but $\mathrm{Cay}(\pi)$ is a retraction in ***D***.

Indeed, the set of morphisms $\mathbf{SgC}(\mathbb{Z}_6, \mathbb{N}_0) = \{c_0\}$, the constant mapping onto 0. Therefore, π cannot be a retraction in **SgC**. But consider $g' : \mathrm{Cay}(\mathbb{Z}_6, \overline{C}) \to \mathrm{Cay}(\mathbb{N}_0, C)$ defined by $g'(\overline{n}) = n$. Then g' is a morphism in $\boldsymbol{D}$ satisfying $\mathrm{Cay}(\pi)g' = \mathrm{id}_{\mathrm{Cay}(\mathbb{Z}_6,\overline{C})}$. So $\mathrm{Cay}(\pi)$ is a retraction in $\boldsymbol{D}$.

Example 11.1.8 (Coretractions). Take $S = (\{2,4\},\cdot) \subseteq (\mathbb{Z}_6,\cdot) = (\{0,1,\ldots,5\},\cdot)$. Note that $S \cong (\mathbb{Z}_2,+)$, i. e., we have a group. Consider the natural embedding $i : S \to \mathbb{Z}_6$. Then $i \in \mathbf{SgC}((S,S),(\mathbb{Z}_6,S))$, and i is not a coretraction in **SgC**. Otherwise, we would have to have $g : \mathbb{Z}_6 \to S$ in **SgC** such that $gi = \mathrm{id}_S$, the identity mapping of S in **SgC**. No such g exists, since S does not have a zero. On the other hand, $\mathrm{Cay}(i)$ is a coretraction in $\boldsymbol{D}$ in the following way. Define $f : \mathbb{Z}_6 \to S$ with $f(2) = 2$ and $f(n) = 4$ for all $2 \neq n \in \mathbb{Z}_6$. Then $\mathrm{Cay}(f) \in \boldsymbol{D}(\mathrm{Cay}(\mathbb{Z}_6,S),\mathrm{Cay}(S,S))$ and $\mathrm{Cay}(f)\,\mathrm{Cay}(i) = \mathrm{id}_{\mathrm{Cay}(S,S)}$.

Corollary 11.1.9. *The functor* Cay *does not reflect retractions or coretractions.*

Does Cay produce strong homomorphisms?

This question seems quite natural; however, it has not been answered definitively. Recall that comorphisms reflect edges and that strong homomorphisms preserve and reflect edges, i. e., they are comorphisms which are also homomorphisms.

Proposition 11.1.10. *Suppose that $f \in \mathbf{SgC}((S,C),(T,D))$ is injective. Then* $\mathrm{Cay}(f)$ *is a strong homomorphism in $\boldsymbol{D}$ if and only if $(f(s),f(s')) \in \mathrm{Cay}(T,D)$ implies $f(s') = f(s)f(c)$ for some $c \in C$.*

Proof. By the definition of a strong homomorphism, if $(f(s),f(s')) \in E(\mathrm{Cay}(T,D))$, then $(s,s') \in E(\mathrm{Cay}(S,C))$. Thus there exists $c \in C$ such that $s' = sc$, and so $f(s') = f(s)f(c)$, which gives the necessity.

Assume to the contrary that $(f(s),f(s')) \in E(\mathrm{Cay}(T,D))$. Then $f(s') = f(s)f(c)$ for some $c \in C$ by hypothesis, and hence $s' = sc$ since f is injective. □

Corollary 11.1.11. *Take $f \in \mathbf{SgC}((S,C),(T,D))$. If f is injective and $f(C) = D$, then* $\mathrm{Cay}(f)$ *is a strong homomorphism in $\boldsymbol{D}$.*

Let $f \in \mathbf{SgC}((S,C),(T,D))$. The following examples show that the conditions "f is injective," "f is surjective," and "$f(C) = D$" are not necessary, while "$f(C) = D$" and "$f^{-1}(D) = C$" are not sufficient, for $\mathrm{Cay}(f)$ to be a strong homomorphism in $\boldsymbol{D}$.

Example 11.1.12. Take $S = (\mathbb{Z}_6,\cdot) = (\{0,1,\ldots,5\},\cdot)$ and define $f : S \to S$ by

$$f(0) = 0, \quad f(1) = f(5) = 1, \quad f(2) = f(4) = 4, \quad f(3) = 3.$$

Take the subsets $C = \{1,5\}$ and $D = \{1\}$ of S. Then $f \in \mathbf{SgC}((S,C),(S,D))$. Now $E(\mathrm{Cay}(S,C))$ contains all loops and the edges $\{(1,5),(2,4),(4,2),(5,1)\}$, while $E(\mathrm{Cay}(S,D))$ contains all loops only. It is easy to check that Cay(f) is a strong homomorphism. Clearly, f is neither injective nor surjective.

Example 11.1.13. Let T be a three-element set with the following multiplication table:

	1	2	3
1	1	2	2
2	2	1	1
3	2	1	1

Clearly, this is a semigroup. Take the subsemigroup $S = \{1,2\}$ of T and set $C := \{2\}$, $D := \{2,3\}$. Then $i : S \hookrightarrow T$, the natural embedding of S into T, belongs to $\mathbf{SgC}((S,C),(T,D))$. Now we get the following edge sets for the respective Cayley graphs:

$$E(\mathrm{Cay}(S,C)) = \{(1,2),(2,1)\}, \quad E(\mathrm{Cay}(T,D)) = \{(1,2),(2,1),(3,1)\}$$

and Cay(i) is a strong homomorphism in $\mathbf{D}$. But $f(C) \neq D$.

Example 11.1.14. Consider $C = L_2 = \{a,b\} \subseteq L_2^0 = S$, i. e., S is the two-element left zero semigroup with zero adjoint. Then

$$E(\mathrm{Cay}(S,C)) = \{(0,0),(a,a),(b,b)\}.$$

Take $D = \{a\}$ and the mapping $f : S \to S$ defined by $f(0) = 0$ and $f(a) = f(b) = a$. Then $f \in \mathbf{SgC}((S,C),(S,D))$.

It is clear that $f(C) = D$ and $f^{-1}(D) = C$. However, Cay(f) is not a strong homomorphism in $\mathbf{D}$ since $(\mathrm{Cay}(f(a)), \mathrm{Cay}(f(b))) = (a,a) \in E(\mathrm{Cay}(S,D))$ but $(a,b) \notin E(\mathrm{Cay}(S,C))$.

11.2 Products and equalizers

Categorical products

Now we turn to the categorical product (the cross product in the category $\mathbf{D}$ of directed graphs) and equalizers (see Chapters 3 and 4). Observe that equalizers are special pullbacks; cf. Remark 3.2.9.

To be precise, we should identify products and other categorical concepts in the category $\mathbf{SgC}$.

Lemma 11.2.1. *Let $\{(S_i,C_i)\}_{i\in I}$ be a family of objects in category $\mathbf{SgC}$. Then $((\prod_{i\in I} S_i, \prod_{i\in I} C_i),(p_i)_{i\in I})$ is the product of $\{(S_i,C_i)\}_{i\in I}$ in $\mathbf{SgC}$, where $\prod_{i\in I} S_i$ and $\prod_{i\in I} C_i$ are Cartesian products of $(S_i)_{i\in I}$ and $(C_i)_{i\in I}$, respectively, and $p_i : (\prod_{i\in I} S_i, \prod_{i\in I} C_i) \to (S_i,C_i)$, $i \in I$, are the canonical projections.*

Proof. Clearly, the p_i, $i \in I$, are morphisms in $\mathbf{SgC}$. For any $(T,D) \in \mathbf{C}$ and any family $(q_i) \in \mathbf{C}((T,D),(S_i,C_i))_{i\in I}$, define $q : (T,D) \to (\prod_{i\in I} S_i, \prod_{i\in I} C_i)$ by $q(t) = (q_i(t))_{i\in I}$, $t \in T$. Then q is the unique morphism in $\mathbf{SgC}$ such that $p_i q = q_i$ for all $i \in I$. □

Theorem 11.2.2. *The functor* Cay *preserves and reflects (multiple) products, i. e., for* $(S, C), (T, D) \in \mathbf{SgC}$ *we have*

$$\mathrm{Cay}(S \times T, C \times D) = \mathrm{Cay}(S, C) \times \mathrm{Cay}(T, D),$$

where $\times$ *on the right-hand side denotes the cross product in* $\mathbf{D}$.

Proof.

$$\begin{aligned}
&\mathrm{Cay}(S, C) \times \mathrm{Cay}(T, D) \\
&= \{\{(x, y), (x', y')\} \mid (x, x') \in \mathrm{Cay}(S, C),\ (y, y') \in \mathrm{Cay}(T, D)\} \\
&= \{(s, t), (sc, td) \mid (s, t) \in S \times T,\ (c, d) \in C \times D\} \\
&= \mathrm{Cay}(S \times T, C \times D).
\end{aligned}$$

It is clear that this can be generalized to multiple products. □

Note that the generating system $C \times D$ of $S \times T$ is not minimal in general. However, Theorem 11.2.2 will be of use in the subsection "Right groups on the torus" of Section 13.2.

Application 11.2.3. We can use this result to determine the Cayley graphs of right groups and left groups, which in the category $\mathbf{SgC}$ have the form of a product between a group and an n-element right zero semigroup $R_n = \{r_1, \ldots, r_n\}$ or left zero semigroup $L_n = \{l_1, \ldots, l_n\}$.

Consider the right group $\mathbb{Z}_2 \times R_2 = \{(0, r_1), (0, r_2), (1, r_1), (1, r_2)\}$. Then $\mathrm{Cay}(\mathbb{Z}_2 \times R_2, \{1, r_2\})$ has the form $\mathrm{Cay}(\mathbb{Z}_2, \{1\}) \times \mathrm{Cay}(R_2, \{r_2\})$, which is the cross product in the category $\mathbf{D}$. (Here and in later pictures, we will write vertices in Cartesian products as xy instead of as (x, y).)

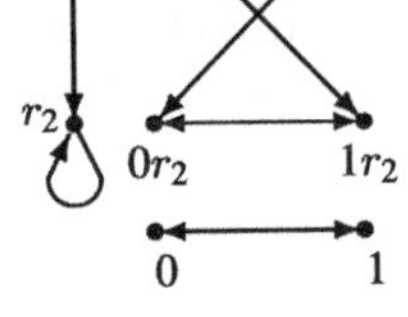

Consider the left group $L_2 \times \mathbb{Z}_2 = \{(l_1, 0), (l_2, 0), (l_1, 1), (l_2, 1)\}$. Then $\mathrm{Cay}(L_2 \times \mathbb{Z}_2, \{l_2, 1\})$ has the form $\mathrm{Cay}(L_2, \{l_1\}) \times \mathrm{Cay}\,(\mathbb{Z}_2, \{1\})(= \mathrm{Cay}(L_2, \{l_1\} \times \mathrm{Cay}(\mathbb{Z}_2, \{1\})))$, which is the cross product in the category $\mathbf{D}$. It is depicted below:

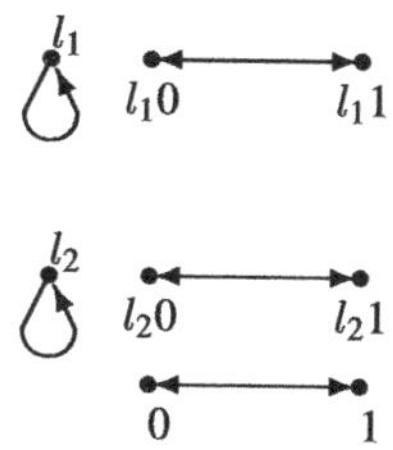

We will resume our discussion of Cayley graphs of left and right groups in Sections 11.3 and 13.2.

Project 11.2.4. Observe that the preservation of products together with the preservation of injective and surjective mappings leads to the preservation of so-called *subdirect products*; see, e. g., [Petrich/Reilly 1999]. This, in turn, opens up many possible avenues of characterizing Cayley graphs of completely regular semigroups. Some steps in this direction are presented in what follows.

Equalizers

Now we identify equalizers in the category $\boldsymbol{SgC}$ which also are not a surprise.

Lemma 11.2.5. *Consider a situation $f, g : (S, C) \rightrightarrows (S', C')$ in $\boldsymbol{SgC}$, which we called an equalizer situation. If $T = \{s \in S \mid f(s) = g(s)\}$ and $D = \{c \in C \mid f(c) = g(c)\} \neq \emptyset$, then $(T, D) \subseteq (S, C)$ with the natural embedding i is the equalizer of f and g in $\boldsymbol{SgC}$; i. e., $fi = gi$, and whenever $fh = gh$, there exists a unique h' with $ih' = h$.*

Proof. Suppose that $D = \{c \in C \mid f(c) = g(c)\} \neq \emptyset$. If $((E, A), h)$ satisfies $fh = gh$, then $h(E) \subseteq T$, $h(A) \subseteq D$ and $h' = h : (E, A) \to (T, D)$ is the unique morphism such that $ih' = h$, where i is the natural embedding. □

Now we show that the functor Cay does not preserve equalizers.

Example 11.2.6. Consider again the semigroup $(\mathbb{Z}_6, \cdot) = (\{0, 1, \ldots, 5\}, \cdot)$ from Examples 11.1.12 and 11.1.8, and define $f : \mathbb{Z}_6 \to \mathbb{Z}_6$ by $f(z) = z^2$ for all $z \in \mathbb{Z}_6$. Let $C = \{0, 5\}$ and $C' = \{0, 1, 5\}$, both subsets of $\mathbb{Z}_6$. Then $f \in \boldsymbol{SgC}((\mathbb{Z}_6, C), (\mathbb{Z}_6, C'))$ and $(T, D) = (\{0, 1, 3, 4\}, \{0\})$ with the natural embedding i is the equalizer of f and $\mathrm{id}_{(\mathbb{Z}_6, C)}$ in $\boldsymbol{SgC}$ by Lemma 11.2.5.

Let $V = \{v\}$ be the one-point digraph in $\boldsymbol{D}$ with one loop, i. e., $E = \{(v, v)\}$. Define $h : V \to \mathbb{Z}_6$ with $h(v) = 3$. Then h is a morphism in $\boldsymbol{D}$ such that $\mathrm{Cay}(f)h = \mathrm{id}_{\mathrm{Cay}(\mathbb{Z}_6, C)}\, h$. Now every morphism from V to $\mathrm{Cay}(T, D)$ in $\boldsymbol{D}$ must map v onto 0, and thus there exists only one such morphism h^*. But $h^* \neq h$. Hence $\mathrm{Cay}(T, D)$ with the embedding is not the equalizer of f and $\mathrm{id}_{\mathbb{Z}_6}$ in $\boldsymbol{D}$, i. e., $(\mathrm{Cay}(T, D), \mathrm{Cay}(i)) \neq \mathrm{Eq}_{\boldsymbol{D}}(\mathrm{Cay}(f), \mathrm{id}_{\mathrm{Cay}(\mathbb{Z}_6, C)})$.

The following can be proved; however, rather than go into details here, we refer to the original literature mentioned in the introduction to this chapter.

Exerceorem 11.2.7. In general, the functor Cay does not preserve or reflect equalizers, and consequently it does not preserve nor reflect pullbacks.

Other product constructions

We now consider box products, boxcross products and lexicographic products of graphs. Remember that in the literature these products have many alternative names.

The box product is, categorically speaking, the tensor product in the category $\mathbf{D}$; see Theorem 4.3.5. Since relatively little is known about the tensor product in the category of semigroups, it does not make sense to talk about preservation of tensor products in this context.

Because of the structure of the coproduct in the category of semigroups (cf. Theorem 4.1.2), we cannot say anything about the preservation of coproducts by the Cay functor either.

Overall, the results in this subsection cannot be seen as preservation properties in the categorical sense.

Remark 11.2.8. It is easy to see that $\mathrm{Cay}(S, C) \oplus \mathrm{Cay}(S, C') = \mathrm{Cay}(S, C \bigcup C')$, where $\oplus$ is the edge sum.

Theorem 11.2.9. *Let $\square$ and $\boxtimes$ denote the box product and boxcross product, respectively. Then for monoids S and T with subsets C and D and identities 1_S and 1_T, we have*
(1) $\mathrm{Cay}(S \times T, (\{1_S\} \times D) \bigcup (C \times \{1_T\})) = \mathrm{Cay}(S, C) \,\square\, \mathrm{Cay}(T, D)$;
(2) $\mathrm{Cay}(S \times T, (\{1_S\} \times D) \bigcup (C \times \{1_T\}) \bigcup (C \times D)) = \mathrm{Cay}(S, C) \boxtimes \mathrm{Cay}(T, D)$.

Proof. (1) We have

$$\begin{aligned}
&E(\,\mathrm{Cay}\,(S \times T, (\{1_S\} \times D) \bigcup (C \times \{1_T\})))\\
&\quad = \{\{(s,t),(s,td)\} \mid (s,t) \in S \times T, d \in D\}\\
&\qquad \bigcup \{\{(s,t),(sc,t)\} \mid (s,t) \in S \times T, c \in C\}\\
&\quad = E(\mathrm{Cay}(S, C) \,\square\, \mathrm{Cay}(T, D)).
\end{aligned}$$

(2) Denote by $\oplus$ the edge sum of graphs. Then

$$\begin{aligned}
&\mathrm{Cay}(S, C) \boxtimes \mathrm{Cay}(T, D)\\
&= (\mathrm{Cay}(S, C) \square \mathrm{Cay}(T, D)) \oplus (\mathrm{Cay}(S, C) \times \mathrm{Cay}(T, D))\\
&= \mathrm{Cay}(S \times T, (\{1_S\} \times D) \bigcup (C \times \{1_T\})) \oplus \mathrm{Cay}(S \times T, C \times D)\\
&= \mathrm{Cay}(S \times T, (\{1_S\} \times D) \bigcup (C \times \{1_T\}) \bigcup (C \times D)).
\end{aligned}$$

□

In the article *Cayley graphs and interconnection networks* by M.-C. Heydemann in [Hahn/Sabidussi 1997] pp. 167–224, the statements of Theorems 11.2.2 and 11.2.9 are contained in the case where S and T are groups. Moreover, for the lexicographic product of graphs, it is stated there, that $\mathrm{Cay}(A,C)[\mathrm{Cay}(A',C')] \cong \mathrm{Cay}(A\times A',(C\times A')\bigcup(1_A\times C'))$, where A and A' are groups and 1_A is the identity of A. Generalized to the situation of semigroups, we have the following.

Theorem 11.2.10. *Let S be a monoid, T a semigroup, and C and D subsets of S and T, respectively. Then*

$$\mathrm{Cay}(S\times T,(C\times T)\bigcup(\{1_S\}\times D)) = \mathrm{Cay}(S,C)[\mathrm{Cay}(T,D)]$$

if and only if $tT = T$ for any $t \in T$, i. e., T has no proper right ideals, i. e., T is right simple.

Proof. We have

$$\begin{aligned}
&E(\mathrm{Cay}\,(S\times T,(C\times T)\bigcup(\{1_S\}\times D)))\\
&= \{\{((s,t),(s,t)(c,t')) = ((s,t),(sc,t'))\} \mid (s,t)\in S\times T,\ (c,t')\in C\times T\}\\
&\quad \bigcup\{\{((s,t),(s,t)(1_S,d)) = ((s,t),(s,td))\} \mid (s,t)\in S\times T,\ (1_S,d)\in\{1_S\}\times D\},
\end{aligned}$$

and

$$\begin{aligned}
E(\mathrm{Cay}(S,C)[\mathrm{Cay}(T,D)]) &= \{\{(s,t),(sc,t')\} \mid (s,sc)\in E(\mathrm{Cay}(S,C)),\ t,t'\in T\}\\
&\quad \bigcup\{\{(s,t),(s,td)\} \mid s\in S,\ (t,td)\in E(\mathrm{Cay}(T,D))\}.
\end{aligned}$$

If $\mathrm{Cay}(S\times T,(C\times T)\bigcup(\{1_S\}\times D)) = \mathrm{Cay}(S,C)[\mathrm{Cay}(T,D)]$, then for any $t,t'\in T$ and $\{(s,t),(sc,t')\}\in E(\mathrm{Cay}(S,C)[\mathrm{Cay}(T,D)])$, where $(s,sc)\in E(\mathrm{Cay}(S,C))$, we have $t' = tx$ for some $x\in T$. So $T\subseteq tT$ and then $T = tT$ for any $t\in T$.

For the converse, suppose that $tT=T$ for any $t\in T$. Then for any arc $\{(s,t),(s',t')\}$ in $\mathrm{Cay}(S,C)[\mathrm{Cay}(T,D)]$, either $s = s'$ and $t' = td$ for some $d\in D$ or $s' = sc$ for some $c\in C$ and $t,t'\in T$. But for any $t,t'\in T$, there exists $y\in T$ such that $t' = ty$ by assumption. Therefore, $\{(s,t),(s',t')\}$ is an arc of $\mathrm{Cay}(S\times T,(C\times T)\bigcup(\{1_S\}\times D))$, and so

$$\mathrm{Cay}(S,C)[\mathrm{Cay}(T,D)] \subseteq \mathrm{Cay}(S\times T,(C\times T)\bigcup(\{1_S\}\times D)).$$

The converse inclusion is obvious. □

Remark 11.2.11. A formal description of the relation between graphs and subgraphs which are subdivisions, with the help of the Cay functor on semigroups with generators, seems to be difficult.

In $\mathrm{Cay}(\mathbb{Z}_6,\{1\})$, e. g., we find a subdivision of K_3 corresponding to $\mathrm{Cay}(\{0,2,4\},\{2\})$ as a subgraph. But subdivision is not a categorical concept. And there is no inclusion between $\{0,2,4\}\times\{2\}$ and $\mathbb{Z}_6\times\{1\}$.

11.3 Characterizations of Cayley graphs

Definition 7.3.1 defines an uncolored Cayley graph $\mathrm{Cay}(S, C)$ for a semigroup S and a connection set $C \subseteq S$. Recall that the Cayley graph $\mathrm{Cay}(S, C)$ has the vertex set S and that (x, y), with $x, y \in S$, is an arc if there exists an element $c \in C$ such that $xc = y$. A digraph G is called a ***semigroup digraph*** or ***digraph of a semigroup*** if there exists a semigroup S and a connection set $C \subseteq S$ such that G is isomorphic to the Cayley graph $\mathrm{Cay}(S, C)$. We speak of S ***semigroup digraphs*** if we want to consider various subsets $C \subseteq S$ and the corresponding Cayley graphs $\mathrm{Cay}(S, C)$.

In Corollary 7.7.11, we characterized connected group digraphs as those connected digraphs that have a regular acting subgroup of the automorphism group.

Question. What chances do we have of generalizing this result to (certain) monoids or semigroups?

For monoids, in Theorem 7.3.7, we showed that if C is a generating set of monoid M, then M is isomorphic to the color endomorphism monoid of the corresponding colored Cayley graph. In general, not much is known about characterizations of (connected) monoid digraphs or semigroup digraphs. In the semigroup setting even the colored version, i. e., a generalization of Theorem 7.3.7 is open. An easy observation is the following.

Remark 11.3.1. Every semigroup graph is what we call ***(k-)out-regular***; i. e., all vertices have the same outdegree k. Here, multiple arcs are counted with their multiplicities and also loops are counted.

The following proposition shows that the class of monoid graphs is quite big.

Proposition 11.3.2. *Every connected, 1-out-regular digraph is a monoid graph.*

Proof. Let $G = (V, E)$ be a connected, 1-out-regular digraph. We construct a monoid M with V as the set of elements. Since every vertex in G has outdegree 1, the digraph contains a unique directed cycle C (possibly a loop). Pick $e \in M$ to be a vertex of maximum distance to C. Since we have outdegree 1, the monoid must have a unique element a, such that $G = \mathrm{Cay}(M, \{a\})$. Thus, all vertices reachable by e are of the form a^k for a certain possibly not unique k. Let us indeed denote $k \odot l := \min\{m \mid a^m = a^{k+l}\}$. Also, more generally the element reached from a vertex x by a path of length k is xa^k. For a vertex x, we define $r(x) = \min\{k \mid \exists \ell : a^{k+\ell} = xa^\ell\}$. Note that $k \geq 0$ by the choice of e of maximum distance to C. Moreover, note that $r(xa^k) = r(x) \odot k$. We can now define $xy := xa^{r(y)}$, with the exception that $ey := y$. First, note that $xa^k = xa^{r(a^k)} = xa^k$, i. e., indeed the element at (forward) distance k of vertex x is xa^k. In particular, $G = \mathrm{Cay}(M, \{a\})$. It remains to show that this operation is associative. We have

$$(xy)z = (xa^{r(y)})z = (xa^{r(y)})a^{r(z)} = xa^{r(y)+r(z)} = xa^{r(y) \odot r(z)} = x(ya^{r(z)}) = x(yz).$$

We refer to Figure 11.1 for an illustration. □

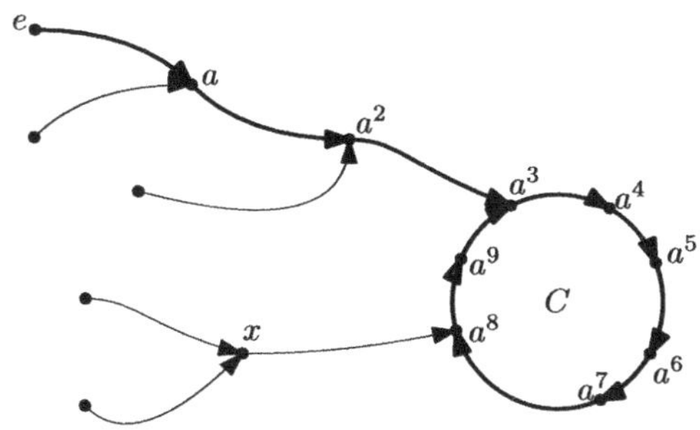

Figure 11.1: An example for Proposition 11.3.2. Here, $r(x) = 7$.

Can Proposition 11.3.2 be generalized to disconnected graphs? It is easy to see that connected 2-out-regular graphs are not monoid digraphs in general. Are they semigroup digraphs?

Cayley graphs of right and left groups

In this subsection, we characterize Cayley graphs of so-called right and left groups, following S. Arworn, U. Knauer and N. N. Chiangmai [3].

The results follow quite easily from the fact that Cay preserves and reflects products; see Theorem 11.2.2 and Application 11.2.3. We will use Remark 11.2.8 in the following form.

Lemma 11.3.3. *Let S be a semigroup, and let C be a subset of S. Then* $\mathrm{Cay}(S,C) = \bigoplus_{c\in C}\mathrm{Cay}(S,\{c\})$.

Here, we will not give proofs, as the results follow in a straightforward manner from the structures of the semigroups under consideration and the preservation of products. They will be illustrated by examples later.

First, we characterize right group digraphs.

Theorem 11.3.4. *The graph (V,E) is a right group digraph, i.e., $(V,E) \cong \mathrm{Cay}(A\times R_k, C)$ with $2 \le k \in \mathbb{N}$, $A\times R_k$ a right group and $C \subseteq A\times R_k$, if and only if*

$$(V,E) \supseteq \bigcup_{i=1}^{k}(V_i,E_i),$$

such that for $u_i, v_i \in V_i$ and $i,j \in \{1,\dots,k\}$,

(1) $\bigcup_{i=1}^{k}(V_i,E_i)$ *is the disjoint union of k group digraphs $(V_1,E_1),\dots,(V_k,E_k)$ with*
   (a) $V_i = A\times\{r_i\}$ *and*
   (b) $(u_i,v_i) = ((u,r_i),(v,r_i)) \in E_i \Leftrightarrow \exists\,(g,r_i)\in C$ *with $ug = v$, and*

(2) $(u_j,v_i) \in E \Leftrightarrow (u_i,v_i)\in E_i$.

Remark 11.3.5. A set of generators C for the right group $A \times R_k$ will always have the property that $p_2(C) = R_k$. Here, p_2 is the second projection.

Example 11.3.6. In Application 11.2.3, we saw that

$$\mathrm{Cay}(\mathbb{Z}_2 \times R_2, \{(1, r_2)\}) = \mathrm{Cay}(\mathbb{Z}_2, \{1\}) \times \mathrm{Cay}(R_2, \{r_2\}).$$

Below we have

$$\mathrm{Cay}(\mathbb{Z}_3 \times R_2, \{(1, r_1)\}) = \mathrm{Cay}(\mathbb{Z}_3, \{1\}) \times \mathrm{Cay}(R_2, \{r_1\}).$$

In both cases, the cross product of the Cayley graphs of the group and the right zero semigroup is visible:

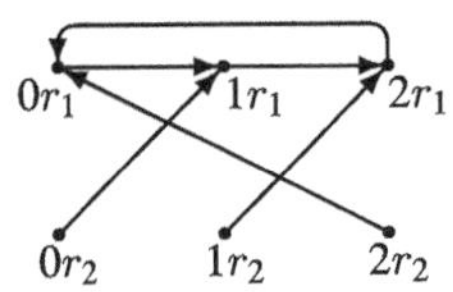

The left analogue of Theorem 11.3.4 is the following.

Theorem 11.3.7. *The graph (V, E) is a left group digraph, i. e., $(V, E) \cong \mathrm{Cay}(L_k \times A, C)$ with $2 \leq k \in \mathbb{N}$, $L_k \times A$ a left group and $C \subseteq L_k \times A$, if and only if*

$$(V, E) = \bigcup_{i=1}^{k} (V_i, E_i),$$

is the vertex disjoint union of k group digraphs $(V_1, E_1), \ldots, (V_k, E_k)$ such that, for $u_i, v_i \in V_i$ and $i \in \{1, \ldots, k\}$,
(a) *$V_i = \{l_i\} \times A$ and*
(b) *$(u_i, v_i) = ((l_i, u), (l_i, v)) \in E_i \Leftrightarrow \exists\, q \in \{1, \ldots k\}$ with $(l_q, g) \in C$ and $ug = v$.*

Remark 11.3.8. A set of generators C for the left group $L_k \times A$ will always have the property that $p_1(C) = L_k$. Here, p_1 is the first projection. Observe that the elements $c \in C$ with the same second component produce parallel arcs in $\mathrm{Cay}(L_k \times A, C)$. That is, for $(l_j, v), (l_k, v) \in C$ we get $(l_i, u)(l_j, v) = (l_i, u)(l_k, v) = (l_i, uv)$, i. e., we have two arcs $((l_i, u), (l_i, uv))$. From this, it is also clear that $(u_i, v_j) = ((l_i, u), (l_j, v)) \notin E$ if $i \neq j$.

A direct consequence of Theorem 11.3.7 is the following.

Corollary 11.3.9. *A digraph is a left group digraph if and only if it is a vertex disjoint union of copies of a group digraph.*

Example 11.3.10. In Application 11.2.3, we also saw that

$$\mathrm{Cay}(L_2 \times \mathbb{Z}_2, \{(l_2, 1)\}) = \mathrm{Cay}(L_2, \{l_2\}) \times \mathrm{Cay}(\mathbb{Z}_2, \{1\}).$$

Below we have

$$\mathrm{Cay}(L_2 \times \mathbb{Z}_3, \{(l_1, 1)\}) = \mathrm{Cay}(L_2, \{l_1\}) \times \mathrm{Cay}(\mathbb{Z}_3, \{1\}).$$

In both cases, the cross product of the Cayley graphs of the group and the left zero semigroup is visible:

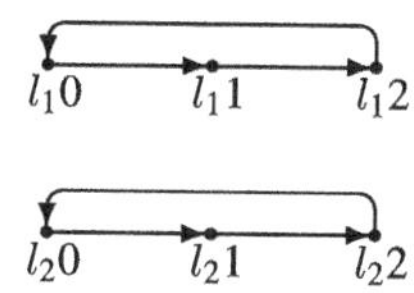

Cayley graphs of strong semilattices of semigroups

In this subsection, we investigate strong semilattices of semigroups and specialize the results to strong semilattices of groups, i. e., Clifford semigroups, and of right or left groups. Compare also Sayan Panma, N. Na Chiangmai, Ulrich Knauer, and Srichan Arworn [72].

We will illustrate the results with an applications section.

For a finite strong semilattice of semigroups $S_\xi = (S_\xi; \circ_\xi)$, $\xi \in Y$, where Y is the (meet) semilattice, we will use the notation $(\bigcup_{\xi \in Y} S_\xi; *)$, as introduced directly before Theorem 9.1.6, or simply $\bigcup_{\xi \in Y} S_\xi$.

In Lemma 11.3.11, we restrict our attention to a one-element connection set $\{c\}$. It says in particular, that in a strong semilattice Y of semigroups, no arcs go from a lower to a higher semigroup, where "higher" and "lower" are with respect to the partial order in Y. Furthermore, if there exists one arc from a higher to a lower semigroup, then there are arcs from every point of the higher semigroup to the lower semigroup.

Lemma 11.3.11. *Consider the strong semilattice of semigroups $(\bigcup_{\xi \in Y} S_\xi; *)$, and take $c \in S_{\xi_0}$ for some $\xi_0 \in Y$. Then the Cayley graph $\mathrm{Cay}(\bigcup_{\xi \in Y} S_\xi, \{c\})$ contains $|Y|$ disjoint semigroup subdigraphs $(S_\xi, E_\xi) \cong \mathrm{Cay}(S_\xi, C_\xi)$, such that*

(a) *If $\xi_0 \geq \xi$, then $C_\xi := \{f_{\xi_0,\xi}(c)\}$;*

(b) *if $\xi_0 < \xi$, then*
$(x_\xi, y_{\xi_0}) \in E(\mathrm{Cay}(\bigcup_{\xi \in Y} S_\xi, \{c\})) \Leftrightarrow (f_{\xi,\xi_0}(x_\xi), y_{\xi_0}) \in E(\mathrm{Cay}(S_{\xi_0}, \{c\}))$, *i. e.,* $y_{\xi_0} = f_{\xi,\xi_0}(x_\xi)c = x_\xi * c$;

(c) *if $\xi_0 \parallel \xi$, set $\alpha := \xi_0 \wedge \xi$, then*
$(x_\xi, y_\alpha) \in E(\mathrm{Cay}(\bigcup_{\xi \in Y} S_\xi, \{c\})) \Leftrightarrow (f_{\xi,\alpha}(x_\xi), y_\alpha) \in E(\mathrm{Cay}(S_\alpha, \{f_{\xi_0,\alpha}(c)\}))$ *for* $y_\alpha \in S_\alpha$. *That is,* $y_\alpha = f_{\xi,\alpha}(x_\xi) f_{\xi_0,\alpha}(c) = x_\xi * c$, *in particular,* $C_\alpha := \{f_{\xi_0,\alpha}(c)\}$.

If $\bigcup_{\xi \in Y} S_\xi$ is a strong chain of semigroups, then Assertion (c) is empty.

Proof. This is clear from the definition of $*$ in $(\bigcup_{\xi \in Y} S_\xi; *)$. Compare the definition of a strong semilattice of semigroups in Section 9.1 just before Theorem 9.1.6. □

Geometric representations of the corresponding Cayley graphs are in Examples 11.3.22, 11.3.23, and 11.3.24.

Next, we extend Lemma 11.3.11 to any finite connection set.

Proposition 11.3.12. *Consider* $(\bigcup_{\xi\in Y} S_\xi; *)$, *and take* $C = \{c_1, \dots, c_k\} \subseteq \bigcup_{\xi\in Y} S_\xi$. *Then we get the Cayley graph* $\mathrm{Cay}(\bigcup_{\xi\in Y} S_\xi, C) = \bigoplus_{c_i\in C} \mathrm{Cay}(\bigcup_{\xi\in Y} S_\xi, \{c_i\})$.

Proof. This follows from Lemma 11.3.11 together with Lemma 11.3.3. □

We now describe the structure of Cayley graphs of a strong semilattice of semigroups with a given one-element connection set. We illustrate the results in Example 11.3.22 for a strong semilattice (containing strong chains) of (right) groups.

Theorem 11.3.13. *Consider the strong semilattice of semigroups* $S = (\bigcup_{\xi\in Y} S_\xi; *)$, *and take a connection set* $C = \bigcup_{\xi\in Y} C_\xi$ *with* $C_\xi \subseteq S_\xi$. *Then the vertex set of* $\mathrm{Cay}(S, C)$ *consists of the union* $\bigcup_{\xi\in Y} S_\xi$ *and for all* $\xi' \geq \xi$ *and* ξ'' *with* $\xi' \wedge \xi'' = \xi$ *there are edges of the form*

- $(x, xf_{\xi',\xi}(c))$ *with* $x \in S_\xi$ *and* $c \in C_{\xi'}$ *the "intra-edges;"*
- $(x, f_{\xi',\xi}(x)f_{\xi'',\xi}(c))$ *with* $x \in S_{\xi'}$ *and* $c \in C_{\xi''}$ *the "inter-edges."*

A way of seeing Theorem 11.3.13 is that $\mathrm{Cay}(S, C)$ is the disjoint union of semigroup digraphs $\bigcup_{\xi\in Y} \mathrm{Cay}(S_\xi, \bigcup_{\xi'\geq\xi} f_{\xi',\xi}(C_{\xi'}))$ whose edges are the intraedges together with some interedges going from $S_{\xi'}$ to S_ξ for $\xi' \geq \xi$.

Generating connection sets

In order for a Cayley graph to be a good representation of an algebraic structure, it is sensible to ask its connection set to be a generating set. A first result concerning generating sets is whether an algebraic substructure can be detected within a semigroup digraph.

Proposition 11.3.14. *Let* S *be a semigroup,* $C \subseteq S$ *such that* $\mathrm{Cay}(S, C)$ *is strongly connected, and* $A < S$ *a subgroup. Then there is a generating system* C' *of* A *such that* $\mathrm{Cay}(A, C')$ *is a contraction from* $\mathrm{Cay}(S, C)$.

Proof. For any $g \in A$ and $(s, sc) \in E(\mathrm{Cay}(S, C))$ by definition, we have $(gs, gsc) \in E(\mathrm{Cay}(S, C))$. Hence, $g \in \mathrm{End}(\mathrm{Cay}(S, C))$. Clearly, g^{-1} is the inverse endomorphism of g. Thus, $g \in \mathrm{Aut}(\mathrm{Cay}(S, C))$. To see that A acts fixed point-free, suppose $gs = s$ for some $s \in S$. Now for every out-neighbor sc of s we have $gsc = sc$, i. e., sc is also a fixed point of g. Choose a directed path P from s to some $h \in A$. We obtain $gh = h$, which implies $g = e$. Hence the left action of A on $\mathrm{Cay}(S, C)$ is strictly fixed point-free. Thus, we may apply Lemma 7.7.18 and obtain some generating system, C' of G such that $\mathrm{Cay}(A, C')$ is a contraction of $\mathrm{Cay}(S, C)$. □

We invite the reader to find an example, showing that the above fails for subsemigroups.

We now turn our attention to generating sets of right groups. For $C \subseteq A \times R_k$ we denote the projections of C on the respective factors by $p_1(C) := \{g \in A \mid \exists j \in \{1,\dots,k\}$ with$(g,r_j) \in C\}$ and $p_2(C) := \{r_j \in R_k \mid \exists g \in A$ with $(g,r_j) \in C\}$. We start with the following lemma, which is basic for right groups, but does not hold for general products of groups and semigroups.

Lemma 11.3.15. *Let $C \subseteq S = A \times R_k$, then the following are equivalent:*
(i) *C generates S;*
(ii) *$p_1(C)$ generates A and $p_2(C)$ generates R_k;*
(iii) $\mathrm{Cay}(S,C)$ *is strongly connected.*

Proof. Clearly, (i) implies (ii). Now, we show (ii)⇒(iii): Take $s = (g, r_i)$, $t = (h, r_j) \in S$. Multiply s from the right by a sequence of elements of C in order to obtain (h, r_ℓ) for some $\ell \in \{1,\dots,n\}$. Now, take some $(f, r_j) \in C$ and multiply it order of f many times to (h, r_ℓ) from the right. We obtain a directed path from s to t. This yields that $\mathrm{Cay}(S,C)$ is strongly connected.

To see, (iii)⇒(i), let $s \in S$. Any directed path in $\mathrm{Cay}(S,C)$ from a vertex $c \in C$ to s corresponds to a word of elements of C generating s. By strong connectivity, such a path exists. □

As an exercise, we recommend to show that the above fails for general (products of) semigroups.

Next, we will study generating systems of strong semilattices of semigroups and in articular of Clifford semigroups. As a first step, we have to consider semilattices on their own. If the poset Y is a meet semilattice, then it is well known that the unique minimal generating system of Y is given by the set of meet irreducibles, i. e.,

$$M(Y) = \{\mu \in Y \mid \eta_1 \wedge \ldots \wedge \eta_k = \mu \implies \mu \in \{\eta_1, \ldots, \eta_k\}\}.$$

In the Hasse diagram of Y, the set $M(Y)$ corresponds to those elements with at most one upward edge.

The following is a useful and easy to see fact.

Remark 11.3.16. In a meet semilattice Y, we have $\xi < \eta$ if and only if $\{\zeta \in M(Y) | \zeta \geq \eta\} \subsetneq \{\zeta \in M(Y) | \zeta \geq \xi\}$. Or in other words, if and only if $\uparrow\eta \cap M(Y) \subsetneq \uparrow\xi \cap M(Y)$, where $\uparrow\xi = \{\zeta | \zeta \geq \xi\}$.

We collect some properties of the arcs of $\mathrm{Cay}(Y, M(Y))$ without proof.

Lemma 11.3.17. *If (η, ξ) is an arc of $\mathrm{Cay}(Y, M(Y))$, then $\xi \leq \eta$. If $\xi \prec \eta$, then (η, ξ) is an arc of $\mathrm{Cay}(Y, M(Y))$. If $\xi \in M(Y)$, then (η, ξ) is an arc of $\mathrm{Cay}(Y, M(Y))$ if and only if $\xi \leq \eta$ or $\eta \leq \xi$.*

Theorem 11.3.18. *Let $S = \bigcup_{\xi \in Y} S_\xi$ be a strong semilattice of semigroups. A subset $C = \bigcup_{\xi \in Y} C_\xi$ with $C_\xi \subseteq S_\xi$ is a generating set of S if and only if*
(a) *$\bigcup_{\xi' \geq \xi} f_{\xi',\xi}(C_{\xi'})$ is a generating set of S_ξ for all $\xi \in Y$; and*
(b) *$C_\xi \neq \emptyset$ for all meet-irreducibles $\xi \in M(Y)$.*

Proof. "⇐": Take C satisfying the two conditions of the theorem and $x \in S_\xi$. Since the meet-irreducibles are a generating system of Y, we can write $\xi = \xi_1 \wedge \ldots \wedge \xi_k$ for $\xi_1, \ldots, \xi_k \in M(Y)$. By (b), we can take $x_i \in C_{\xi_i}$ for all $1 \leq i \leq k$ and get that $x' = x_1 * \ldots * x_k \in S_\xi$. By (a), we can find $c_1, \ldots, c_\ell \in \bigcup_{\xi' \geq \xi} f_{\xi',\xi}(C_{\xi'})$ such that $x' * c_1 * \ldots * c_\ell = x$.

"⇒": It is easy to see that both properties are necessary: If (b) is not satisfied, it is not possible to generate elements within S_ξ for $C_\xi = \emptyset$ and $\xi \in M(Y)$. If (a) is violated for some $\xi \in Y$, then the corresponding S_ξ cannot be generated entirely. □

Together with Theorem 11.3.13, Lemma 11.3.15, and observing that structure homomorphisms yield graph morphisms, we get a description of Cayley graphs of strong semilattices of right groups with generating connection sets.

Corollary 11.3.19. *Let $S = \bigcup_{\xi \in Y} S_\xi$ be a strong semilattice of right groups and generating set $C = \bigcup_{\xi \in Y} C_\xi$ with $C_\xi \subseteq S_\xi$. We have that*
(a) *for all $\xi \in Y$ the strongly connected graph $\mathrm{Cay}(S_\xi, \bigcup_{\xi' \geq \xi} f_{\xi',\xi}(C_{\xi'}))$ is a strong subgraph of $\mathrm{Cay}(S, C)$;*
(b) *the digraph $\mathrm{Cay}(Y, M(Y))$ can be obtained by contracting the strong subgraph of $\mathrm{Cay}(S, C)$ induced by S_ξ into a single vertex for all $\xi \in Y$;*
(c) *for $\beta \geq \alpha$ the structure homomorphism $f_{\beta,\alpha}$ gives a graph morphism taking $\mathrm{Cay}(S_\beta, \bigcup_{\xi \geq \beta} f_{\xi,\beta}(C_\xi))$ to $\mathrm{Cay}(S_\alpha, \bigcup_{\xi \geq \alpha} f_{\xi,\alpha}(C_\xi))$.*

We apply this to two special types of strong semilattices of semigroups.

Construction 11.3.20. We describe how to construct the Cayley graph of the following special strong semilattice $S = \bigcup_{\xi \in Y} S_\xi$ of semigroups S_ξ, with generating set $C = \bigcup_{\xi \in Y} C_\xi$ with $C_\xi \subseteq S_\xi$ for $\xi \in Y$ and $Y = \{\alpha < \beta, \alpha < \gamma\}$. We write $S = S_\beta \cup S_\gamma \xrightarrow{f_\beta, f_\gamma} S_\alpha$. From Corollary 11.3.19, we get that
- C_β and C_γ generate the graphs induced by S_β and S_γ, respectively;
- $C_\alpha \cup f_{\beta,\alpha}(C_\beta) \cup f_{\gamma,\alpha}(C_\gamma)$ generates the graph of S_α, the lower graph of S.

Moreover,
- for every $c_\beta \in C_\beta$, we have an arc from every $y \in S_\gamma$ to $f_{\gamma,\alpha}(y) f_{\beta,\alpha}(c_\beta) \in S_\alpha$;
- for every $c_\gamma \in C_\gamma$, we have an arc from every $x \in S_\beta$ to $f_{\beta,\alpha}(x) f_{\gamma,\alpha}(c_\gamma) \in S_\alpha$;
- for every $c \in C_\alpha \backslash (f_{\beta,\alpha}(C_\beta) \cup f_{\gamma,\alpha}(C_\gamma))$, we have an arc from every $x \in S_\beta$ to $f_{\beta,\alpha}(x)c \in S_\alpha$, and from every $y \in S_\gamma$ to $f_{\gamma,\alpha}(y)c \in S_\alpha$.

Note that C_α is not necessarily a generating system of S_α, it may even be empty. This is not a problem, since α is not meet-irreducible in the semilattice $Y = \{\alpha < \beta, \gamma\}$.

This situation of Construction 11.3.20 is exemplified for the Cayley graph of $S_\beta^{\{1_\beta\}} \cup S_\gamma^{\{1_\gamma\}} \xrightarrow{f_{\beta,\alpha}(1_\beta)=0_\alpha, f_{\gamma,\alpha}(1\gamma)=2_\alpha} S_\alpha^{\{1_\alpha\}}$ in Example 11.3.25.

And the Cayley graph of $S_\beta^{\{1_\beta\}} \cup S_\gamma^{\{1_\gamma\}} \xrightarrow{f_{\beta,\alpha}(1_\beta)=0_\alpha, f_{\gamma,\alpha}(1\gamma)=1_\alpha} S_\alpha^{\emptyset}$ is $\mathrm{Cay}(S, \{1_\beta\}) \oplus \mathrm{Cay}(S, \{1_\gamma\})$ in Example 11.3.22.

Construction 11.3.21. We turn our attention to the strong semilattice of the form $S = S_\beta \xrightarrow{f} S_\alpha$ of semigroups S_α, S_β, with generating set $C := C_\alpha \cup C_\beta$ with $C_\xi \subseteq S_\xi$, $\xi = \alpha, \beta$.

The Cayley graph $\mathrm{Cay}(S, C)$ of S with respect to C is constructed using Corollary 11.3.19 as follows:

- C_β generates the graph of S_β, the upper graph, i. e., $\mathrm{Cay}(S_\beta, C_\beta)$
- $C_\alpha \cup f_{\beta,\alpha}(C_\beta)$ generates the graph of S_α, the lower graph, i. e., $\mathrm{Cay}(S_\alpha, C_\alpha \cup f_{\beta,\alpha}(C_\beta))$
- for every $c \in C_\alpha$, we have an arc from every $x \in S_\beta$ to $f_{\beta,\alpha}(x)c \in S_\alpha$.

Again, C_α is not necessarily a generating system of S_α, but now $C_\alpha \neq \emptyset$, since α is meet-irreducible in the semilattice $Y = \{\alpha < \beta\}$.

In this situation, we use the notation

$$S_\beta^{C_\beta} \xrightarrow{f_{\beta,\alpha}} S_\alpha^{C_\alpha}.$$

In Section 13.3, we give many more examples of Cayley graphs of Clifford semigroups which are of the form $S = A_\beta \xrightarrow{f_{\beta,\alpha}} A_\alpha$, where A_β, A_α are groups.

Examples of strong semilattices of (right or left) groups

Now we apply Theorem 11.3.13 and Construction 11.3.20 to strong semilattices of (right or left) groups, by giving several examples.

Example 11.3.22. Now consider the strong semilattice of groups, i. e., the Clifford semigroup $S = \bigcup_{\xi \in Y} S_\xi$, with semilattice $Y = \{\alpha < \beta, \gamma\}$. The defining homomorphisms are the identity mapping (from S_γ) and the constant mapping c_0 onto the identity 0_α (from S_β), as indicated in Diagram (a). There we have the Clifford semigroup $\mathbb{Z}_2 \cup \mathbb{Z}_2 \xrightarrow{c_0, id_{\mathbb{Z}_2}} \mathbb{Z}_2$.

We give the Cayley graphs $\mathrm{Cay}(S, C)$ for all six different one-element connection sets C, as shown in the Diagrams (b)–(g) below. Moreover, we give $\mathrm{Cay}(S, \{1_\beta, 1_\gamma\})$, which is obtained by taking the edge sum $\mathrm{Cay}(S, \{1_\beta\}) \oplus \mathrm{Cay}(S, \{1_\gamma\})$ of the graphs in (b) and (d). Compare Remark 11.2.8 and Lemma 11.3.3. Note that $\{1_\beta, 1_\gamma\}$ is a generating system of S. The respective Cayley graph has the connection set $\{0_\alpha\} \cup \{c_0(1_\beta)\} \cup$

$\{id_{\mathbb{Z}_2}(1_\gamma)\} \cup \{1_\beta\} \cup \{1_\gamma\} = \{0_\alpha, 1_\alpha, 1_\beta, 1_\gamma\}$. Here, $C_0 = \{0_\alpha\}$, compare Construction 11.3.20.

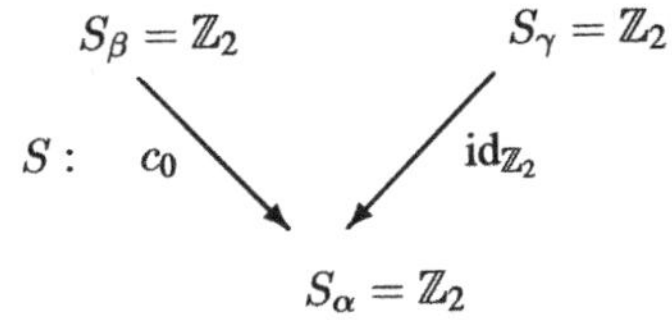

Diagram (a).

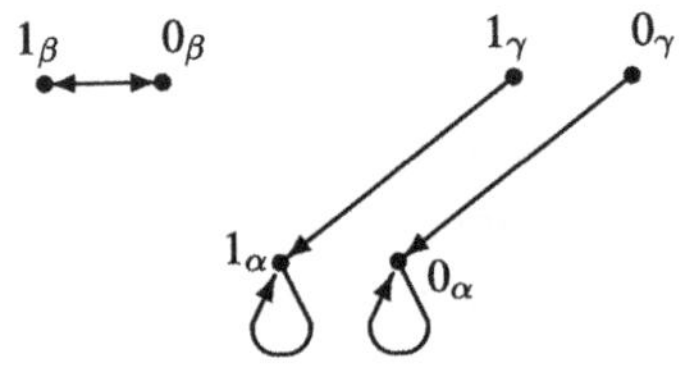

Diagram (b). $\mathrm{Cay}(S, \{1_\beta\})$

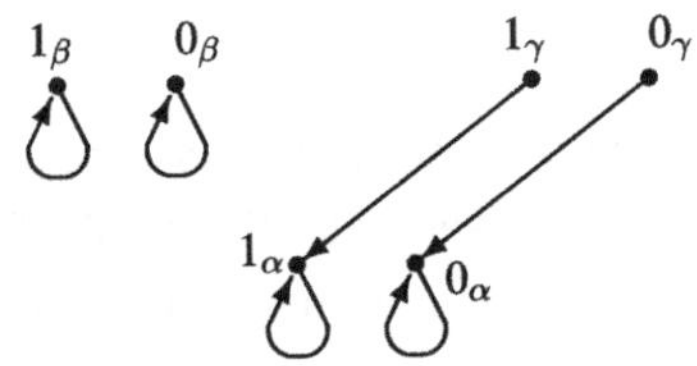

Diagram (c). $\mathrm{Cay}(S, \{0_\beta\})$

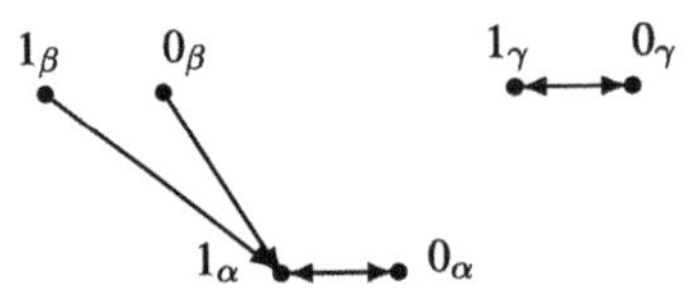

Diagram (d). $\mathrm{Cay}(S, \{1_\gamma\})$

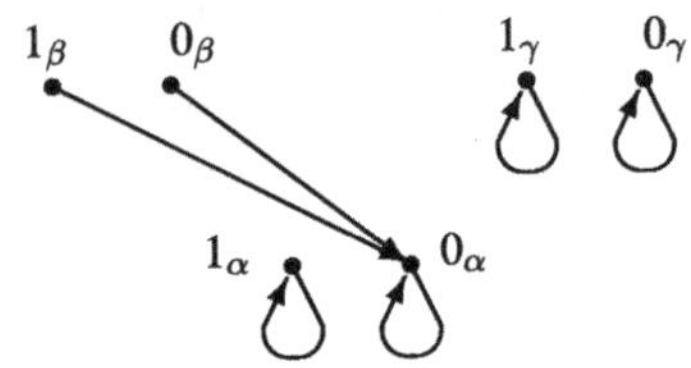

Diagram (e). $\mathrm{Cay}(S, \{0_\gamma\})$

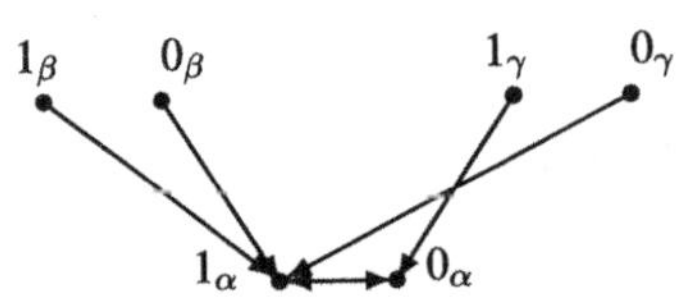

Diagram (f). $\mathrm{Cay}(S, \{1_\alpha\})$

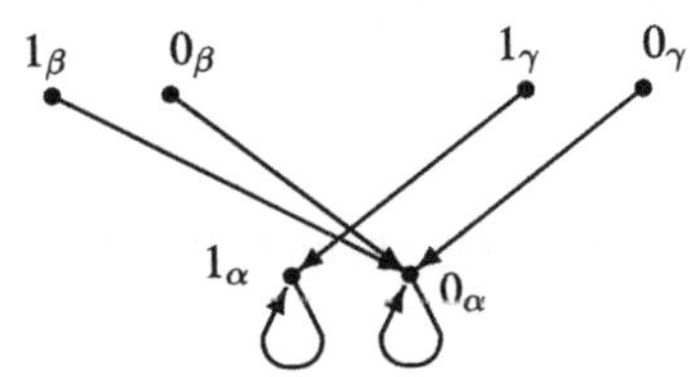

Diagram (g). $\mathrm{Cay}(S, \{0_\alpha\})$

In the next diagrams, we again write xy for the pair (x, y).

Example 11.3.23. For $\xi = \alpha, \beta$, let

$$S_\xi = \mathbb{Z}_2 \times R_2 = \{(0_\xi, r_1), (0_\xi, r_2), (1_\xi, r_1), (1_\xi, r_2)\},$$
$$S_\gamma = \mathbb{Z}_3 \times R_2 = \{(0_\gamma, r_1), (0_\gamma, r_2), (1_\gamma, r_1), (1_\gamma, r_2), (2_\gamma, r_1), (2_\gamma, r_2)\}.$$

As defining homomorphisms, take

$$f_{\beta,\alpha} = \mathrm{id}_{\mathbb{Z}_2} \times \mathrm{id}_{R_2} : S_\beta \to S_\alpha, \text{ in particular, } f_{\beta,\alpha}((1_\beta, r_2)) = (1_\alpha, r_2);$$
$$f_{\gamma,\alpha} = c_{0_\alpha} \times \mathrm{id}_{R_2} : S_\gamma \to S_\alpha, \text{ in particular, } f_{\gamma,\alpha}((1_\gamma, r_1)) = (0_\alpha, r_1).$$

Then $S = \bigcup_{\xi \in Y} S_\xi$ is a strong semilattice of right groups; see Figures (a)–(c) below.

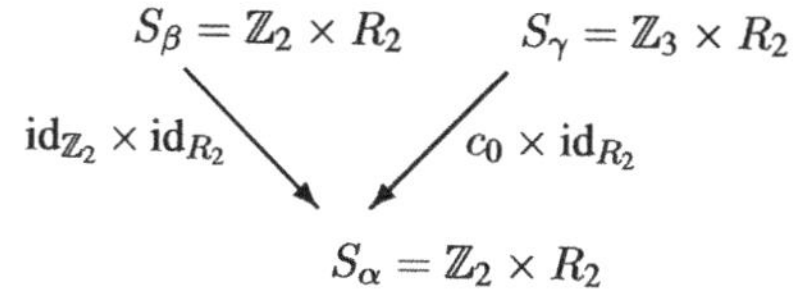

Diagram (a).

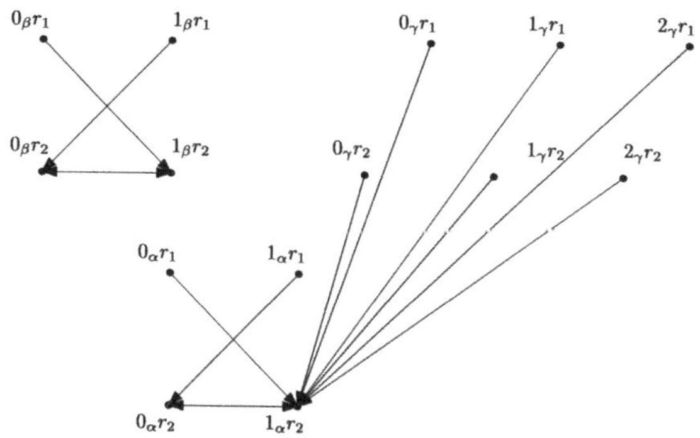

Diagram (b). Cay$(S, \{(1_\beta, r_2)\})$

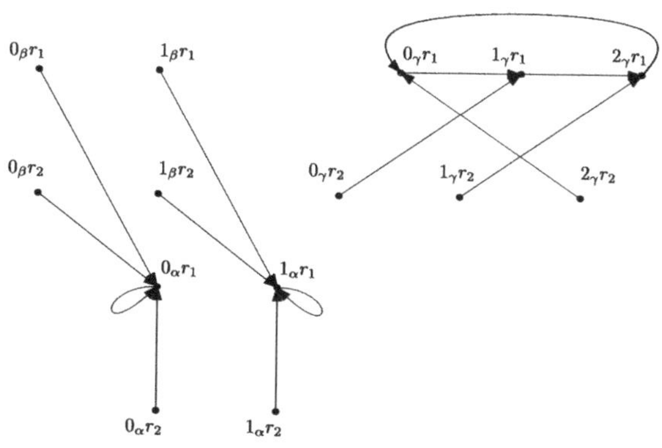

Diagram (c). Cay$(S, \{(1_\gamma, r_1)\})$

Example 11.3.24. For $\xi = \alpha, \beta$, let

$$S_\xi = L_2 \times \mathbb{Z}_2 = \{(l_1, 0_\xi), (l_2, 0_\xi), (l_1, 1_\xi), (l_2, 1_\xi)\},$$
$$S_\gamma = L_2 \times \mathbb{Z}_3 = \{(l_1, 0_\gamma), (l_2, 0_\gamma), (l_1, 1_\gamma), (l_2, 1_\gamma), (l_1, 2_\gamma), (l_2, 2_\gamma)\}.$$

As defining homomorphisms take

$$f_{\beta,\alpha} = \mathrm{id}_{L_2} \times \mathrm{id}_{\mathbb{Z}_2} : S_\beta \to S_\alpha, \quad f_{\beta,\alpha}((l_2, 1_\beta)) = (l_2, 1_\alpha);$$
$$f_{\gamma,\alpha} = \mathrm{id}_{L_2} \times c_0 : S_\gamma \to S_\alpha, \quad f_{\gamma,\alpha}((l_1, 1_\gamma)) = (l_1, 0_\alpha).$$

See the Diagrams (a)–(c) below.

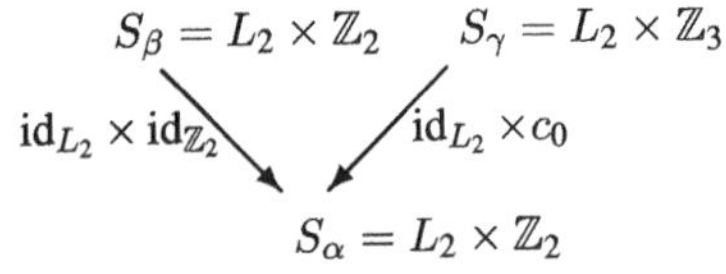

Diagram (a).

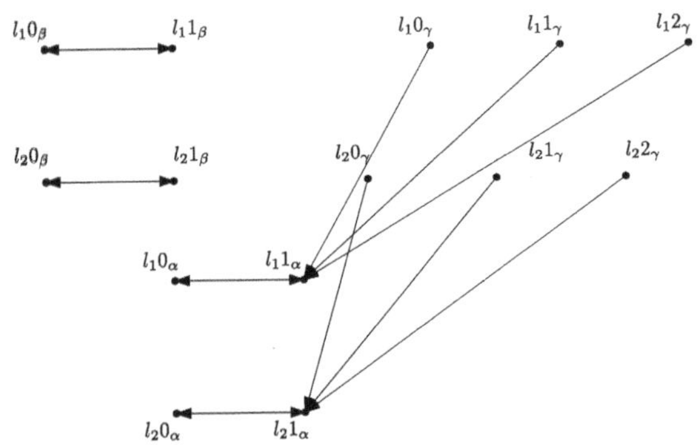

Diagram (b). $\mathrm{Cay}(S, \{(l_2), 1_\beta\})$

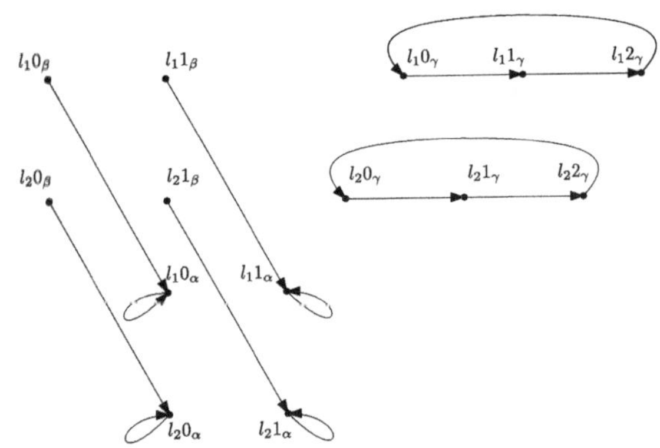

Diagram (c). $\mathrm{Cay}(S, \{(l_1), 1_\gamma\})$

Example 11.3.25. Now consider the following situation:

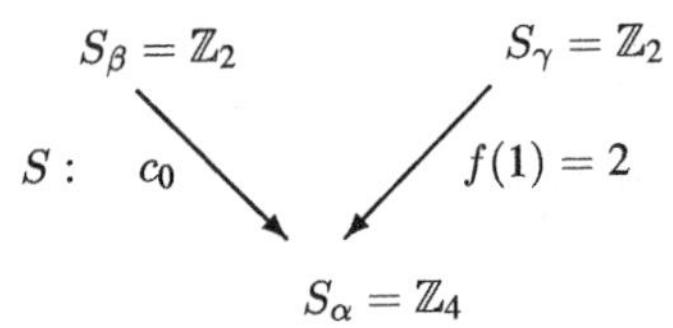

Here, $\{1_\alpha, 1_\beta, 1_\gamma\}$ is a generating system for the Clifford semigroup. According to Construction 11.3.20, we see that $C_\alpha = \{1_\alpha\}$, $c_0(1_\beta) = 0_\alpha, f(1_\gamma) = 2 = 1_\beta * 1_\gamma$. The elements of the corresponding connection set for the Cayley graph are $1_\alpha, 0_\alpha, 2_\alpha, 1_\beta, 1_\gamma$. The lower

Cayley graph of A_α is $\vec{C}_4$ with edges $\{0,2\}$ and $\{1,3\}$ and loops everywhere. The upper Cayley graphs are K_2. In between there are arcs from $0_\beta, 1_\beta$ to 1_α by 1_α and to 2_α by $2_\alpha = f(1_\gamma)$ as well as from 0_γ to 0_α by $0_\alpha = c_0(1_\beta)$ and to 1_α by 1_α, and from 1_γ to 2_α by $0_\alpha = c_0(1_\beta)$ and to 3_α by 1_α as indicated in Construction 11.3.20. So the underlying graph on the vertices of $S_\alpha \cup S_\gamma$ contains a $K_{3,3}$ minor.

11.4 Comments

In this chapter, we started with the categorical viewpoint and discussed the Cayley construction as a functor, namely the Cay functor. For this, it was necessary to consider a new category, the category ***SgC*** of semigroups with connection sets.

Here, several known categorical constructions have to be identified; they do not differ much from the category ***Sgr*** of semigroups with semigroup homomorphisms.

The so-called preservation properties of the Cay functor we applied only to the categorical products, i. e., the Cartesian product in the new category ***SgC*** and the cross product in the category ***D*** of digraphs.

Besides the completely regular semigroups and their semilattices studied in this chapter, it would be interesting to investigate other completely regular semigroups; cf. [Petrich/Reilly 1999]. Without further preparation, several semigroups are accessible in a straightforward manner from what we have already discussed.

When investing Cayley graphs of so-called rectangular semigroups, which have the form $L_m \times S \times R_n$, where again, L_m is a left zero semigroup, R_n is a right zero semigroup and S is a semigroup, it will turn out that $\mathrm{Cay}(L_m \times S \times R_n) \cong \bigcup_{i=1\ldots m} \mathrm{Cay}(S \times R_n)$. For this, see Example 11.3.24. So it suffices to study $\mathrm{Cay}(S \times R_n)$ in this case. Analogously, the study of Cayley graphs of so-called rectangular bands, which have the form $L_m \times R_n$, reduces to the study of $\mathrm{Cay}(R_n)$. If S is a group, then $L_m \times S \times R_n$ is called a rectangular group. Then their Cayley graphs are unions of $\mathrm{Cay}(S \times R_n)$, i. e., of Cayley graphs of right groups.

Then we can consider Cayley graphs of strong semilattices of these; cf. Example 11.3.23.

As noted in Remark 11.2.4, the Cayley graphs of certain completely regular semigroups might be constructed from the Cayley graphs of their components when using the preservation of subdirect products by the functor Cay.

Another, sort of dual point of view is to fix the image of the functor Cay and study its preimage. This is studied with respect to transitive graphs in Chapter 12 and graphs of given genus in Chapter 13. Whenever the directed Cayley graph is transitive, acyclic, and has loops at all vertices, it corresponds to a poset and one can speak of Cayley posets. This is class is studied in [24].

12 Vertex transitive Cayley graphs

In this chapter, we take up the problem of automorphism vertex transitivity from Sections 7.7 and 8.7; we also touch briefly on endomorphism vertex transitivity. Again, results will be applied to special semigroups, and we calculate and present pictures for many examples.

Recall Lemma 7.7.18, which says that a group digraph $G = (V, E)$ has the property that there exists a subgroup $U \subseteq \mathrm{Aut}(G)$ with regular left action on G; i. e., for any two vertices $x, y \in V$ there exists exactly one $s \in U$ such that $s(x) = y$. This means that G is U-vertex transitive, with strictly fixed point-free action of U on G, compare Definition 7.7.6. These properties will have to be interpreted in the context of semigroups if we have only semigroup digraphs. So, for instance, End-vertex transitive, where the endomorphisms act from the left, means that the semigroup is left simple or, in other words, *left solvable*.

The presentation in this chapter mainly follows Sayan Panma [69]. Compare also Sayan Panma, Ulrich Knauer, and Srichan Arworn [70, 71].

12.1 Vertex transitivity

Recall that a digraph $G = (V, E)$ is said to be Aut(G)-*vertex transitive* or simply *vertex transitive* if for any two vertices $x, y \in V$, there exists an automorphism $\varphi \in \mathrm{Aut}(G)$ such that $\varphi(x) = y$. More generally, a subset $A \subseteq \mathrm{End}(G)$ is said to act vertex transitively on G (or we say that G is A-vertex transitive) if for any two vertices $x, y \in V$ there exists an endomorphism $\varphi \in A$ such that $\varphi(x) = y$. Compare Definition 7.7.1.

Now let S be a semigroup and let $C \subseteq S$. We denote the automorphism group and the endomorphism monoid of Cay(S, C) by Aut(S, C) and End(S, C), respectively. Recall that an element $\varphi \in \mathrm{End}(S, C)$ is said to be color preserving if $xa = y$ implies $\varphi(x)a = \varphi(y)$ for $x, y \in S$ and $a \in C$; see Definition 7.3.4. We write ColEnd(S, C) and ColAut(S, C) for the color preserving endomorphisms and automorphisms of Cay(S, C), respectively.

The following facts are well known and quite obvious.

Lemma 12.1.1. *Let $G = (V, E)$ be a finite vertex transitive digraph. Then the indegree $\overrightarrow{d}(v)$ is the same for each vertex v and is equal to the outdegree $\overleftarrow{d}(v)$ of v.*

Lemma 12.1.2. *Let $G = (V, E)$ be a finite digraph and let $G_1, G_2, \ldots, G_n$ be the connected components of G. Then G is vertex transitive if and only if the following conditions hold:*
(a) *$G_1, G_2, \ldots, G_n$ are isomorphic; and*
(b) *G_i is* Aut(G_i)*-vertex transitive for all $i \in \{1, 2, \ldots, n\}$.*

https://doi.org/10.1515/9783110617368-012

In this part, we first obtain results on transitivity properties of strong semilattices of semigroups. We take up the discussion from Section 7.7; refer, in particular, to Definition 7.7.1.

Lemma 12.1.3. *Denote by* $S = (\bigcup_{\xi \in Y} S_\xi, \beta)$ *a finite strong semilattice of semigroups with a maximal element* $\beta \in Y$, *and take* $\emptyset \neq C \subseteq S$. *Then, for all* $v \in S_\beta$, *the indegrees of* v *in* $\mathrm{Cay}(S_\beta, C \bigcap S_\beta)$ *and in* $\mathrm{Cay}(S, C)$ *are equal.*

Proof. Take $v \in S_\beta$. Then by Lemma 11.3.11 there is no $\alpha \neq \beta$ such that (x_α, v) is an arc in $\mathrm{Cay}(S, C)$. Therefore, the indegrees of v in $\mathrm{Cay}(S_\beta, C \bigcap S_\beta)$ and in $\mathrm{Cay}(S, C)$ are equal. □

An immediate consequence is the following.

Lemma 12.1.4. *Let* $S = (\bigcup_{\xi \in Y} S_\xi, \beta)$ *with a maximal element* $\beta \in Y$, *and take* $\emptyset \neq C \subseteq S$. *If* $\mathrm{Cay}(S, C)$ *is* $\mathrm{Aut}(S, C)$*-vertex transitive, then* $C \subseteq S_\beta = \{v_1, \ldots, v_n\}$.

Proof. Assume that $C \bigcap S_\beta \neq C$. Consider the following two cases.

Case 1. If $C \bigcap S_\beta = \emptyset$, then in $\mathrm{Cay}(S_\beta, C \bigcap S_\beta)$ we have $\overleftarrow{d}(v) = 0$ for all $v \in S_\beta$. Since β is maximal in Y, in $\mathrm{Cay}(S, C)$ we get $\overleftarrow{d}(v) = 0$ for all $v \in S_\beta$ by Lemma 12.1.3. Because $C \neq \emptyset$, in $\mathrm{Cay}(S, C)$ we get $\overrightarrow{d}(v) \geq 1$ for all $v \in S_\beta$. Hence $\mathrm{Cay}(S, C)$ cannot be $\mathrm{Aut}(S, C)$-vertex transitive by Lemma 12.1.1.

Case 2. If $C \bigcap S_\beta \neq \emptyset$, then in $\mathrm{Cay}(S_\beta, C \bigcap S_\beta)$ we have $\sum_{i=1}^n \overrightarrow{d}(v_i) = \sum_{i=1}^n \overleftarrow{d}(v_i)$. By Lemma 12.1.3, $\sum_{i=1}^n \overleftarrow{d}(v_i)$ in $\mathrm{Cay}(S_\beta, C \bigcap S_\beta)$ and $\sum_{i=1}^n \overleftarrow{d}(v_i)$ in $\mathrm{Cay}(S, C)$ are equal. Since $C \bigcap S_\beta \neq \emptyset$ and $C \bigcap S_\beta \neq C$, there exists $a \in C \setminus (C \bigcap S_\beta)$, say $a \in S_\gamma$ for some $\gamma \in Y$ and $\gamma \neq \beta$. Therefore, $(v_i, v_i a)$ is an arc in $\mathrm{Cay}(S, C)$ where $v_i \in S_\beta$ and $v_i a \in S_{\gamma \wedge \beta}$, and thus $\sum_{i=1}^n \overrightarrow{d}(v_i)$ in $\mathrm{Cay}(S_\beta, C \bigcap S_\beta)$ is less than $\sum_{i=1}^n \overrightarrow{d}(v_i)$ in $\mathrm{Cay}(S, C)$. Hence, in $\mathrm{Cay}(S, C)$, $\sum_{i=1}^n \overleftarrow{d}(v_i) < \sum_{i=1}^n \overrightarrow{d}(v_i)$. Then there exists $v \in S_\beta$ such that $\overleftarrow{d}(v) \neq \overrightarrow{d}(v)$. By Lemma 12.1.1, $\mathrm{Cay}(S, C)$ is not $\mathrm{Aut}(S, C)$-vertex transitive. □

The following lemma is an immediate consequence of the above; it gives a necessary condition for $\mathrm{Aut}(S, C)$-vertex transitive Cayley graphs of strong semilattices of semigroups.

Lemma 12.1.5. *Let* $S = \bigcup_{\xi \in Y} S_\xi$ *and* $\emptyset \neq C \subseteq S$. *If* $\mathrm{Cay}(S, C)$ *is* $\mathrm{Aut}(S, C)$*-vertex transitive, then* Y *has the maximum* μ *with* $C \subseteq S_\mu$.

The first example in the next section shows that the conditions of Lemma 12.1.5 are not sufficient for $\mathrm{Cay}(S, C)$ to be $\mathrm{Aut}(S, C)$-vertex transitive.

12.2 Application to strong semilattices of right groups

We now study strong semilattices of right groups with automorphism vertex transitive Cayley graphs. We start with an example which illustrates Lemma 12.1.5.

Theorem 12.2.4 in this section shows that the Cayley graph of a strong semilattice of right groups is ColAut(S, C)-vertex transitive only if it is a strong semilattice of groups.

For the most part, we do not give proofs but rather illustrate the result with some examples.

Example 12.2.1. Let $S_\alpha = R_3 = \{r_1, r_2, r_3\}$, $S_\beta = \mathbb{Z}_2 \times R_2 = \{(0_\beta, r_1), (0_\beta, r_2), (1_\beta, r_1), (1_\beta, r_2)\}$, and $S_\gamma = R_2 = \{r_1', r_2'\}$. Let the defining homomorphism be as shown in figure (a) below, i. e., $f_{\beta,\alpha} = p_2$ and $f_{\gamma,\alpha}$ is the inclusion with $f_{\gamma,\alpha}(r_1') = r_1$ and $f_{\gamma,\alpha}(r_2') = r_2$.

Then $S = \bigcup_{\xi \in Y} S_\xi$ is a strong semilattice of right groups.

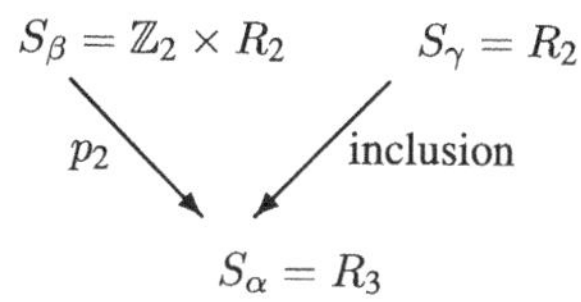

Diagram (a).

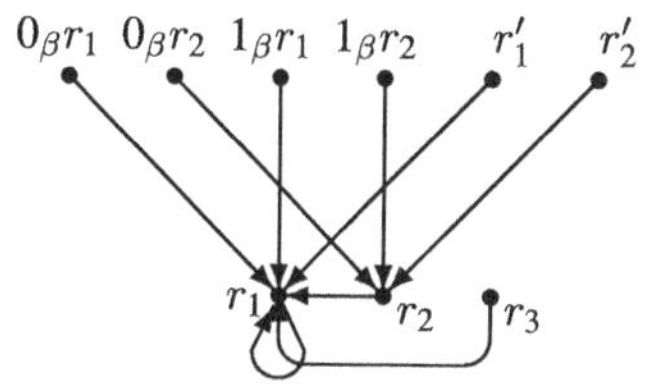

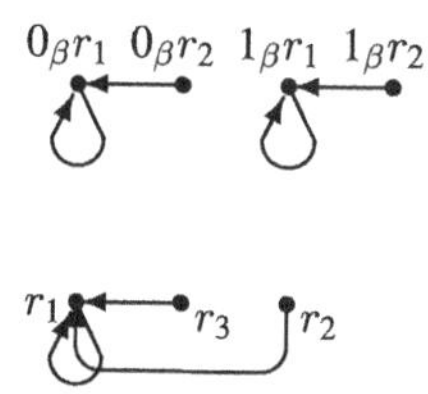

Diagram (b). Cay$(S, \{r_1\})$ Diagram (c). Cay$(S_\alpha \bigcup S_\beta, \{(0_\beta, r_1)\})$

In Diagram (b) above, the semilattice does not have a maximum, and we see that Cay(S, C) is not Aut(S, C)-vertex transitive.

If we take $Y = \{\alpha, \beta\}$, i. e., remove r_1', r_2' and the corresponding arcs from Diagram (b), then Y has a maximum but $C = \{r_1\} \not\subseteq S_\beta$. The resulting picture shows that Cay(S, C) is not Aut(S, C)-vertex transitive either.

In Diagram (c), the conditions of Lemma 12.1.5 are satisfied, but we see that Cay$(S_\alpha \bigcup S_\beta, C)$ is still not Aut$(S_\alpha \bigcup S_\beta, C)$-vertex transitive. Here, Condition (c) of Theorem 12.2.8 is not fulfilled.

Lemma 12.2.2. *Take a finite right group $A \times R_r$, and take $\emptyset \neq C \subseteq A \times R_r$. Then $\langle C \rangle = \langle p_1(C) \rangle \times p_2(C) \subseteq A \times R_r$, where p_1 and p_2 are the projections from $A \times R_r$.*

Proof. It is clear that $\langle p_2(C) \rangle = p_2(C)$ and thus $\langle C \rangle \subseteq \langle p_1(C) \rangle \times p_2(C)$. For the converse implication, take $(g, r_i) \in \langle p_1(C) \rangle \times p_2(C)$. Then because of finiteness, $(g, r_i)^s = (1_A, r_i)$ for some power s, and $g = g_1 \ldots g_t \in \langle p_1(C) \rangle$. Then $(g_1, r_i) \in C$ and thus $(g, r_i) \in \langle C \rangle$. □

Lemma 12.2.3. *Take a finite strong semilattice Y with maximum μ of right groups*

$$\bigcup_{\xi\in Y} S_\xi = \bigcup_{\xi\in Y} (A_\xi \times R_{n_\xi}, \mu),$$

with groups A_ξ and right zero semigroups $R_{n_\xi} = \{r_1, \ldots, r_{n_\xi}\}$. Let $\emptyset \neq C \subseteq S_\mu$.

Then, for all $s \in S_\xi$, we have $|s\langle C\rangle| = |f_{\mu,\xi}(\langle C\rangle)|$.

Proof. Take $s_\alpha = (g, r_i^\alpha) \in S_\alpha$. Since $\langle C\rangle$ is a right group and a subsemigroup of S_m, we have $f_{\mu,\alpha}(\langle C\rangle) = \langle A'_\alpha\rangle \times R'_{n_\alpha}$, where $A'_\alpha \subseteq A_\alpha$ and $R'_{n_\alpha} \subseteq R_{n_\alpha}$, by Lemma 12.2.2. Then $|s_\alpha\langle C\rangle| = |(g, r_i^\alpha) f_{\mu,\alpha}(\langle C\rangle)| = |(g, r_i^\alpha)(\langle A'_\alpha\rangle \times R'_{n_\alpha})| = |g\langle A'_\alpha\rangle \times r_i^\alpha R'_{n_\alpha}| = |\langle A'_\alpha\rangle \times R'_{n_\alpha}| = |f_{\mu,\alpha}(\langle C\rangle)|$. □

ColAut(S, C)-vertex transitivity

The following result is clear from the structure of Cayley graphs of right groups; compare with Theorem 11.3.4 and also $\mathrm{Cay}(S_\beta, \{(0_\beta, r_1)\})$ in Diagram (b) of Example 12.2.6.

Theorem 12.2.4. *Take a finite right group $S = A \times R_r$, and let $\emptyset \neq C \subseteq S$. If $\mathrm{Cay}(S, C)$ is $\mathrm{ColAut}(S, C)$-vertex transitive, then it is $\mathrm{Aut}(S, \{a\})$-vertex transitive for any $\{a\} \in C$, and S is a group, i. e., $|R_r| = 1$.*

Corollary 12.2.5. *Take a finite strong semilattice Y, with maximum μ, of right groups*

$$S = \bigcup_{\xi\in Y} S_\xi = \bigcup_{\xi\in Y} (A_\xi \times R_{n_\xi}, \mu),$$

with groups A_ξ and right zero semigroups $R_{n_\xi} = \{r_1, \ldots, r_{n_\xi}\}$.

If the Cayley graph $\mathrm{Cay}(S, C)$ is $\mathrm{ColAut}(S, C)$-vertex transitive, then $|R_{n_\xi}| = 1$ for all $\xi \in Y$, i. e., S_ξ is a group for all $\zeta \in Y$; in other words, S is a Clifford semigroup with identity $1_S = 1_{S_\mu}$.

Diagram (b) of the following example illustrates the situation.

Example 12.2.6. Take

$$S_\alpha = \mathbb{Z}_2 = \{0_\alpha, 1_\alpha\},$$
$$S_\beta = \mathbb{Z}_2 \times R_2 = \{(0_\beta, r_1), (0_\beta, r_2), (1_\beta, r_1), (1_\beta, r_2)\},$$
$$S_\gamma = \mathbb{Z}_2 \times \mathbb{Z}_2 = \{(0,0)_\gamma, (0,1)_\gamma (1,0)_\gamma, (1,1)_\gamma\}$$

with defining homomorphisms

$$f_{\beta,\alpha} = p_1 : S_\beta \to S_\alpha,$$
$$f_{\gamma,\alpha} = p_2 : S_\gamma \to S_\alpha,$$

as shown in Diagram (a) below.

Then $S = \bigcup_{\xi \in Y} S_\xi$ is a strong semilattice of right groups.

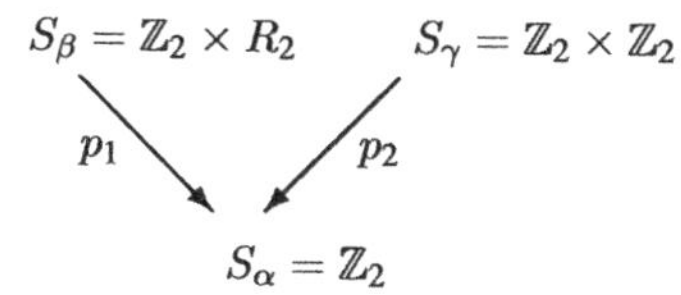

Diagram (a).

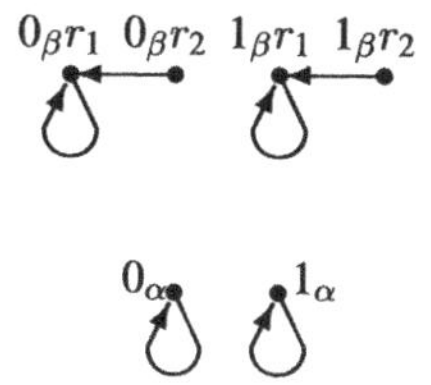

Diagram (b).
$\mathrm{Cay}(S_\alpha \bigcup S_\beta, \{(0_\beta, r_1)\})$

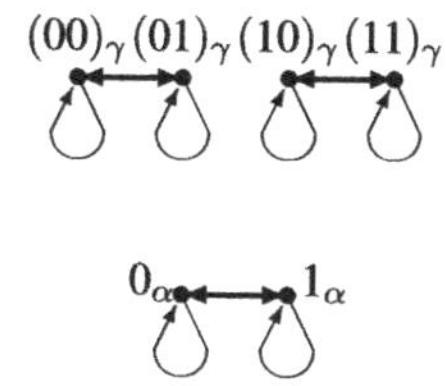

Diagram (c).
$\mathrm{Cay}(S_\alpha \bigcup S_\gamma, \{(0,0)_\gamma, (0,1)_\gamma\})$
$(0,1)_\gamma$: thick line; $(0,0)_\gamma$: thin line

In Diagram (b), we have $S_\alpha \bigcup S_\beta$ with $S_\beta = \mathbb{Z}_2 \times R_2$ such that $|R_2| > 1$, and we see that $\mathrm{Cay}(S_\alpha \bigcup S_\beta, C)$ is not $\mathrm{Aut}(S_\alpha \bigcup S_\beta, C)$-vertex transitive, and thus not $\mathrm{ColAut}(S_\alpha \bigcup S_\beta, C)$-vertex transitive.

Diagram (c) satisfies the conditions of Theorem 12.2.8 for right groups, which actually are groups in the present case. Note that p_2 has different domains here and there. We see that $\mathrm{Cay}(S_\alpha \bigcup S_\gamma, C)$ is $\mathrm{Aut}(S_\alpha \bigcup S_\gamma, C)$-vertex transitive.

This example will be used again for Theorem 12.2.8, whose Conditions (a) and (b) are fulfilled here. We see that $\mathrm{Cay}(S_\alpha \bigcup S_\gamma, C)$ is also $\mathrm{ColAut}(S_\alpha \bigcup S_\gamma, C)$-vertex transitive.

The following result is also clear.

Corollary 12.2.7. *Take a finite strong semilattice with maximum μ of right zero semigroups $S = \bigcup_{\xi \in Y} S_\xi = \bigcup_{\xi \in Y} R_{n_\xi}$, and let $\emptyset \neq C \subseteq S_\mu$. If the Cayley graph* $\mathrm{Cay}(S, C)$ *is* $\mathrm{ColAut}(S, C)$*-vertex transitive, then we have $|R_{n_\xi}| = 1$ for all $\xi \in Y$, i. e., $S = Y$.*

Aut(S, C)-vertex transitivity

Now we consider $\mathrm{Aut}(S, C)$-vertex transitive Cayley graphs of a strong semilattice of right groups.

In the next theorem, we characterize $\mathrm{Aut}(S, C)$-vertex transitive Cayley graphs of strong semilattices of right groups. Note that $\mathrm{Aut}(S, C)$-vertex transitivity is a weaker requirement than $\mathrm{ColAut}(S, C)$-vertex transitivity and, indeed, nontrivial $\mathrm{ColAut}(S, C)$-vertex transitive right groups are possible.

Theorem 12.2.8. *Take a finite strong semilattice Y of right groups*

$$S = \bigcup_{\xi \in Y} S_\xi = \bigcup_{\xi \in Y} (A_\xi \times R_{n_\xi}),$$

with groups A_ξ and right zero semigroups $R_{n_\xi} = \{r_1, \ldots, r_{n_\xi}\}$, and let $\emptyset \neq C \subseteq S$.

Then the Cayley graph Cay(S, C) *is* Aut(S, C)*-vertex transitive if and only if the following conditions hold:*

(a) *Y has a maximum μ with $C \subseteq S_\mu$;*
(b) $|p_2(f_{\mu,\xi}(C))| = |R_{n_\xi}|$ *for all $\xi \in Y$, where p_2 is the second projection;*
(c) *the restrictions of $f_{\mu,\xi}$ to $\langle C\rangle$ are injections for all $\xi \in Y$;*
(d) *the Cayley graph* Cay$(\langle C\rangle, C)$ *is* Aut$(\langle C\rangle, C)$*-vertex transitive.*

Proof. Necessity of Condition (a) comes from Lemma 12.1.5; necessity of (d) is obvious. For (b) and (c), use Lemmas 12.2.2 and 12.2.3. Example 12.2.9 will illustrate the situation, so we omit the rest of the proof. Compare also the Diagrams (c) in Examples 12.2.1 and 12.2.6. □

It is clear that Condition (d) is not really satisfactory since it still requires the proof of Aut-vertex transitivity.

Example 12.2.9. For $Y = \{\alpha, \beta, \gamma\}$ and $\xi \in Y$, take

$$S_\xi = \{(0_\xi, r_1), (0_\xi, r_2), (1_\xi, r_1), (1_\xi, r_2)\} \cong \mathbb{Z}_2 \times R_2,$$

with the defining homomorphisms as indicated in Diagram (a) below.

Then $S = \bigcup_{\xi \in Y} S_\xi$ is a strong semilattice of right groups.

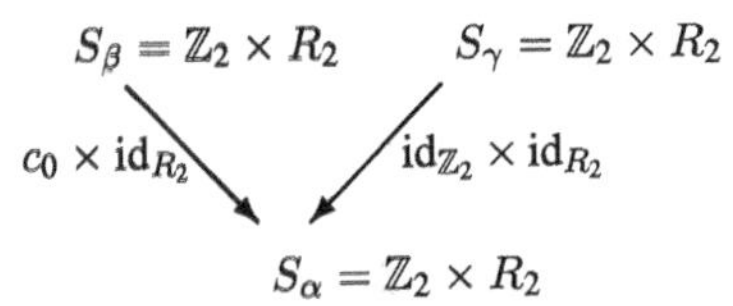

Diagram (a).

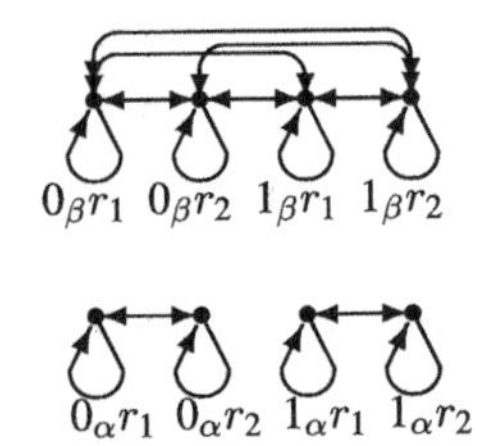

Diagram (b). Cay$(S_\alpha \bigcup S_\beta, S_\beta)$

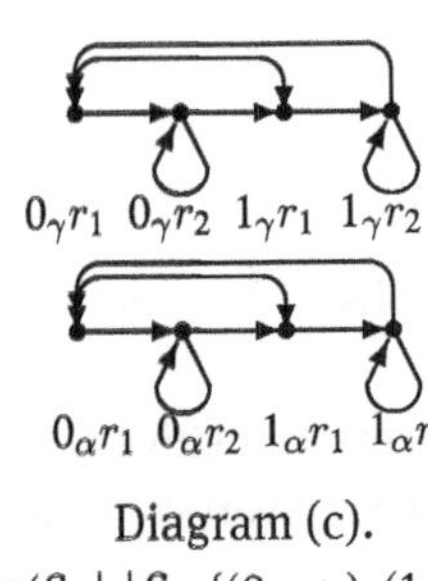

Diagram (c).
Cay$(S_\alpha \bigcup S_\gamma, \{(0_\gamma, r_2), (1_\gamma, r_1)\})$

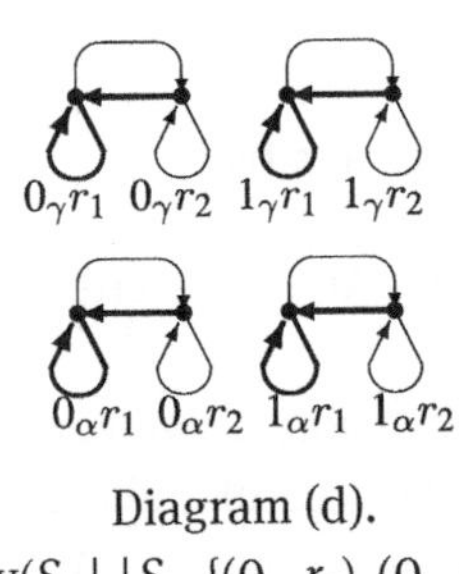

Diagram (d).
Cay$(S_\alpha \bigcup S_\gamma, \{(0_\gamma, r_1), (0_\gamma, r_2)\})$
$(0_\gamma, r_1)$: thick line; $(0_\gamma, r_2)$: thin line

In Diagram (b) above, Conditions (a), (b), and (d) from Theorem 12.2.8 are satisfied, but not Condition (c), and we see that $\mathrm{Cay}(S_\alpha \bigcup S_\beta, C)$ is not $\mathrm{Aut}(S_\alpha \bigcup S_\beta, C)$-vertex transitive. The upper component is the Cayley graph from Condition (d).

In Diagram (c), Conditions (a), (b), and (c) from Theorem 12.2.8 are satisfied, but not Condition (d), and we see that $\mathrm{Cay}(S_\alpha \bigcup S_\gamma, C)$ is not $\mathrm{Aut}(S_\alpha \bigcup S_\gamma, C)$-vertex transitive. The upper component is the Cayley graph from Condition (d).

In Diagram (d), Conditions (a), (b), (c), and (d) from Theorem 12.2.8 are satisfied, and we see that $\mathrm{Cay}(S_\alpha \bigcup S_\gamma, C)$ is $\mathrm{Aut}(S_\alpha \bigcup S_\gamma, C)$-vertex transitive but not $\mathrm{ColAut}(S_\alpha \bigcup S_\gamma, C)$-vertex transitive. The upper left component is the Cayley graph from Condition (d).

Parallel to Lemma 12.2.4 and Corollaries 12.2.5 and 12.2.7, we specialize the preceding theorem.

Corollary 12.2.10. *Take $S = \bigcup_{\xi \in Y} R_{n_\xi}$, a strong semilattice of right zero semigroups, and $\emptyset \neq C \subseteq S$.*

Then the Cayley graph $\mathrm{Cay}(S, C)$ is $\mathrm{Aut}(S, C)$-vertex transitive if and only if the following conditions hold:

(a) *Y has the maximum μ, with $C = R_{n_\mu}$;*
(b) *the defining homomorphisms $f_{\mu,\xi}$ are isomorphisms for all $\xi \in Y$; in particular, $n_\xi = n_\mu$ for all $\xi \in Y$.*

Corollary 12.2.11. *Let R_r be a finite right zero semigroup and $\emptyset \neq C \subseteq R_r$. Then $\mathrm{Cay}(R_r, C)$ is $\mathrm{Aut}(R_r, C)$-vertex transitive if and only if $C = R_r$.*

Corollary 12.2.12. *Let $S = A \times R_r$ be a finite right group, take $\emptyset \neq C \subseteq S$, and let p_2 be the second projection. Then $\mathrm{Cay}(S, C)$ is $\mathrm{Aut}(S, C)$-vertex transitive if and only if the following conditions hold:*

(a) *$p_2(C) = R_r$;*
(b) *the Cayley graph $\mathrm{Cay}(\langle C \rangle, C)$ is $\mathrm{Aut}(\langle C \rangle, C)$-vertex transitive.*

Proof. This is a direct consequence of Theorem 12.2.8. □

12.3 Application to strong semilattices of left groups

Here, we consider left groups instead of right groups. In Theorem 12.3.4, we characterize $\mathrm{Aut}(S, C)$-vertex transitive and $\mathrm{ColAut}(S, C)$-vertex transitive Cayley graphs of strong semilattices of left groups.

As a left dual of Lemma 12.2.2, we have the following.

Lemma 12.3.1. *Let $S = L_l \times A$ be a finite left group, where A is a group, $L_l = \{l_1, \dots, l_l\}$ a left zero semigroup, and $\emptyset \neq C \subseteq S$. Then $\langle C \rangle = p_1(C) \times \langle p_2(C) \rangle$ is a left group contained in S.*

Note that the result of the following lemma is not left dual to Lemma 12.2.3 in the direct sense.

Lemma 12.3.2. *Take a finite strong semilattice Y with maximum μ of left groups*

$$\bigcup_{\xi\in Y} S_\xi = \bigcup_{\xi\in Y}(L_{n_\xi}\times A_\xi,\mu),$$

with groups A_ξ and left zero semigroups $L_{n_\xi} = \{l_1,\dots,l_{n_\xi}\}$. Let $\emptyset \neq C \subseteq S_\mu$ and denote by p_2 the second projection.

Then, for all $s\in S_\xi$, we have $|s\langle C\rangle| = |p_2(f_{\mu,\xi}(\langle C\rangle))|$.

Proof. Let $s = (l_i^\alpha, g_\alpha)\in S_\alpha$. Since $\langle C\rangle$ is a left subgroup of S_m, we have $f_{m,\alpha}(\langle C\rangle) = \langle L'_{n_\alpha}\times A'_\alpha\rangle$ where $A'_\alpha \subseteq A_\alpha$ and $L'_{n_\alpha}\subseteq L_{n_\alpha}$, by Lemma 12.3.1. Therefore, $|s\langle C\rangle| = |(l_i^\alpha, g_\alpha)f_{m,\alpha}(\langle C\rangle)| = |(l_i^\alpha,g_\alpha)(L'_{n_\alpha}\times\langle A'_\alpha\rangle)| = |l_i^\alpha L'_{n_\alpha}\times g_\alpha\langle A'_\alpha\rangle| = |\{l_i^\alpha\}\times\langle A'_\alpha\rangle| = |\langle A'_\alpha\rangle| = |p_2(f_{m,\alpha}(\langle C\rangle))|$. □

We state without proof Theorem 2.1 from A. V. Kelarev and C. E. Praeger [44]. Note that the original paper uses left action for the construction of the Cayley graph. For our purpose, we have changed this to right action and specialized the statement to finite semigroups.

Theorem 12.3.3. *Take a semigroup S and a subset C. Then* Cay(S, C) *is* ColAut(S, C)*-vertex transitive if and only if:*

1. *$Sa = S$ for all $a\in C$;*
2. *$\langle C\rangle$ is a left group; and*
3. *$|s\langle C\rangle|$ is independent of the choice of $s\in S$.*

Note that Condition (1) means that S is left simple if $\langle C\rangle = S$.

Theorem 12.3.4. *Take a finite strong semilattice Y of left groups*

$$S = \bigcup_{\xi\in Y} S_\xi = \bigcup_{\xi\in Y}(L_{n_\xi}\times A_\xi),$$

with groups A_ξ and left zero semigroups $L_{n_\xi} = \{l_1,\dots,l_{n_\xi}\}$ and let $\emptyset\neq C\subseteq S$.

Then the following conditions are equivalent:

(i)
   (a) *Y has the maximum μ with $C\subseteq S_\mu$; and*
   (b) *$|p_2(\langle C\rangle)| = |p_2(f_{\mu,\xi}(\langle C\rangle))|$ for all $\xi\in Y$.*

(ii) Cay(S, C) *is* ColAut(S, C)*-vertex transitive.*

(iii) Cay(S, C) *is* Aut(S, C)*-vertex transitive.*

Proof. (i) ⇒ (ii): Assuming (i), we prove Statements (1), (2), and (3) of Theorem 12.3.3; Assertion (ii) then follows:

(1) Take $a \in C$ and $\alpha \in Y$. Since $C \subseteq S_\mu$, we have that $f_{\mu,\alpha}(a) = (l,g) \in S_\alpha$ for some $g \in A_\alpha$ and $l \in L_{n_\alpha}$. Thus

$$\begin{aligned} S_\alpha a &= (L_{n_\alpha} \times A_\alpha) f_{\mu,\alpha}((a)) = (L_{n_\alpha} \times A_\alpha)(l,g) \\ &= L_{n_\alpha} l \times A_\alpha g = L_{n_\alpha} \times A_\alpha = S_\alpha. \end{aligned}$$

Therefore, $Sa = (\bigcup_{\alpha\in Y} S_\alpha)a = \bigcup_{\alpha\in Y}(S_\alpha a) = \bigcup_{\alpha\in Y} S_\alpha = S$.

(2) Since $C \subseteq S_\mu$, we obtain from Lemma 12.3.1 that $\langle C\rangle$ is a left group.

(3) Let $s, s' \in S$. Then $s \in S_\alpha$ and $s' \in S_\beta$ for some $\alpha, \beta \in Y$. By Lemma 12.3.2, we have $|s\langle C\rangle| = |p_2(f_{\mu,\alpha}(\langle C\rangle))|$ and $|s'\langle C\rangle| = |p_2(f_{\mu,\beta}(\langle C\rangle))|$. From ($b$), we then obtain $|s\langle C\rangle| = |s'\langle C\rangle|$.

(ii) $\Rightarrow$ (iii) is obvious.

(iii) $\Rightarrow$ (i): We know from Lemma 12.1.5 that (a) is necessary. So it remains to prove that the Cayley graph $\mathrm{Cay}(S,C)$ is not $\mathrm{Aut}(S,C)$-vertex transitive if there exists $\beta \in Y$ such that $|p_2(\langle C\rangle)| \neq |p_2(f_{\mu,\beta}(\langle C\rangle))|$. We leave this as an exercise. □

Example 12.3.5 will illustrate the situation; see also Example 12.4.11.

Example 12.3.5. For $Y = \{\alpha,\beta,\gamma\}$ and $\xi \in Y$ take $S_\xi = \{(l_1,0_\xi),(l_2,0_\xi),(l_1,1_\xi),(l_2,1_\xi)\} \cong L_2 \times \mathbb{Z}_2$, with defining homomorphisms $f_{\beta,\alpha} := \mathrm{id}_{L_2} \times c_0 : S_\beta \to S_\alpha$ and $f_{\gamma,\alpha} := \mathrm{id}_{L_2} \times \mathrm{id}_{\mathbb{Z}_2} : S_\gamma \to S_\alpha$.

Then $S = \bigcup_{\xi\in Y} S_\xi$ is a strong semilattice of left groups.

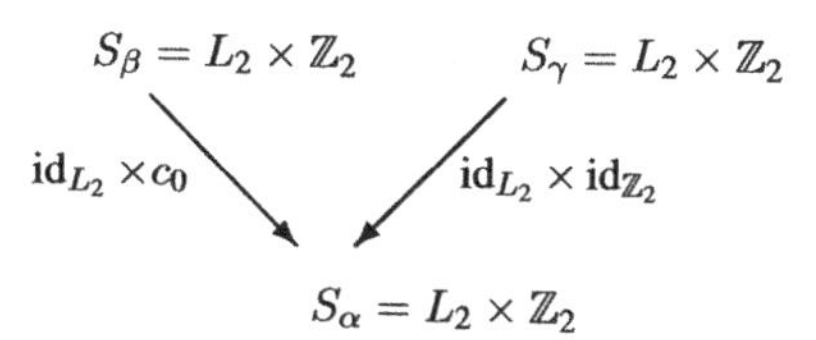

Diagram (a).

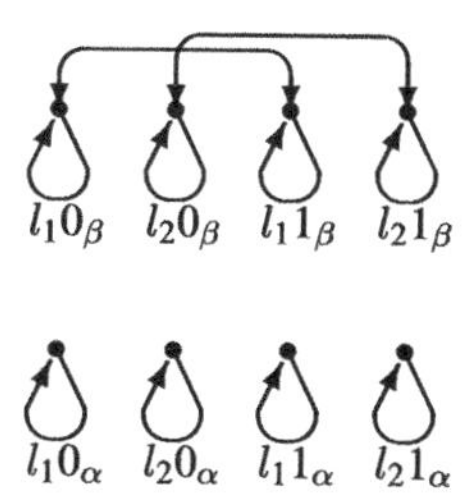

Diagram (b).
$\mathrm{Cay}(S_\alpha \bigcup S_\beta, \{(l_1,0_\beta),(l_2,1_\beta)\})$

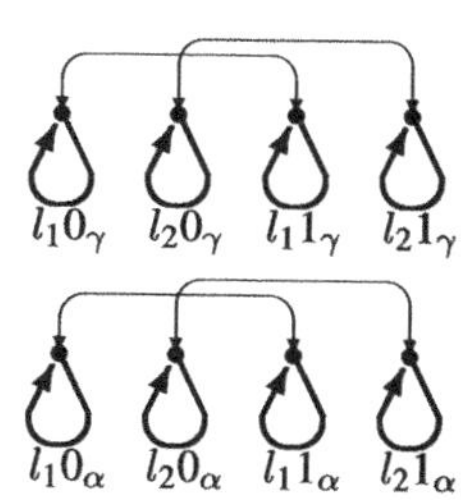

Diagram (c).
$\mathrm{Cay}(S_\alpha \bigcup S_\gamma, \{(l_1,0_\gamma),(l_1,1_\gamma)\})$
$(l_1,0_\gamma)$: thick line; $(l_1,1_\gamma)$: thin line

In Diagram (b), Condition (a) from Theorem 12.3.4 is satisfied but not Condition (b), and we see that $\mathrm{Cay}(S_\alpha \bigcup S_\beta, C)$ is not $\mathrm{Aut}(S_\alpha \bigcup S_\beta, C)$-vertex transitive.

In Diagram (c), Conditions (a) and (b) from Theorem 12.3.4 are both satisfied, and we see that $\mathrm{Cay}(S_\alpha \bigcup S_\gamma, C)$ is $\mathrm{ColAut}(S_\alpha \bigcup S_\gamma, C)$-vertex transitive, and thus also $\mathrm{Aut}(S_\alpha \bigcup S_\gamma, C)$-vertex transitive.

Again, we specialize the preceding result to strong semilattices of left zero semigroups.

Corollary 12.3.6. *Take a finite strong semilattice of left zero semigroups $S = \bigcup_{\xi \in Y} S_\xi = \bigcup_{\xi \in Y} L_{n_\xi}$, with $\emptyset \neq C \subseteq S$. Then the following conditions are equivalent:*
(i) *Y has the maximum μ with $C \subseteq L_{n_m}$.*
(ii) *$\mathrm{Cay}(S, C)$ is $\mathrm{ColAut}(S, C)$-vertex transitive.*
(iii) *$\mathrm{Cay}(S, C)$ is $\mathrm{Aut}(S, C)$-vertex transitive.*

Proof. (i) ⇒ (ii): Since $|p_2(\langle C\rangle)| = |p_2(f_{m,\xi}(\langle C\rangle))| = 1$, we get from Theorem 12.3.4 that $\mathrm{Cay}(S, C)$ is $\mathrm{ColAut}(S, C)$-vertex transitive, if we interpret S_ξ as $L_{n_\xi} \times \{e\}$.

(ii) ⇒ (iii) is obvious.

(iii) ⇒ (i) follows from Theorem 12.3.4. □

Corollary 12.3.7. *Let $S = L_l \times G$ be a finite left group, and $\emptyset \neq C \subseteq S$. Then the Cayley graph $\mathrm{Cay}(S, C)$ is always $\mathrm{ColAut}(S, C)$-vertex transitive, and thus $\mathrm{Aut}(S, C)$-vertex transitive.*

Application to Clifford semigroups

Here, we treat groups as a special case of right groups; of course, they could also be considered as special left groups.

Theorem 12.3.8. *Take the finite strong semilattice $S = \bigcup_{\xi \in Y} A_\xi$ of groups A_ξ, and let $\emptyset \neq C \subseteq S$. Then the following conditions are equivalent:*
(i)
  (a) *Y has the maximum μ, with $C \subseteq A_\mu$; and*
  (b) *the restrictions of $f_{\mu,\xi}$ to $\langle C\rangle$ are injections for all $\xi \in Y$.*
(ii) *$\mathrm{Cay}(S, C)$ is $\mathrm{ColAut}(S, C)$-vertex transitive.*
(iii) *$\mathrm{Cay}(S, C)$ is $\mathrm{Aut}(S, C)$-vertex transitive.*

Proof. We know that a right group $S_\xi = A_\xi \times R_{n_\xi}$ is a group if $|R_{n_\xi}| = 1$, and for all $C \subseteq A_\xi$ we know that $\mathrm{Cay}(\langle C\rangle, C)$ is $\mathrm{Aut}(\langle C\rangle, C)$-vertex transitive because $\langle C\rangle$ is a subgroup of the group A_ξ. By Theorem 12.2.8, we have the equivalence between (i) and (iii). By Theorem 12.3.4 specialized to groups, we get everything. □

As an example, consider Diagram (c) in Example 12.2.6.

12.4 End′(S, C)-vertex transitive Cayley graphs

In this section, we give some preliminary results on strong semilattices of semigroups with $\mathrm{End}'(S, C)$-vertex transitive Cayley graphs or with $\mathrm{ColEnd}'(S, C)$-vertex transitive Cayley graphs. Here,

$$\mathrm{End}'(S, C) := \mathrm{End}(S, C) \setminus \mathrm{Aut}(S, C),$$
$$\mathrm{ColEnd}'(S, C) := \mathrm{ColEnd}(S, C) \setminus \mathrm{ColAut}(S, C).$$

Evidently, $\mathrm{ColAut}(S, C) \subseteq \mathrm{Aut}(S, C) \subseteq \mathrm{End}(S, C)$, $\mathrm{ColEnd}'(S, C) \subseteq \mathrm{End}'(S, C) \subseteq \mathrm{End}(S, C)$, and $\mathrm{ColEnd}(S, C) \subseteq \mathrm{End}(S, C)$.

It seems natural that there exist Cayley graphs which are Aut-vertex transitive and not End′-vertex transitive. But there also exist Cayley graphs which are Aut-vertex transitive and at the same time End′-vertex transitive (see Theorem 12.4.10), or not Aut-vertex transitive but End′-vertex transitive; cf. Example 12.4.11.

First, we state and prove a lemma from A. V. Kelarev and C. E. Praeger [44].

Lemma 12.4.1. *Let S be a semigroup and C a subset of S.*
If $\mathrm{Cay}(S, C)$ is $\mathrm{ColEnd}(S, C)$-vertex transitive, then $Sc = S$ for every $c \in C$.
If $\mathrm{Cay}(S, C)$ is $\mathrm{End}(S, C)$-vertex transitive, then $SC = S$.

Proof. Take $s \in S$ and $c \in C$. Then here exists $f \in \mathrm{End}(S, C)$ with $f(sc) = s$. Since (s, sc) is an edge, $(f(s), f(sc))$ is also an edge. Hence $f(sc) = f(s)c'$ for some $c' \in C$. Thus $s = f(s)c' \in SC$; so $SC = S$ and the second statement holds. In the first case, we may assume that $f \in \mathrm{ColEnd}(S, C)$, whence $c' = c$, and so $Sc = S$, i. e., the first statement holds. □

Lemma 12.4.2. *Let $S = (\bigcup_{\xi \in Y} S_\xi, \beta)$ be a finite strong semilattice of semigroups, $\beta \in Y$ a maximal element of Y, and $\emptyset \neq C \subseteq S$. If $\mathrm{Cay}(S, C)$ is $\mathrm{ColEnd}(S, C)$-vertex transitive, then $C \subseteq S_\beta$.*

Proof. Suppose there exists $c \in C \setminus S_\beta$, say $c \in S_\gamma$, with $\gamma \neq \beta$. As β is maximal, we have $\alpha \wedge \gamma \neq \beta$ for all $\alpha \in Y$. Now $S_\alpha c = f_{\alpha,\alpha\wedge\gamma}(S_\alpha) f_{\gamma,\alpha\wedge\gamma}(c) \subseteq S_{\alpha\wedge\gamma} \neq S_\beta$, and thus $S_\alpha c \bigcap S_\beta = \emptyset$ for all $\alpha \in Y$. This implies that $Sc \bigcap S_\beta = \bigcup_{\alpha \in Y} S_\alpha = \emptyset$, and hence $Sc \neq S$. Now Lemma 12.4.1 implies that $\mathrm{Cay}(S, C)$ is not $\mathrm{ColEnd}(S, C)$-vertex transitive. □

Lemma 12.4.3 is an immediate consequence; it gives two necessary conditions for $\mathrm{ColEnd}(S, C)$-vertex transitive Cayley graphs of strong semilattices of semigroups. The conditions are identical to those in Lemma 12.1.5, but the proofs of the lemmas used, namely Lemmas 12.4.2 and 12.1.4, are different.

Lemma 12.4.3. *Let $S = \bigcup_{\xi \in Y} S_\xi$ be a finite strong semilattice of finite semigroups, and let $\emptyset \neq C \subseteq S$. If $\mathrm{Cay}(S, C)$ is $\mathrm{ColEnd}(S, C)$-vertex transitive, then Y has the maximum μ with $C \subseteq S_\mu$.*

The following observations are relatively straightforward for nonconnected Aut(D)-vertex transitive graphs.

Lemma 12.4.4. *Let $D = (V, E)$ be a finite digraph, let $f \in \text{Aut}(D)$, and let D_1 and D_2 be components of D. If $f(x) = y$ for some $x \in D_1$ and $y \in D_2$, then $f(D_1) = D_2$.*

Lemma 12.4.5. *Let D be a nonconnected digraph. If D is* Aut(D)*-vertex transitive, then it is* End$'$(D)*-vertex transitive.*

Lemma 12.4.6. *Let S be a semigroup, let C be a nonempty subset of S, and let $D_1, D_2, \dots, D_n$ be components of* Cay(S, C)*. Then* Cay(S, C) *is* ColAut(S, C)*-vertex transitive if and only if the components of* Cay(S, C) *are color isomorphic and each component is color automorphism vertex transitive.*

Lemma 12.4.7. *Let S be a semigroup and C a nonempty subset of S. If* Cay(S, C) *is a nonconnected digraph and is* ColAut(S, C)*-vertex transitive, then it is* ColEnd$'$(S, C)*-vertex transitive.*

The next theorem gives some descriptions of End$'$(S, C)-vertex transitive Cayley graphs and of ColEnd$'$(S, C)-vertex transitive Cayley graphs.

Theorem 12.4.8. *Let $S = \bigcup_{\xi \in Y} S_\xi$ be a finite strong semilattice of semigroups, with $|Y| > 1$, and let $\emptyset \neq C \subseteq S$.*
1. *If* Cay(S, C) *is* Aut(S, C)*-vertex transitive, then it is* End$'$(S, C)*-vertex transitive. In this case, $C \bigcap S_\beta \neq \emptyset$ for all maximal $\beta \in Y$.*
2. *If* Cay(S, C) *is* ColAut(S, C)*-vertex transitive, then it is* ColEnd$'$(S, C)*-vertex transitive. In this case, Y has the maximum μ and $C \subseteq S_\mu$.*

Proof. (1) Let Cay(S, C) be Aut(S, C)-vertex transitive. By Lemma 12.1.5, we obtain that Y has the maximum μ and $C \subseteq S_\mu$. It is clear that Cay(S, C) $= \dot{\bigcup}_{\xi \in Y}$ Cay($S_\xi, f_{\mu,\xi}(C)$). Now $|Y| > 1$ implies that Cay(S, C) is not connected. Therefore, Cay(S, C) is End$'$(S, C)-vertex transitive by Lemma 12.4.5.

Suppose now that Cay(S, C) is End$'$(S, C)-vertex transitive. Assume that $C \bigcap S_\beta = \emptyset$ for some maximal element $\beta \in Y$. Choose $s \in S_\beta$. We will show that $s \notin SC$. If $s \in SC$, then $s = tc$ for some $t \in S$ and $c \in C$. Hence $t \in S_\gamma$ and $a \in S_\xi$ for some γ and $\xi \neq \beta$ in Y. Therefore, $s = tc = f_{\gamma,\gamma\wedge\xi}(t) f_{\xi,\gamma\wedge\xi}(c)$. Since $s \in S_\beta$, we get $\gamma \wedge \xi = \beta$, and hence $\beta < \xi$. Thus we obtain a contradiction, because β is a maximal element in Y. Hence $s \notin SC$ and so $SC \neq S$. By Lemma 12.4.1, we get that Cay(S, C) is not End(S, C)-vertex transitive, and thus also not End$'$(S, C)-vertex transitive.

(2) The proof is similar to that for (1), but using Lemma 12.4.7. The second part follows from Lemma 12.4.3. □

In Diagram (d) of Example 12.2.9, we have that Cay(S, C) is Aut(S, C)-vertex transitive, so now we see that it is End$'$(S, C)-vertex transitive.

In Diagram (c) of Example 12.3.5, we have Cay(S, C) is ColAut(S, C)-vertex transitive, so also ColEnd′(S, C)-vertex transitive, Y has the maximum $\mu = \gamma$, and $\emptyset \neq C \subseteq S_\gamma$.

From Example 12.4.11 Diagram (b), Cay(S, C) is End′(S, C)-vertex transitive, and we see that $C \bigcap S_\beta \neq \emptyset$ for all maximal $\beta \in Y$.

Corollary 12.4.9. *Let $S = \bigcup_{\xi \in Y}(A_\xi \times \{r_1, \ldots, r_{n_\xi}\})$ be a finite strong semilattice of right groups, and take $\emptyset \neq C \subseteq S$. Then the following hold:*
(1) *If* Cay(S, C) *is* ColEnd′(S, C)*-vertex transitive, then:*
  (a) *Y has the maximum μ with $C \subseteq S_m$; and*
  (b) *$n_\xi = 1$ for all $\xi \in Y$, i. e., S is a Clifford semigroup.*
(2) *If* Cay(S, C) *is* End′(S, C)*-vertex transitive, then $p_2(C \bigcap S_\beta) = \{r_1, \ldots, r_{n_\beta}\}$ for all maximal $\beta \in Y$.*

In the next theorem, we consider Cayley graphs of strong semilattices of left groups with a one-element connection set.

Theorem 12.4.10. *Let $S = \bigcup_{\xi \in Y}(L_{n_\xi} \times A_\xi)$ be a finite strong semilattice of left groups, i. e., A_ξ are groups and L_{n_ξ} are left zero semigroups. Take $C = \{c\} \subseteq S$. Then* Cay$(S, \{c\})$ *is* End′$(S, \{c\})$*-vertex transitive if and only if it is* Aut$(S, \{c\})$*-vertex transitive.*

Example 12.4.11 will illustrate the situation. We present the Cayley graph of a strong semilattice of semigroups which is End′(S, C)-vertex transitive but not Aut(S, C)-vertex transitive. Note that the connection set has more than one element.

Example 12.4.11. Consider the semilattice $Y = \{\alpha < \beta, \gamma\}$. For $\xi = \alpha, \beta$, take $S_\xi = \mathbb{Z}_2 = \{0_\xi, 1_\xi\}$, $S_\gamma = \{(l_1, 0_\gamma), (l_2, 0_\gamma), (l_1, 1_\gamma), (l_2, 1_\gamma)\} \cong L_2 \times \mathbb{Z}_2$ and the defining homomorphisms $f_{\beta,\alpha} = \mathrm{id}_{\mathbb{Z}_2}$ and $f_{\gamma,\alpha} = p_2$ as indicated. Then $S = \bigcup_{\xi \in Y} S_\xi$ is a strong semilattice of semigroups.

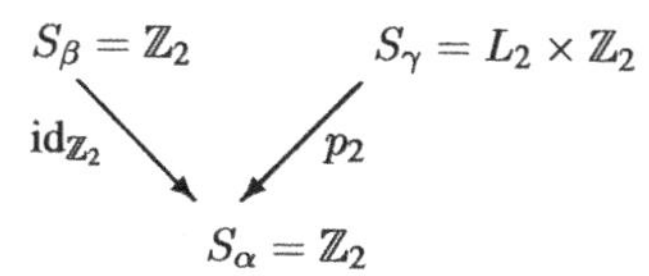

Diagram (a).

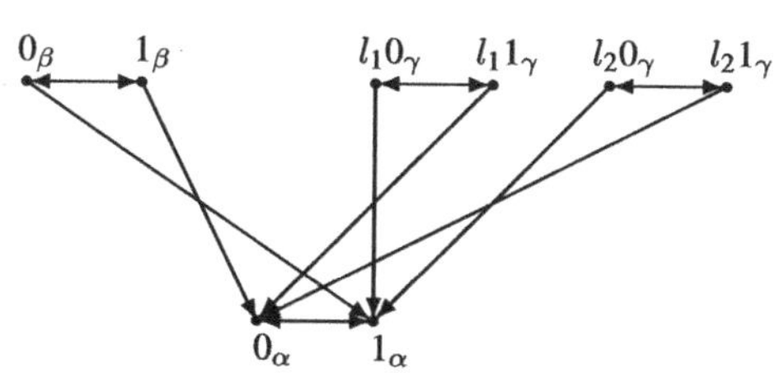

Diagram (b). Cay$(S, \{1_\beta, (l_2, 1_\gamma)\})$

In Diagram (b), we see that $\mathrm{Cay}(S, C)$ is $\mathrm{End}'(S, C)$-vertex transitive but not $\mathrm{Aut}(S, C)$-vertex transitive.

Example 12.4.12. Consider again the strong semilattice from Example 12.3.5, with one-element connection sets, which is $\mathrm{Aut}(S, C)$-vertex transitive and $\mathrm{End}'(S, C)$-vertex transitive for both chains of left groups contained but not overall.

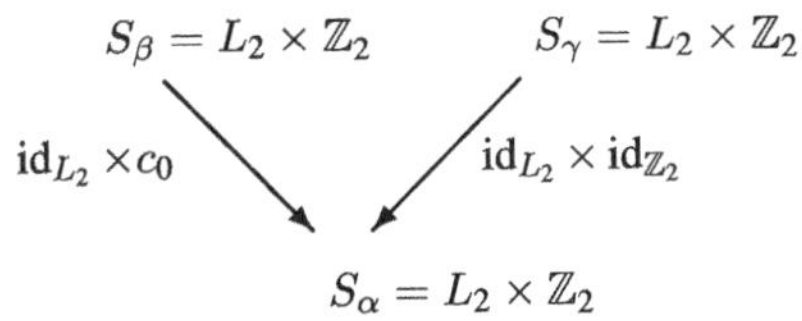

Diagram (a).

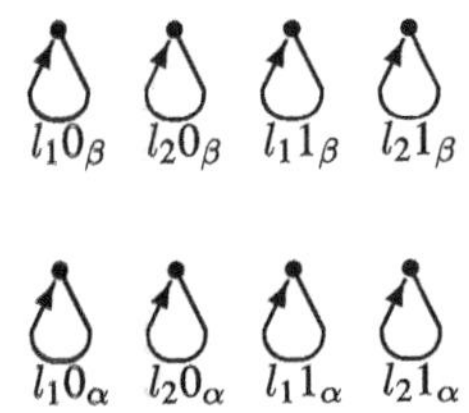

Diagram (c).
$\mathrm{Cay}(S_\alpha \bigcup S_\beta, \{(l_1, 0_\beta)\})$

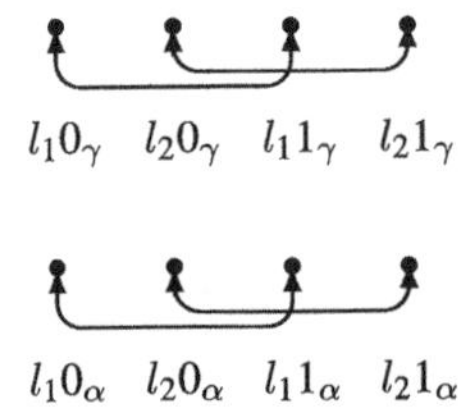

Diagram (b).
$\mathrm{Cay}(S_\alpha \bigcup S_\gamma, \{(l_1, 1_\gamma)\})$

12.5 Comments

Much can be done on End′-vertex transitivity. Note that End′ = End, if we do not have Aut-vertex transitivity. One can get started by considering examples and proceed by trial and error. Then one can develop hypotheses and go on to construct proofs. The problems arising, range in difficulty from exercises to thesis projects on various levels.

As was already mentioned in Section 11.4, it would be interesting to study other completely regular semigroups, cf. [Petrich/Reilly 1999]. And then we can consider Cayley graphs of strong semilattices of these, and so on.

Here we have already used the basic paper A. V. Kelarev and C. E. Preager [44]. There the authors give a characterization of transitive Cayley graphs of semigroups.

For $\mathrm{ColAut}(S, C)$-vertex transitive Cayley graphs of bands and of completely simple semigroups compare Z. Jiang [42].

For $\mathrm{Aut}(S, C)$-vertex transitive Cayley graphs of bands and for $\mathrm{ColAut}(S, C)$-vertex transitive Cayley graphs of rectangular bands compare S. Fan and Y. Zeng [19].

Completely 0-simple semigroups are Aut-vertex transitive only in the trivial cases, see Shoufeng Wang and Yinghui Li [88].

Which of them are End-vertex transitive?

13 Embeddings of Cayley graphs—genus of semigroups

It is known that each group can be defined in terms of generators and relations, and that corresponding to each such (nonunique) presentation there is a unique graph, called the Cayley graph of the presentation. A "drawing" of this graph gives a "picture" of the group from which certain properties of the group can be determined. The same principle can be used for other algebraic systems. So we will say that algebraic systems with a given system of generators are *planar* or *toroidal* if the respective Cayley graphs can be embedded in the plane or on the torus.

If $\mathrm{Cay}(S, C)$ is planar, for some generating set C of S, we call S a ***planar semigroup***. If a nonplanar graph can be embedded on the torus, i. e., on the orientable surface of genus 1, it is said to be ***toroidal***.

It is clear that when considering embeddings in surfaces, directions, colors, and multiplicities of the edges and loops in the Cayley graph are not important. This means we can consider the Cay functor as going from the category ***SgC*** to the category ***Gra***; more formally, we apply a suitable forgetful functor after Cay. In this chapter, we investigate the genus of semigroups, and we concentrate on genus 0 and 1. An investigation of arbitrary semigroups seems hopeless, since their number is growing rapidly with the number of elements (Theorem 9.1.7). As there are only few types of planar groups, we focus on semigroups which are close to groups. The biggest class of these are so-called completely regular semigroups. They are unions of groups. So one target could be a description of planar completely regular semigroups with the help of planar groups. But we are still quite far from this. What we do here is to discuss the question for very special unions of groups, namely right groups and certain Clifford semigroups.

13.1 The genus of a group

For convenience, we first repeat some standard group notation.

$(\mathbb{Z}_n, +)$ with $\mathbb{Z}_n = \{0, \ldots, n-1\}$ the **cyclic group** with n elements, addition modulo n. Note that $\mathbb{Z}_m \times \mathbb{Z}_n \cong \mathbb{Z}_{mn}$ if and only if $gcd(m, n) = 1$. In particular, $\mathbb{Z}_2 \times \mathbb{Z}_2 \ncong \mathbb{Z}_4$.

D_n the **dihedral group**. The elements of D_n are the symmetries of the n-gon with the vertices $1, \ldots, n$, that is $|D_n| = 2n$. Note that $\mathbb{Z}_2 \times D_n \cong D_{2n}$, if n is odd. If n is even, $\mathbb{Z}_2 \times D_n \ncong D_{2n}$, since the numbers of elements of order 2 are different. For example, all elements of $\mathbb{Z}_2 \times D_2 \cong \mathbb{Z}_2 \times \mathbb{Z}_2 \times \mathbb{Z}_2$ have order 2; in D_4, there are 2 elements which do not have order 2. The rotation by 1 is denoted by $(1 \ldots n)$. For elements of order 2 like, we will mostly use just a, b, or c.

https://doi.org/10.1515/9783110617368-013

A_n, S_n the **alternating group** and the **symmetric group**, respectively, on the n points $1, \dots, n$. Here, we deal only with $n \leq 5$. For their elements, we use the cycle notation.

The identity element is denoted by e or e_A for all groups A except for $\mathbb{Z}_n$, where we rather use 0.

Definition 13.1.1. For a finite group A, the ***genus*** of A is defined to be the minimum of all genera of Cayley graphs $\mathrm{Cay}(A, C)$ with generating sets $C \subseteq A$. We will call such C a ***genus-minimal generating set***.

In analogy with Definition 13.1.1, we formulate the following.

Definition 13.1.2. The ***genus of a finite semigroup*** S is defined to be the genus of a Cayley graph of S with a genus-minimal generating set. We write $\gamma(S)$.

Clearly, when considering genus we may ignore edge-directions, multiple edges, and loops. We call the resulting simple undirected graph the *underlying graph* of $\mathrm{Cay}(S, C)$ and denote it by $\underline{\mathrm{Cay}}(S, C)$.

Example 13.1.3. Here, we give several plane representations of planar groups, Figure 13.1. Many more follow in Figures 13.2, pages 263, 13.3, pages 263, 13.4, pages 264, 13.5, pages 264, 13.6, pages 265, 13.7, pages 265, 13.8, pages 266. Figure 13.9, page 266, represents a toroidal group.

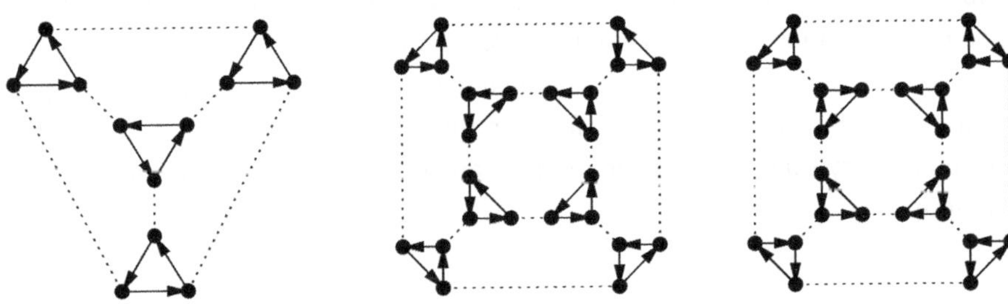

Figure 13.1: From left to right: $\mathrm{Cay}(A_4, \{(12)(34), (123)\})$ (solid: (123), dotted: $(12)(34)$), truncated tetrahedron underlying, $\mathrm{Cay}(S_4, \{(123), (34)\})$ (solid: (123), dotted: (34)), truncated cube underlying, and $\mathrm{Cay}(\mathbb{Z}_2 \times A_4, \{(0, (123)), (1, (12)(34))\})$ (solid: $(0, (123))$, dotted: $(1, (12)(34))$), truncated cube underlying. Note that $\mathrm{Cay}(S_4, \{(123), (34)\})$ and $\mathrm{Cay}(\mathbb{Z}_2 \times A_4, \{(0, (123)), (1, (12)(34))\})$ are not isomorphic but $\underline{\mathrm{Cay}}(S_4, \{(123), (34)\})$ and $\underline{\mathrm{Cay}}(\mathbb{Z}_2 \times A_4, \{(0, (123)), (1, (12)(34))\})$ are isomorphic.

Remark 13.1.4. Definition 13.1.1 says that finding the genus of a group A amounts to finding a generating set C for A such that the genus of $\mathrm{Cay}(A, C)$ is minimal. Note that this does not mean that the generating set is minimal in the number of elements: A planar representation of $\mathbb{Z}_2 \times A_5$ can be obtained with three generators (compare Figure 13.8, p. 266), but not with two generators, although $\mathbb{Z}_2 \times A_5$ can be generated

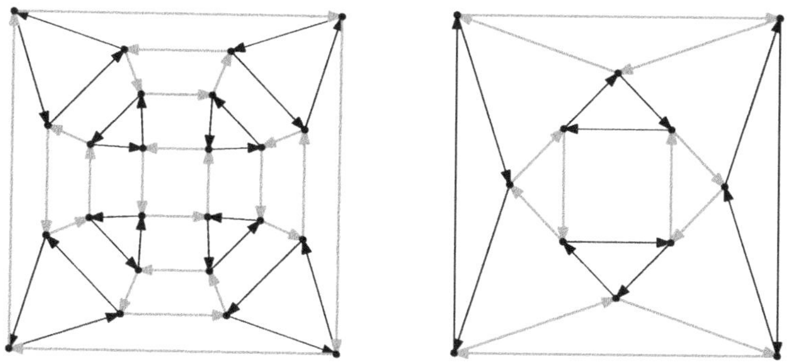

Figure 13.2: Plane Cayley graphs Cay(S_4, {(1234), (123)}), rhombicuboctahedron underlying and Cay(A_4, {(123)(234)}), cuboctahedron underlying.

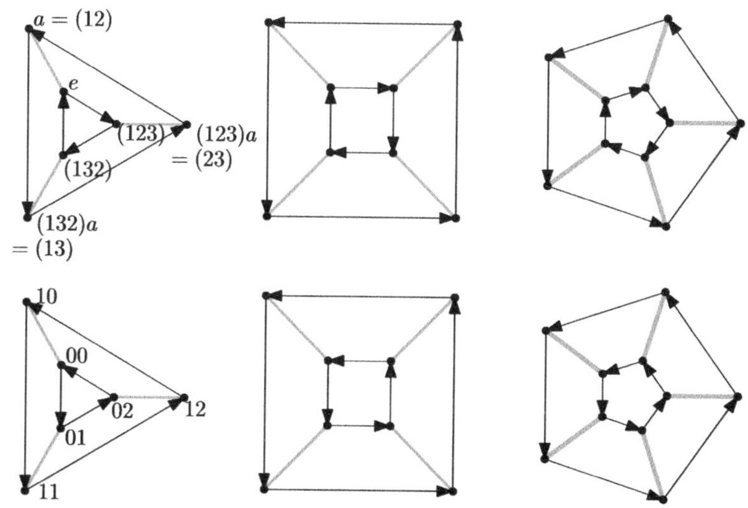

Figure 13.3: The plane Cayley graphs of Cay(D_n, {a, (1, ..., n)}) with $a^2 = e$ and Cay($\mathbb{Z}_2 \times \mathbb{Z}_n$, {(1, 0), (0, 1)}), $n = 3, 4, 5$. The underlying graphs are prisms.

by two elements; cf. [Gross/Tucker 1987]. Neither does it mean that the generating set (and thus the Cayley graph) is unique.

So it is clear that the form of the geometric presentation depends on the set of generators C chosen for the Cayley graph. Take the four-dimensional cube Q_4. This gives the Cayley graph Cay($(\mathbb{Z}_2)^4$, {(1, 0, 0, 0), (0, 1, 0, 0), (0, 0, 1, 0), (0, 0, 0, 1)}), and this is nonplanar. Apparently, it has genus 1; see Figure 13.9, page 266.

Theorem 13.1.5 ([59]). *The finite group A is planar if and only if $A = B_1 \times B_2$ with $B_1 \in \{\mathbb{Z}_1, \mathbb{Z}_2\}$ and $B_2 \in \{\mathbb{Z}_n, D_n, S_4, A_4, A_5 \mid n \in \mathbb{N}\}$.*

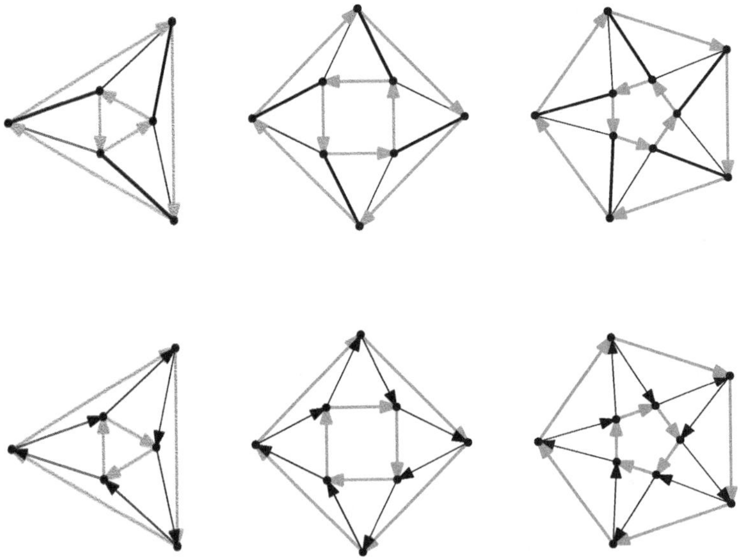

Figure 13.4: The plane Cayley graphs of Cay(D_n, $\{a, b, (1, \dots, n)\}$) with $a^2 = b^2 = e$, first line, and Cay($\mathbb{Z}_{2n}$, $\{1, 2\}$), second line, $n = 3, 4, 5$. The underlying graphs are antiprisms.

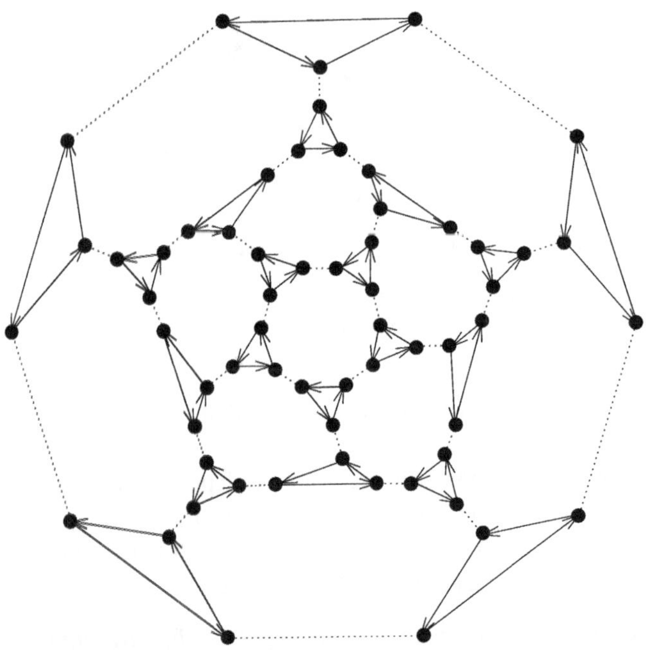

Figure 13.5: The plane Cayley graph Cay(A_5, $\{(23)(45), (124)\}$) (solid: (124), dotted: (23)(45)), truncated dodecahedron underlying.

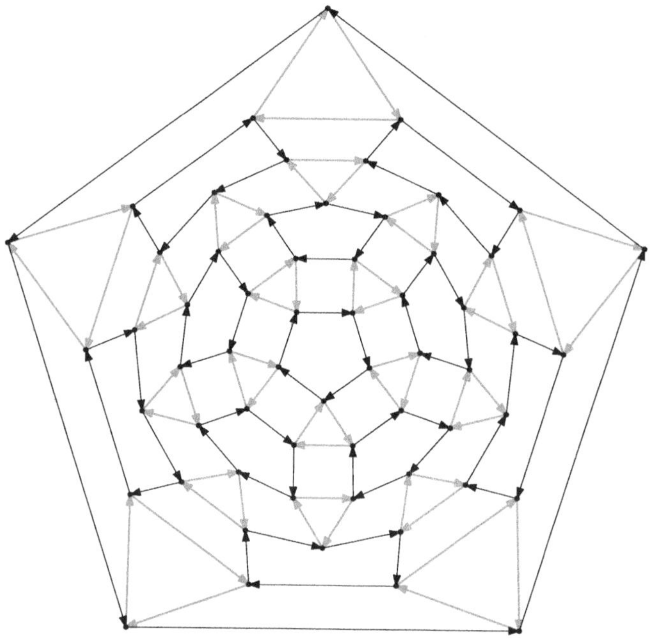

Figure 13.6: The plane Cayley graph $\mathrm{Cay}(A_5, \{(12345), (124)\})$, rhombicosidodecahedron underlying.

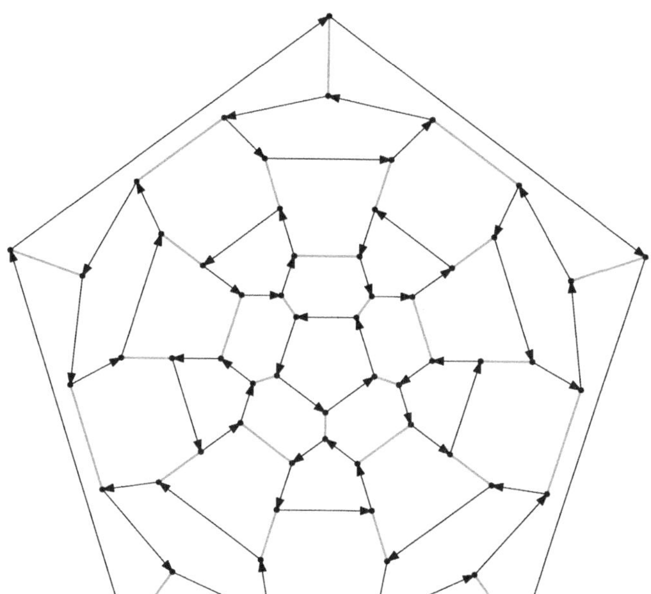

Figure 13.7: The plane Cayley graph $\mathrm{Cay}(A_5, \{(12345), (23)(45)\})$, truncated icosahedron underlying.

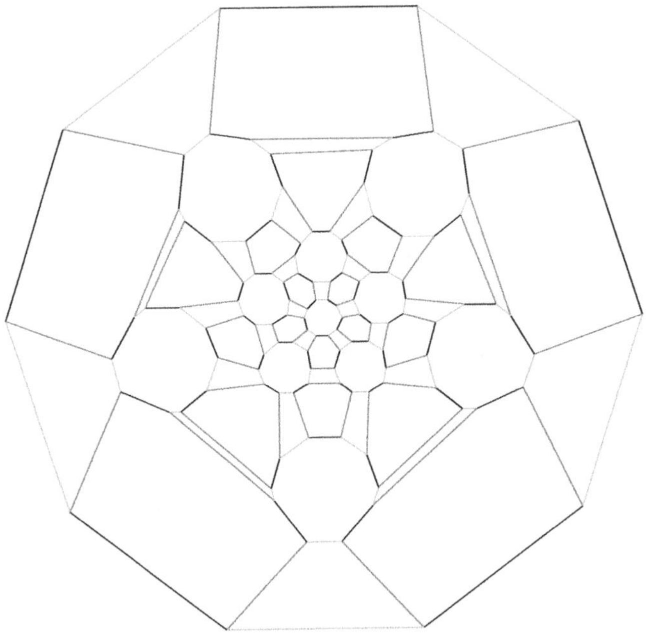

Figure 13.8: The plane Cayley graph $\mathrm{Cay}(Z_2 \times A_5, \{(1,(12)(35)),(1,(24)(35)),(1,(23)(45))\})$, (rhombi)truncated icosidodecahedron underlying.

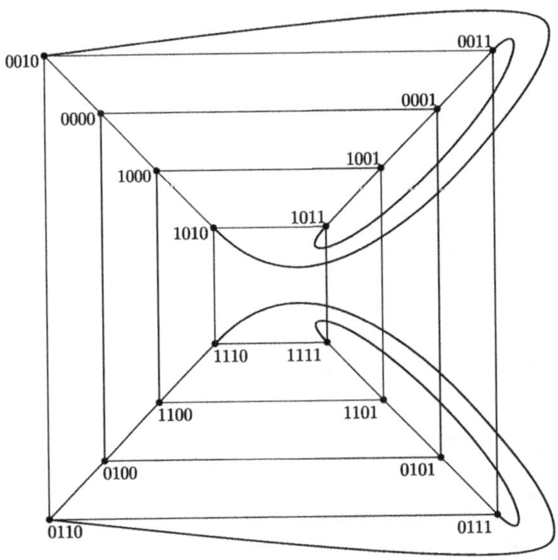

Figure 13.9: A toroidal embedding of Q_4 the Cayley graph $\mathrm{Cay}((\mathbb{Z}_2)^4, \{(1,0,0,0),(0,1,0,0),(0,0,1,0),(0,0,0,1)\})$.

Table 13.1: The planar groups, their planar generating systems, where $a, b \in D_n$ are of order 2, and the three-dimensional solids these graphs are the graphs of.

Group	Generators	$\|V\|$	$\|E\|$	Geometric realization, Figure
$\mathbb{Z}_n$	1	n	n	$\vec{C}_n$
$\mathbb{Z}_4$	$1, 2$	4	6	(tetrahedron underlying)
$\mathbb{Z}_{2n}$	$1, 2$	$2n$	$4n$	$\vec{C}_n$-antiprism, 13.4, p. 264
$\mathbb{Z}_2 \times \mathbb{Z}_2 \cong D_2$	$(1,0), (0,1)$	4	4	2-prism = square
$\mathbb{Z}_2 \times \mathbb{Z}_4 (\not\cong \mathbb{Z}_8)$	$(1,0), (0,1)$	8	12	$\vec{C}_4$-prism (cube underlying), 13.3, p. 263
$\mathbb{Z}_2 \times \mathbb{Z}_n$	$(1,0), (0,1)$	$2n$	$3n$	$\vec{C}_n$-prism, 13.3, p. 263
$\mathbb{Z}_2 \times \mathbb{Z}_2$	$(1,0), (0,1), (1,1)$	4	6	tetrahedron
$D_3 (\cong S_3)$	$(123), (12)$	6	9	$(\overleftarrow{C}_3\text{-}\vec{C}_3)$-prism, 13.3 p. 263
	$(123), (12), (23)$	6	12	(octahedron underlying)
D_4	$(1234), a$	8	12	$(\overleftarrow{C}_4\text{-}\vec{C}_4)$-prism, 13.3, p. 263
D_n	a, b	$2n$	$2n$	$2n$-gon
	$(1 \ldots n), a$	$2n$	$3n$	$(\overleftarrow{C}_n\text{-}\vec{C}_n)$-prism. 13.3, p. 263
$\mathbb{Z}_2 \times D_n$	$(1,e), (0,a), (0,b)$	$4n$	$6n$	$2n$-prism, 13.3, p. 263
D_{2n}	$a, b, w; (aw)^2 = e$	$4n$	$6n$	$2n$-prism, 13.3, p. 263
				(neglecting directions)
A_4	$(123), (12)(34)$	12	18	truncated tetrahedron, 13.1, p. 262
	$(123), (234)$		24	cuboctahedron, 13.2, p. 263
	$(123), (234), (13)(24)$		30	icosahedron
				(neglecting directions)
$\mathbb{Z}_2 \times A_4$	$(0, (123)), (1, (12)(34))$	24	36	truncated cube, 13.1, p. 262
				(neglecting directions)
S_4	$(123), (34)$	24	36	truncated cube, 13.1, p. 262
	$(12), (23), (34)$		36	truncated octahedron
	$(12), (1234)$		36	truncated octahedron
	$(123), (1234)$		48	rhombicuboctahedron, 13.2, p. 263
	$(1234), (123), (34)$		60	snub cuboctahedron
$\mathbb{Z}_2 \times S_4$	$(1, (12)), (0, (23)), (0, (34))$	48	72	(rhombi)truncated cuboctahedron
				(neglecting directions)
A_5	$(124), (23)(45)$	60	90	truncated dodecahedron, 13.5
	$(12345), (23)(45)$		90	truncated icosahedron, 13.7, p. 265
	$(12345), (124)$		120	rhombicosidodecahedron, 13.6, p. 265
	$(12345), (124), (23)(45)$		150	snub icosidodecahedron
$\mathbb{Z}_2 \times A_5$	$(1, (12)(35)),$ $(1, (24)(35)), (1, (23)(45))$	120	180	(rhombi)truncated icosidodecahedron, 13.8, p. 266

The theorem is in: H. Maschke [59]; see also [Halin 1980], [Gross/Tucker 1987] and [White 2001]. In Table 13.1, we make the results of Maschke's theorem more precise by giving the planar generators and their corresponding three-dimensional solids. It turns out that as underlying graphs we get ***Archimedean graphs***, i. e., graphs of ***Archimedean solids***.

Some of these results can be found in the following references:
- https://en.wikipedia.org/wiki/Archimedean_graph
- http://garsia.math.yorku.ca/~zabrocki/posets/phedron4/per4outlinec.jpg.
- http://www.antiquark.com/math/permutahedron_4.gif (figure of S_4 generated by the transpositions $(12), (23), (34)$).
- http://www.jaapsch.net/puzzles/cayley.htm (figures of solids for all planar groups with all possible generator sets except for $\mathbb{Z}_2 \times A$).
- T. Roman [Roman 1978].
- [Grossmann/Magnus 1964].

Remark 13.1.6. We see that the only outer planar groups are $\mathbb{Z}_2$ and $\mathbb{Z}_1$, whose Cayley graphs are K_2 and K_1, and $\mathbb{Z}_n^{\{1\}}$ and $D_n^{\{a,b\}}$, where a, b are of order 2. Their Cayley graphs are cycles: $\vec{C}_n$ and C_{2n}, respectively.

Various results and questions about genera

Remark 13.1.7. There are many results on the genus of groups. We cite some, mostly from [White 2001], which seem specially interesting and/or surprising.

The finite Abelian planar groups are exactly $\mathbb{Z}_n$, $\mathbb{Z}_2 \times \mathbb{Z}_{2n}$, $(\mathbb{Z}_2)^3$.

The group Q of quaternions is the smallest group with genus 1.

$(\mathbb{Z}_2)^4$ has genus 1 (see Figure 13.9, p. 266), $(\mathbb{Z}_2)^5$ has genus 5, $(\mathbb{Z}_3)^3$ has genus 7, S_5 has genus 4.

There is exactly one group of genus 2: the automorphism group of the generalized Petersen graph $G(8, 3)$—also know as Möbius-Kantor graph, see Figure 1.2, page 7, which has 96 elements, with generators and relations as follows $[a, b, c; a^2 = b^2 = c^2 = (ab)^2 = (bc)^3 = (ac)^8 = b(ac)^4b(ac)^4 = e]$.

The smallest groups (by order) with unknown genus are non-Abelian with 32 elements.

There are at most finitely many groups of genus g, if $g > 1$.

Viera K. Proulx gives a characterization of groups of genus 1: [74].

There exists a characterization of graphs of nonorientable genus 1, which could be called ***Möbius graphs*** or ***projective graphs***, by forbidden subdivided subgraphs, cf. Theorem 11–31 in [White 2001]. The list of these graphs contains 103 graphs.

How could this be used for the investigation of Groups of nonorientable genus 1?

There is a question "in the other direction": The graphs of all Archimedean and Platonic solids are Cayley graphs of a group with minimal generating system with three exceptions: the dodecahedron, the icosidodecahedron, and the antiprisms (in particular the tetrahedron). The antiprism is the Cayley graph of a group with non-minimal generating system though, e. g., $\mathrm{Cay}(\mathbb{Z}_{2n}, \{1, 2\})$. The other two are not even this. Are they underlying graphs of directed Cayley graphs of semigroups (with minimal generating system)? It has been shown in Yifei Hao, Xing Gao, and Yanfeng Luo [30] that the dodecahedron graph is an induced subgraph of a Brandt semigroup Cayley graph.

Remark 13.1.8. The scope of the questions can be extended to include infinite groups, see, e. g., [Geogakopoulos 2017]. There a complete description of the planar cubic Cayley graphs is given.

In [White 2001], Chapter 14, there is an interesting discussion of the genus of field graphs, along with many questions. Since finite fields $GF(p^r)$ have the additive group $\mathbb{Z}_{p^r}$ for a prime p, they are considered to have one additive and one multiplicative generator. This suggests a definition of their genus. We quote some of the results here:

The finite field $GF(p^r)$ is planar if and only if $p^r = 2, 3, 4, 5, 7, 11$.
The finite field $GF(p^r)$ is toroidal if and only if $p^r = 8, 9, 13, 17$.
The first field with unknown genus has 16 elements.
Which are the planar or toroidal rings $\mathbb{Z}_n$, starting with $n = 6, 10, 12, 14, 15, 16$?

13.2 On the genus of right groups

Recall that a *right zero semigroup* on k elements is the semigroup R_k on the set $\{r_1, \ldots, r_k\}$ such that $r_i r_j := r_j$ for all $i, j \in \{1, \ldots, k\}$. A semigroup S is called a *right group* if it is isomorphic to the product $A \times R_k$ of a group and a right zero semigroup.

Lemma 13.2.1. *For a right group $A \times R_k$, we have $\gamma(A) \leq \gamma(A \times R_k)$.*

Proof. By Lemma 11.3.15, we have that for any generating set C the digraph $\mathrm{Cay}(A \times R_k, C)$ is strongly connected. Thus, by Proposition 11.3.14, we get that there is a generating system C' of A such that $\mathrm{Cay}(A, C')$ is a contraction of $\mathrm{Cay}(A \times R_k, C)$. The claim follows since contraction cannot increase the genus of a graph; cf. Lemma 1.8.1. □

The result in Lemma 13.2.1 justifies the following approach. If we are interested in planar right groups, we need to consider only planar groups. We will see later that such an approach is also valid for other classes of completely regular semigroups. Or, if we are interested in toroidal right groups, we need only consider planar or toroidal groups. We also know that planarity of $\mathrm{Cay}(A \times R_r, C)$, with $C \subseteq A \times R_r$, implies planarity of $\mathrm{Cay}(A, p_1(C))$, where p_1 is the first projection; see Theorem 13.2.7. However, an analogous statement for higher genus remains open.

Figure 13.10, page 270, shows (plane) Cayley graphs of four right groups.

Remark 13.2.2. As before, analogously one considers a *left zero semigroup* $L_k = \{l_1, \ldots, l_k\}$ on k elements such that $l_i l_j := l_i$ for all $i, j \in \{1, \ldots, k\}$. Now consider the *left group* $L_k \times A$ whose generating systems always have the form $L_k \times C$ where C is a generating system of A. The right Cayley graph $\underline{\mathrm{Cay}}(L_k \times A, L_k \times C)$ of a left group $L_k \times A$

consists of k copies of $\underline{\mathrm{Cay}}(A, C)$. Consequently, a left group $L_k \times A$ is planar if and only if the group A is planar, for arbitrary $k \in \mathbb{N}$.

Planar right groups

On the way to characterizing planar right groups, we start with positive results. We mainly use: Kolja Knauer and Ulrich Knauer [49].

We start with a result which generalizes the examples of Figure 13.10.

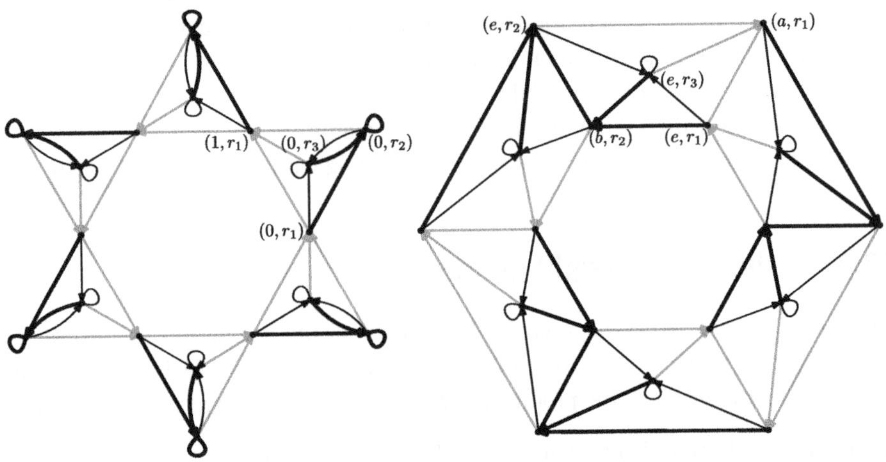

Figure 13.10: Plane Cayley graphs $\mathrm{Cay}(\mathbb{Z}_6 \times R_3, \{(1, r_1), (0, r_2), (0, r_3)\})$ and $\mathrm{Cay}(D_3 \times R_3, \{(a, r_1), (b, r_2), (e, r_3)\})$. The graphs $\mathrm{Cay}(\mathbb{Z}_6 \times R_2, \{(1, r_1), (0, r_2)\})$ and $\mathrm{Cay}(D_3 \times R_2, \{(a, r_1), (b, r_2)\})$ can be obtained by deleting the vertices inside the triangles and inside the quadrangles, respectively.

Lemma 13.2.3. *The right groups $\mathbb{Z}_n \times R_2, \mathbb{Z}_n \times R_3$ and $D_n \times R_2, D_n \times R_3$ are planar.*

Proof. We know that $\underline{\mathrm{Cay}}(\mathbb{Z}_n, \{1\})$ is a cycle. Now $C := \{(1, r_1), (0, r_2), (0, r_3)\}$ is a generating system of $\mathbb{Z}_n \times R_3$. A plane drawing of $\mathrm{Cay}(\mathbb{Z}_6 \times R_3, C)$ is on the left of Figure 13.10, p. 270.

For D_n with two order two generators a, b again $\underline{\mathrm{Cay}}(D_n, \{a, b\})$ is a cycle and $C := \{(a, r_1), (b, r_2), (e, r_3)\}$ is a generating system of $D_n \times R_3$. A plane drawing of $\mathrm{Cay}(D_3 \times R_3, C)$ is shown on the right of Figure 13.10.

Now, for $A \in \{\mathbb{Z}_n, D_n\}$ and $S' := A \times R_2$ note that $C' := C \setminus \{(0, r_3)\}$ and $C' := C \setminus \{(e, r_3)\}$, respectively, is a generating system of S' and $\mathrm{Cay}(S', C')$ is a subgraph of $\mathrm{Cay}(S, D)$. Thus, it is also planar. □

To extend this result to other groups, we recall Lemma 13.2.4. proceed to the following lemma.

Lemma 13.2.4. *Let A be a group with generating system $C = \{a, b\}$ where $a^2 = e$ and $b^2 \neq e$. Suppose that* Cay(A, C) *has an embedding on a surface M such that the semicycles containing a-edges are cycles, i. e., directed, see the left and the middle graph of Figure* 13.1, *page* 262, *in Example* 13.1.3. *Then $\gamma(A \times R_2), \gamma(A \times R_3) \leq \gamma(M)$.*

Proof. Note that the condition is fulfilled in the left and the middle graph but not in the right graph of Figure 13.1, page 262, in Example 13.1.3. We first show the statement for $S := A \times R_3$. As a generating system for S, we use $D := \{(a, r_1), (b, r_2), (e, r_3)\}$. The property of our embedding of Cay(A, C) on M makes it possible to blow up a-edges to rectangles such that incoming b-arcs are attached to one pair of opposite vertices and outgoing b-arcs to an entire side of the rectangle each; see the middle of Figure 13.11. This blown up graph can be completed to the wanted embedding of Cay(S, D). This is shown on the right of Figure 13.11.

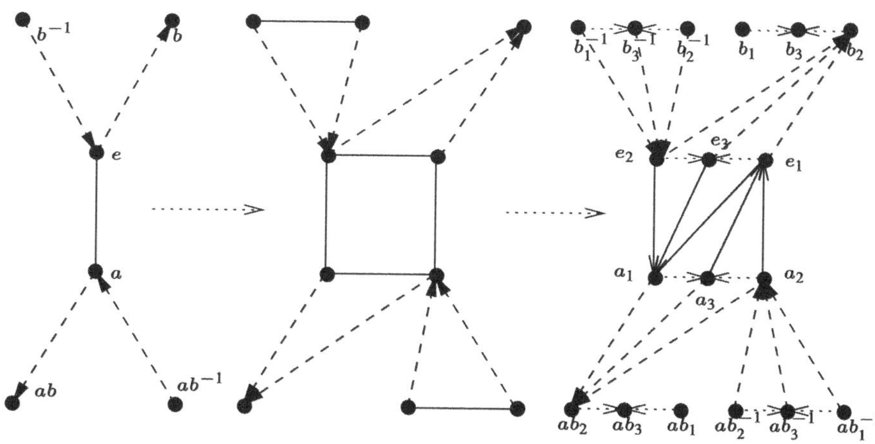

Figure 13.11: Left: Local configuration of b-arcs (dashed) around an a-edge (solid). Middle: Topologically relevant part of the transformation. Right: Completion to the graph of the right group. Arcs corresponding to elements in $A \times \{r_1\}$ are solid, in $A \times \{r_2\}$ dashed, and in $A \times \{r_3\}$ dotted. We write e_1 for (e, r_1), b_2^{-1} for (b^{-1}, r_2), etc.

For $S' := A \times R_2$ note that $D' := D \setminus \{(e, r_3)\}$ is a generating system of S' and Cay(S', D') is a subgraph of Cay(S, D). Thus, Cay(S', D') embeds in M. An example for this construction in the case $S' := A_4 \times R_2$ is depicted in Figure 13.12, page 272.

However, similarly one can see that choosing $D'' = \{(e, r_1), (a, r_2), (b, r_2)\}$, also yields a graph Cay(S', D'') which can be embedded into M. This constructions is exemplified in Figure 13.13, page 273, for $S' := A_4 \times R_2$. □

Theorem 13.2.5. *If $A \in \{\{e\}, \mathbb{Z}_n, D_n, S_4, A_4, A_5\}$, then $A \times R_k$ and $\{e\} \times R_{k'}$ are planar for $k \leq 3$ and $k' \leq 4$.*

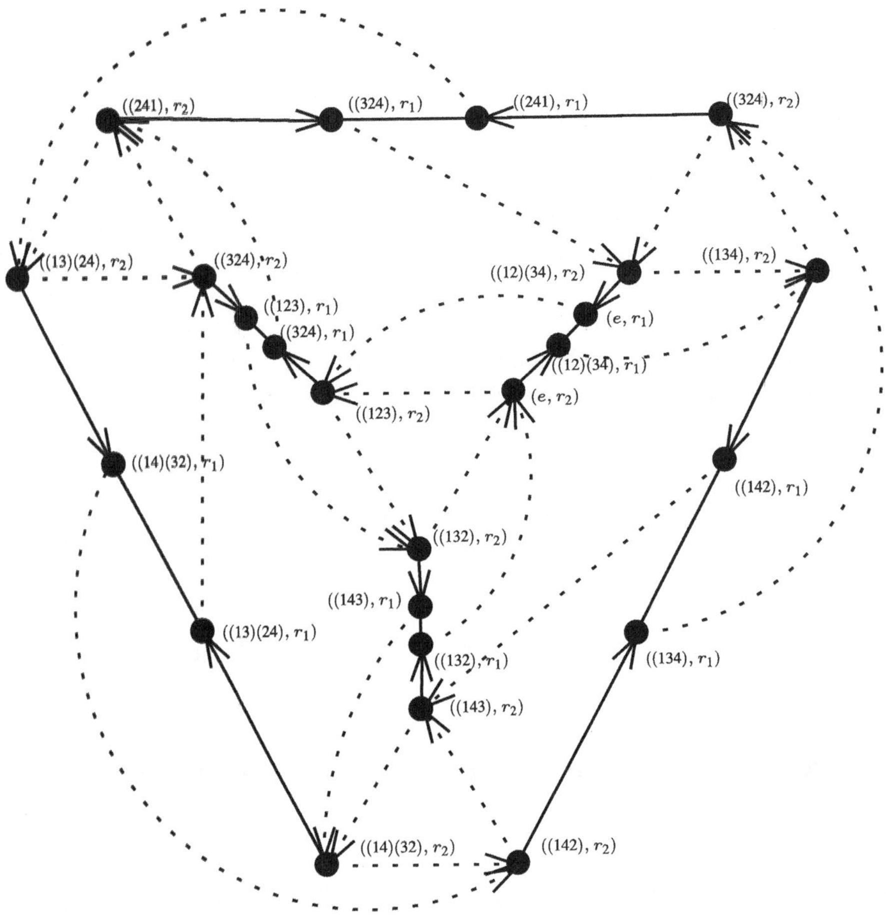

Figure 13.12: The plane Cayley graph $\mathrm{Cay}(A_4 \times R_2, \{((12)(34), r_1), ((123), r_2))\})$ (dotted: $((123), r_2)$, solid: $((12)(34), r_1)$). This exemplifies the principal construction in Lemma 13.2.4 and generalizes to $A_4 \times R_3$.

Proof. For $\mathbb{Z}_n$ and D_n, this was proved in Lemma 13.2.3. For the remaining groups, this follows by their plane drawings provided in Figure 13.1, page 262, and Figure 13.5, page 264, and application of Lemma 13.2.4. Moreover, $\underline{\mathrm{Cay}}(\{e\} \times R_{k'}, R_{k'})$ is isomorphic to the complete graph on k' vertices $K_{k'}$. Together this yields the claim. □

Planar right groups come from planar groups

Lemma 13.2.1 yields that in our characterization we only need to consider planar groups as factors. Unfortunately, Lemma 13.2.1 does not preserve the generating system of the right group we start out from. Specializing to right groups, we can obtain a stronger statement under certain circumstances. For this, we introduce the notation $p_1(C)_j := \{g \in A \mid (g, r_j) \in C\}$.

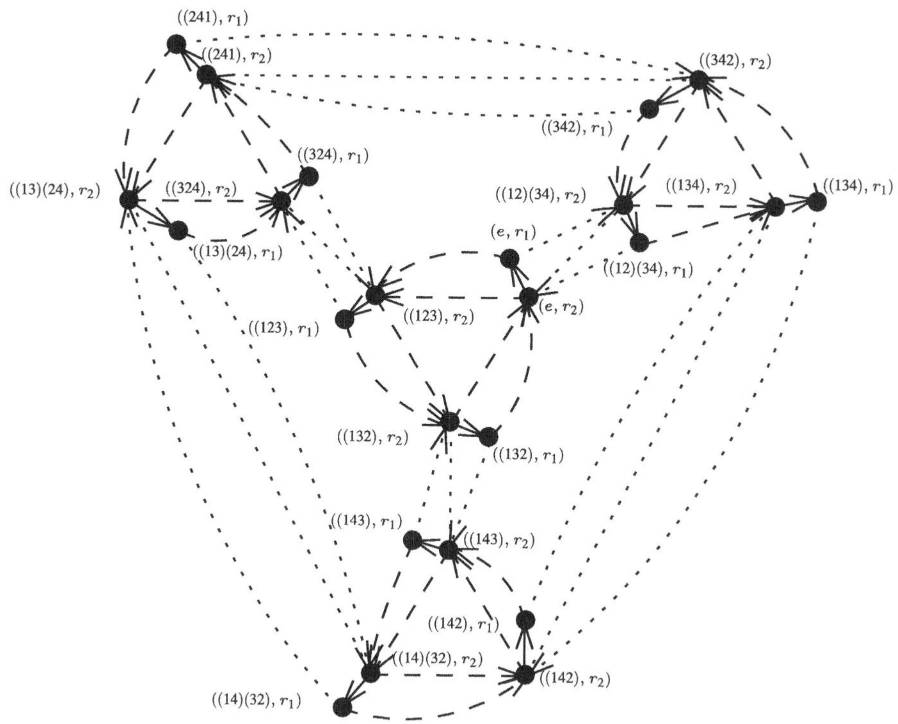

Figure 13.13: The graph $\mathrm{Cay}(A_4 \times R_2, \{(e, r_1), ((12)(34), r_2), (123), r_2))\})$ (dotted: $((12)(34), r_2)$, solid: (e, r_1), dashed: $((123), r_2)$). This exemplifies the alternative construction for $A_4 \times R_2$ mentioned in the proof of Lemma 13.2.4.

We now need Theorem 1.8.4, and as a second ingredient we need a formula for the number of edges of the underlying undirected Cayley graph of the right group $A \times R_k$. For $C \subseteq A \times R_k$ and $a \in A$, we set $c_a := |\{j \in \{1, \dots, k\} \mid (a, r_j) \in C\}|$. Furthermore, set $m := |A|$.

Lemma 13.2.6. *Let $S = A \times R_k$ with generating system C. Then $\underline{\mathrm{Cay}}(S, C)$ has mk vertices and its number of edges is*

$$|E(\underline{\mathrm{Cay}}(S, C))| = m\Big(\Big(\sum_{a \in p_1(C)} c_a k - \frac{c_{a^{-1}}}{2}\Big) - \frac{c_e}{2}\Big) \geq \frac{m}{2}\Big((2k-1) \sum_{a \in p_1(C)} c_a - c_e\Big).$$

Proof. The number of vertices $|S| = |A \times R_k| = mk$; this is clear.

We start by proving the equality for the number of edges. Every element $(a, r_i) \in C$ contributes an outgoing arc at every element of S. But if $(a^{-1}, r_j) \in C$ all arcs of the form $((g, r_j), (ga, r_i))$ are counted twice and there are m of them. Note that this occurs in particular if $a^2 = e$ and also if $i = j$. So, this yields $mkc_a - m\frac{c_{a^{-1}}}{2}$ edges labeled a. In the particular case that $a = e$ additionally at each vertex (g, r_i), a loop can be deleted,

i. e., instead of counting half an edge at each such vertex we count none. This yields the $-m\frac{c_e}{2}$ in the formula.

Together we obtain the claimed equality $|E(\underline{\mathrm{Cay}}(S,C))| = m((\sum_{a\in p_1(C)} c_a k - \frac{c_{a^{-1}}}{2}) - \frac{c_e}{2})$.

Observe now that for fixed c_e the left-hand-side of the formula is minimized if $a^2 = e$ for every $a \in p_1(C)$, i. e., $c_a = c_{a^{-1}}$. This yields the lower bound. □

Theorem 13.2.7. *Let $S = A \times R_k$ and C a generating system such that $\underline{\mathrm{Cay}}(S,C)$ is planar. Then $\underline{\mathrm{Cay}}(A, p_1(C))$ is a minor of $\underline{\mathrm{Cay}}(S,C)$, i. e., in particular planar.*

Proof. The statement is trivial for $k = 1$ so assume $k \geq 2$.

We can now use Lemma 13.2.6 to estimate the number of edges of $\underline{\mathrm{Cay}}(S,C)$. Indeed, since $c_e = 0$ as $e \notin p_1(C)$, we get the first $\geq$, for the second $\geq$ we use that $|p_1(C)_j| > 1$ for all $j \in \{1,\dots,k\}$ and in particular $\sum_{a\in p_1(C)} c_a \geq 2k$ and that $k \geq 2$:

$$|E(\underline{\mathrm{Cay}}(S,C))| \geq \frac{m}{2}(2k-1) \sum_{a\in p_1(C)} c_a \geq (2k-1)mk > 3mk > 3mk - 6.$$

So $3mk - 6$ is a lower bound for $|E(\underline{\mathrm{Cay}}(S,C))|$ whereas it is the upper bound for the number of edges of a planar graph given by Theorem 1.8.5 – a contradiction. □

Nonplanar right groups from planar groups

In this subsection, we show that the right groups $\mathbb{Z}_2 \times H \times R_k$ with $H \in \{\mathbb{Z}_{2m}, D_{2n}, A_4, S_4, A_5 | n \geq 1, m \geq 2\}$ are not planar for $k \geq 2$. Moreover, $\{e\} \times R_k$ is not planar for $k \geq 5$. Since $\mathbb{Z}_2 \times \mathbb{Z}_2 \cong D_2$, $\mathbb{Z}_2 \times \mathbb{Z}_{2m-1} \cong \mathbb{Z}_{4m-2}$ and $\mathbb{Z}_2 \times D_{2n+1} \cong D_{4n+2}$ for all $m \geq 2$ and $n \geq 1$, this is exactly the set of right groups we have to prove to be nonplanar in order to show that the list from Theorem 13.2.5 is complete.

Euler's formula (Theorem 1.8.5) already allows to restrict the size of the right zero semigroup in a planar right group:

Proposition 13.2.8. *If A is nontrivial and $A \times R_k$ is planar, then $k \leq 3$. Moreover, $A \times R_k$ is nonplanar for any A and $k \geq 5$.*

Proof. Let $k \geq 4$ and $A \times R_k$ with A nontrivial, i. e., there is $a \in A$ such that $c_a := |\{j \in \{1,\dots,k\} \mid (a, r_j) \in C\}| > 0$. The lower bound in Lemma 13.2.6 is minimized if $k = 4$ and there is exactly one such $a \in A$ and $c_a = 1$. In this case, we get $m((3k - \frac{3}{2}) + (k - \frac{1}{2}) - \frac{3}{2}) = 12.5m > 12m - 6$ – a contradiction to Euler's formula (Theorem 1.8.5).

Since $\underline{\mathrm{Cay}}(\{e\} \times R_k, R_k) \cong K_k$ and K_k is nonplanar for all $k \geq 5$, we obtain the second part of the statement. □

With some further edge counting, we obtain the following

Proposition 13.2.9. *For $n \geq 1$ and $k = 2, 3$, the right groups $\mathbb{Z}_2 \times D_{2n} \times R_k$, $\mathbb{Z}_2 \times S_4 \times R_k$ and $\mathbb{Z}_2 \times A_5 \times R_k$ are not planar.*

Proof. Let $S = A \times R_k$ be one of the right groups from the statement and set $m = |A|$. Suppose that C is a generating system of S such that $\underline{\mathrm{Cay}}(S, C)$ is planar. By Theorem 13.2.7, we know that there is a generating system $C' \subseteq p_1(C)$ of A such that $\underline{\mathrm{Cay}}(A, C')$ is planar. Comparing with Table 13.1, page 267, we see that all planar generating systems for our choice of A consist of three generators all having order two, say a_1, a_2, a_3.

If (up to relabeling of R_k), we have $C'' := \{(a_1, r_1), (a_2, r_2), (a_3, r_3)\} \subseteq C$ (in particular $k = 3$), then we consider the subgraph $\underline{\mathrm{Cay}}(S, C'')$ of $\underline{\mathrm{Cay}}(S, C)$. By Lemma 13.2.6, we know that $\underline{\mathrm{Cay}}(S, C'')$ has $\geq 7.5m$ edges. Since $p_1(C'')$ contains only order 2 elements and $p_1(C'')$ is a minimal generating system of A we get that $\underline{\mathrm{Cay}}(S, C'')$ is triangle-free. Hence it has at most $2(mk - 2) = 6m - 4$ edges by Theorem 1.8.5—a contradiction in all cases.

If (up to relabeling of R_k), we have $C'' := \{(a_1, r_1), (a_2, r_2), (a_3, r_2)\} \subseteq C$ then we consider the subgraph $\underline{\mathrm{Cay}}(A \times R_2, C'')$ of $\underline{\mathrm{Cay}}(S, C)$. By Lemma 13.2.6, we know that $\underline{\mathrm{Cay}}(A \times R_2, C'')$ has $\geq 4.5m$ edge. As in the previous case, $\underline{\mathrm{Cay}}(A \times R_2, C'')$ is triangle-free and has most $2(mk - 2) = 4m - 4$ edges by Theorem 1.8.5—a contradiction in all cases.

If (up to relabeling of R_k) $C'' := \{(a_1, r_1), (a_2, r_1), (a_3, r_1), (x, r_2)\} \subseteq C$ again we consider the subgraph $\underline{\mathrm{Cay}}(A \times R_2, C'')$ of $\underline{\mathrm{Cay}}(S, C)$. Here we distinguish two subcases:

If $x \neq e$, then by the lower bound in Lemma 13.2.6 we know that $\underline{\mathrm{Cay}}(A \times R_2, C'')$ has at least $6m$ edges. On the other hand, Theorem 1.8.5 gives an upper bound of $6m - 6$ – a contradiction in all cases.

If $x = e$, then $\underline{\mathrm{Cay}}(A \times R_2, C'')$ has $5.5m$ edges and we have to come up with a stronger upper bound than Theorem 1.8.5 for this particular case. Note that in $\underline{\mathrm{Cay}}(A \times R_2, C'')$ every edge may appear in two triangles except for edges of the form $\{(g, r_2), (ga_i, r_1)\}$ for $i = 1, 2, 3$. The latter edges appear only in the triangle $\{(g, r_2), (ga_i, r_1), (g, r_1)\}$ and there are $3m$ of them. We therefore have that the number of triangular faces $|F_3|$ is bounded from above by $\frac{2|E|-3m}{3}$ and there are at least $\frac{3m}{4}$ larger faces. Plugging this into Euler's formula yields $|E| \leq \frac{21}{4}m - 6$, which is less than $5.5m$ – a contradiction in all cases. □

Now we turn to the remaining cases. Here, edge counting does not suffice to prove nonplanarity. Instead we will use Wagner's theorem (see Theorem 1.8.2), i. e., we will find K_5 and $K_{3,3}$ minors which prove nonplanarity. First, we prove a lemma somewhat complementary to Lemma 13.2.4.

Lemma 13.2.10. *Let A be a group with generating system $C' = \{a, b\}$ where a is of order two and b of larger order such that the neighborhood of some a-edge in* Cay(A, C') *looks as depicted in Figure 13.14, page 276. Then $\underline{\mathrm{Cay}}(A \times R_k, C)$ is non-planar if $C' \subseteq p_1(C)$ and $k \geq 2$.*

Proof. We distinguish two cases of what C looks like:

If (up to relabeling of R_k), we have $C'' = \{(a, r_1), (b, r_1)\} \subseteq C$ consider $\mathrm{Cay}(A \times R_2, C'')$. This graph contains $\mathrm{Cay}(A \times R_1, C'') \cong \mathrm{Cay}(A, C')$. Consider the a-edge from Figure 13.14, page 276 say that it connects vertices (e, r_1) (left) and (a, r_1) (right). Add

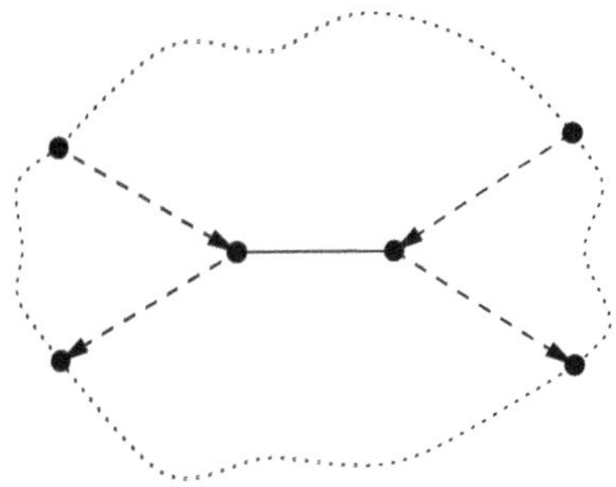

Figure 13.14: The solid edge is the a-edge, the dashed arcs are b-arcs, the dotted curves correspond to paths, vertex-disjoint from all other elements of the figure.

vertices (e, r_2) and (a, r_2) to the picture. The first has an arc to (a, r_1) and to the bottom-left vertex (b, r_1). The second has an arc to (e, r_1) and to the bottom-right vertex (ab, r_1). Contracting these two 2-paths to a single edge each, as well as the left, bottom and right dotted path to single edges and the top dotted path to a single vertex we obtain K_5,; see Figure 13.15.

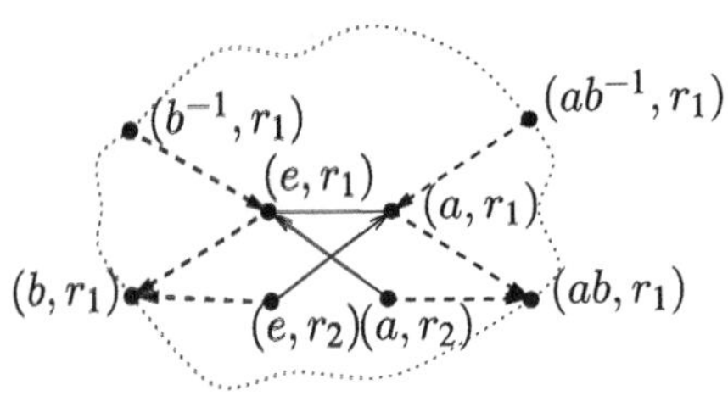

Figure 13.15: Finding K_5.

If (up to relabeling of R_k), we have $C'' = \{(a, r_1), (b, r_2)\} \subseteq C$ again consider Cay($A \times R_2, C''$). Denote by $H \subseteq A$ the elements corresponding to the vertices of Figure 13.14, without the two vertical dotted paths. Take the b-arcs in Cay($H \times \{r_2\}, C''$) and all a-edges Cay($H \times R_2, C''$). As in the first case, we assume without loss of generality that the central a-edge connects from left to right elements e and a in Cay(A, C''). This edge will now be represented by a 3-path (e, r_2), (a, r_1), (e, r_1), (a, r_2).

Also in the dotted paths, we need to replace a-edges by some new paths. We do this in the same way as with the central edge. The paths resulting this way from the dotted paths will again be pairwise disjoint and we draw them dotted in Figure 13.16, page 277.

To obtain a $K_{3,3}$-minor focus on the 3-path (e, r_2), (a, r_1), (e, r_1), (a, r_2) representing the central a-edge in our argument. We include the b-arcs $((e, r_1), (b, r_2))$ and $((a, r_1), (ab, r_2))$ starting from the inner vertices of this 3-path.

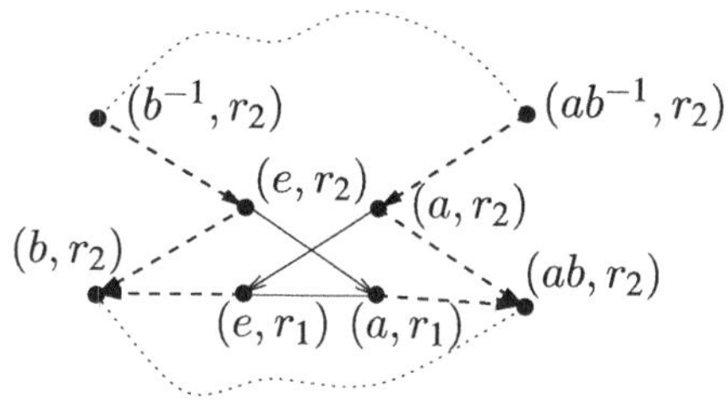

Figure 13.16: Finding $K_{3,3}$.

Contract the dotted path between (b, r_2) and (ab, r_2) and the one between (b^{-1}, r_2) and (ab^{-1}, r_2) to a single edge, respectively. Last, we contract the two arcs $((b^{-1}, r_2), (e, r_2))$ and $((ab^{-1}, r_2), (a, r_2))$. The resulting graph is $K_{3,3}$. □

The lemma yields the following

Proposition 13.2.11. *The right groups $\mathbb{Z}_2 \times \mathbb{Z}_{2n} \times R_k$ and $\mathbb{Z}_2 \times A_4 \times R_k$ with $k, n \geq 2$ are not planar.*

Proof. Let $S = A \times R_k$ be one of the right groups from the statement and suppose that C is a generating system of S such that $\underline{\mathrm{Cay}}(S, C)$ is planar. By Theorem 13.2.7, we know that there is a generating system $C' \subseteq p_1(C)$ of A such that $\underline{\mathrm{Cay}}(A, C')$ is planar. Comparing with Table 13.1, page 267, we see that for $A \in \{\mathbb{Z}_2 \times \mathbb{Z}_{2n}, \mathbb{Z}_2 \times A_4\}$ there is exactly one planar generating system. The preconditions of Lemma 13.2.10 are satisfied for $A = Z_2 \times \mathbb{Z}_{2n}$, which is easy to check directly; see Figure 13.3, page 263, and for $A = \mathbb{Z}_2 \times A_4$ we refer to Figure 13.1, page 262. Thus, the respective Cayley graphs cannot be planar. □

Now we have proved the following.

Theorem 13.2.12. *The right groups $\mathbb{Z}_2 \times H \times R_k$ with $H \in \{\mathbb{Z}_{2m}, D_{2n}, A_4, S_4, A_5\}$, $m \geq 2$, $n \geq 1$, and $k \geq 2$ are not planar.*

Raising the genus

Example 13.2.13. Here, we show directly that $\mathrm{Cay}(\mathbb{Z}_2 \times \mathbb{Z}_4 \times R_2, \{(1, 0, r_1), (0, 1, r_1), (0, 0, r_2)\})$ contains $K_{3,3}$. In Figure 13.17 below, page 278, we start from the Cayley graph $\mathrm{Cay}(\mathbb{Z}_2 \times \mathbb{Z}_4, \{(1, 0), (0, 1)\})$, which is $K_2 \square \vec{C_4}$. This is the graph without the two points $(0, 0, r_2), (1, 0, r_2)$ and the incident arcs. The first step on the way to $\mathrm{Cay}(\mathbb{Z}_2 \times \mathbb{Z}_4 \times R_2, \{(1, 0, r_1), (0, 1, r_1), (0, 0, r_2)\})$ is as follows. We start with the edge $\{(0, 0, r_1), (1, 0, r_1)\}$. Then we insert the arcs

$$((0, 0, r_1), (0, 0, r_2)),\ ((0, 0, r_2), (1, 0, r_1)),\ ((0, 0, r_2), (0, 1, r_1))$$
$$\text{and}\ ((1, 0, r_1), (1, 0, r_2)),\ ((1, 0, r_2), (0, 0, r_1)),\ ((1, 0, r_2), (1, 1, r_1)).$$

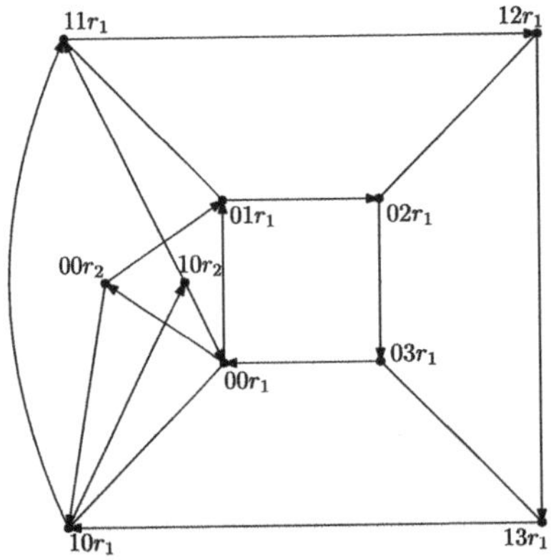

Figure 13.17: On the way to $\mathrm{Cay}(\mathbb{Z}_2 \times \mathbb{Z}_4 \times R_2, \{(1,0,r_1),(0,1,r_1),(0,0,r_2)\})$.

Since in this graph the inner $\overrightarrow{C_4}$ and the outer $\overrightarrow{C_4}$ are directed in the same way, i. e., clockwise, it follows that the 4-semicycles formed with the help of the generator $(1,0) \in \mathbb{Z}_2 \times \mathbb{Z}_4$, which is of order 2, are not directed.

To obtain the Cayley graph $\mathrm{Cay}(\mathbb{Z}_2 \times \mathbb{Z}_4 \times R_2, \{(1,0,r_1),(0,1,r_1),(0,0,r_2)\})$, this construction has to be applied to each of the four edges corresponding to the generator $(1,0,r_1)$ of order 2, surrounding the inner $\overrightarrow{C_4}$. It is clear that already after the first step the graph contains $K_{3,3}$ with the partition $\{(0,0,r_1),(1,0,r_1),(0,1,r_1)\}$, $\{(0,0,r_2),(1,0,r_2),(0,2,r_1)\}$, for example. This procedure applies to all groups $\mathbb{Z}_2 \times \mathbb{Z}_{2n}$ and, similarly, also to the group $\mathbb{Z}_2 \times A_4$. In this case, instead of two $\overrightarrow{C_{2n}}$ we have several $\overrightarrow{C_3}$; see the right graph in Example 13.1.3.

Question. What is the genus of $\mathbb{Z}_2 \times \mathbb{Z}_{2n} \times R_2$ and of $\mathbb{Z}_2 \times A_4 \times R_2$?

Example 13.2.14. We know from Theorem 13.1.5 that the groups $\mathbb{Z}_m \times \mathbb{Z}_n$ with $gcd(m,n) > 1$, $m,n > 2$ are not planar. Here, we see that they have genus 1.

Moreover, also $\mathbb{Z}_m \times \mathbb{Z}_n \times R_2$ where $gcd(m,n) > 1$, $m,n > 2$, has genus 1. In Figure 13.18, page 279, we give a representation of $\mathrm{Cay}(\mathbb{Z}_3 \times \mathbb{Z}_3 \times R_2, \{(1,0,r_1),(0,1,r_1),(0,0,r_2)\})$ on the torus; points with the same label in the square to be identified, loops are omitted. Clearly, this can be generalized to any $\mathrm{Cay}(\mathbb{Z}_m \times \mathbb{Z}_n \times R_2, \{(1,0,r_1),(0,1,r_1),(0,0,r_2)\})$.

Example 13.2.15. The genus of $\mathrm{Cay}(\mathbb{Z}_2 \times D_{2n} \times R_2, \{(1,1_{D_{2n}},r_1),(0,a,r_2),(0,b,r_2)\})$ is $\leq 4n$. Consider $\mathbb{Z}_2 \times D_2 \times R_2$, which apparently has genus ≤ 4; see Figure 13.19, page 279.

Theorem 13.2.16. *Table 13.2 on page 280 shows our results for $A \times R_r$, $r < 4$. Here, a, b, and c are the respective group generators of order 2, as described in the previous theorems and examples. For the elements of A_4 in $\mathbb{Z}_2 \times A_4$, we use the cycle notation.*

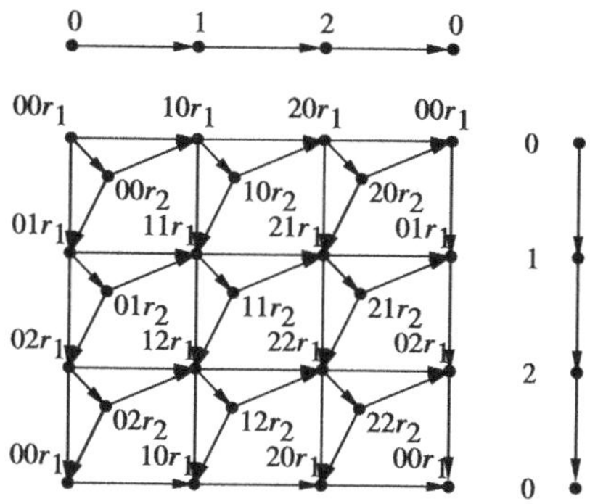

Figure 13.18: A representation of $\mathrm{Cay}(\mathbb{Z}_3 \times \mathbb{Z}_3 \times R_2, \{(1,0,r_1),(0,1,r_1),(0,0,r_2)\})$ on the torus.

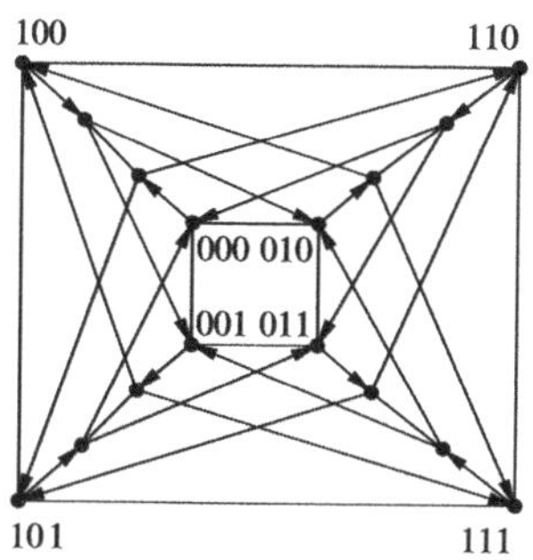

Figure 13.19: $\mathrm{Cay}(\mathbb{Z}_2 \times D_{2n} \times R_2, \{(1, 1_{D_{2n}}, r_1), (0, a, r_2), (0, b, r_2)\})$.

Right groups generated by products on the torus and the plane

We now study the minimal genus among Cayley graphs $\mathrm{Cay}(A \times R_r, C \times R_r)$ taken over all minimum generating sets C of the groups A, i. e., we restrict to generating systems that are Cartesian products of generating systems of the factors. This is natural from a categorical point of view, compare also Theorem 11.2.2.

In this setting, we consider the following question: when does the Cay functor produce a graph of genus 1 or 0, i. e., is toroidal or planar? See Theorem 13.2.24 and Corollary 13.2.21, respectively.

The results in this part come mainly from Kolja Knauer and Ulrich Knauer [48].

As before, we denote by $\times$ the *cross product* for graphs as well as the direct product for semigroups and sets. The following statement comes from Theorem 11.2.2. It is essential for the present considerations.

Corollary 13.2.17. *If A is a group and $C \subseteq A$, then*

$$\mathrm{Cay}(A \times R_r, C \times R_r) = \mathrm{Cay}(A, C) \times K_r^{(r)},$$

where $K_r^{(r)}$ is the complete graph with r loops.

Table 13.2: Our results for $A \times R_r$, $r < 4$. Also see Theorem 13.2.16.

Group A	Genus	Genus preserving generators Geometric realization	Genus raising generators of $A \times R_r$
$\mathbb{Z}_n$	0	$(1,r_1),(0,r_2)$ $(1,r_1),(0,r_2),(0,r_3)$ Both in Figure 13.10, p. 270	$(1,r_1),(0,r_2),(0,r_3),(0,r_4)$ Proposition 13.2.8
D_n	0	$(a,r_1),(b,r_1),(e_{D_n},r_2)$ $(a,r_1),(b,r_2)$ $(a,r_1),(b,r_2),(e_{D_n},r_3)$ Lemma 13.2.3	$(a,r_1),(b,r_1),(e_{D_n},r_2),(e_{D_n},r_3)$
A_4, S_4, A_5 $a^2 = e, b^2 \neq e$ generating	0	$(a,r_2),(b,r_2),(e_{A_4},r_1)$ $(a,r_1),(b,r_2)$ $(a,r_1),(b,r_2),(e_{A_4},r_3)$ Fig. 13.13, p. 273, Fig. 13.12, p. 272	$(a,r_1),(b,r_1),(e_A,r_2),(e_A,r_3)$
$\mathbb{Z}_2 \times \mathbb{Z}_{2n}$	0	Proposition 13.2.11	$(1,0,r_1),(0,1,r_2)$ $(1,0,r_1),(0,1,r_1),(0,0,r_2)$ Example 13.2.13
$\mathbb{Z}_2 \times A_4$	0	Proposition 13.2.11	$(0,(123),r_1),(1,(12)(34),r_2)$ $(0,(123),r_1),(1,(12)(34),r_1)$, $(0,e_{A_4},r_2)$
$\mathbb{Z}_2 \times D_{2n}$	0	Proposition 13.2.9	$(1,1_{D_{2n}},r_1),(0,a,r_2),(0,b,r_2)$ Figure 13.19, p. 279
$\mathbb{Z}_2 \times S_4, \mathbb{Z}_2 \times A_5$	0	Proposition 13.2.9	$(a,r_1),(b,r_2),(c,r_2)$
$\mathbb{Z}_m \times \mathbb{Z}_n$	1	$(1,0,r_1),(0,1,r_1),(0,0,r_2)$	$(1,0,r_1),(0,1,r_1),(0,0,r_2)$, $(0,0,r_3)$
$gcd(m,n)>1, m,n>2$		Figure 13.18, p. 279	

Proof. Note that the formula

$$\mathrm{Cay}(S \times T, C \times T) = \mathrm{Cay}(S, C) \times \mathrm{Cay}(T, D)$$

of Theorem 11.2.2 now turns into

$$\mathrm{Cay}(A \times R_r, C \times R_r) = \mathrm{Cay}(A, C) \times \mathrm{Cay}(R_r, R_r) = \mathrm{Cay}(A, C) \times K_r^{(r)}.$$ □

As far as we know, there do not exist general formulas relating the genus of a cross product of two graphs to the genera of the factors; see, e. g., [Gross/Tucker 1987], [Imrich/Klavžar 2000] or [White 2001].

The following will be useful.

Remark 13.2.18. If a subdivision of a graph H is a subgraph of G, then a subdivision of $H \times K_r^{(r)}$ is a subgraph of $G \times K_r^{(r)}$.

Lemma 13.2.19. *If* Cay(A, C) *is not planar, then* Cay$(A \times R_r, C \times R_r)$ *with $r \geq 2$ cannot be embedded on the torus.*

Proof. Note that $K_{3,3} \times K_2^{(2)} \cong K_{6,6}$ already has genus 4; see [White 2001]. Moreover, the graph $K_5 \times K_2^{(2)}$ has 10 vertices and 40 edges. An embedding on the torus would have 30 faces by the Euler–Poincaré formula (Theorem 1.8.4). Even if all faces were triangles in this graph, this would require 45 edges. So the graph cannot be toroidal. □

Lemma 13.2.20. *Suppose that* $\underline{\text{Cay}}(A, C)$ *contains a $K_{2,2}$ subdivision. If $r \geq 2$, then* Cay$(A \times R_r, C \times R_r)$ *cannot be embedded on the sphere. If $r \geq 3$, then* Cay$(A \times R_r, C \times R_r)$ *cannot be embedded on the torus.*

Proof. If $r \geq 2$, the resulting graph $\underline{\text{Cay}}(A \times R_r, C \times R_r)$ contains a subdivision of $K_{4,4}$, which has genus 1. If $r \geq 3$, the resulting graph contains a subdivision of $K_{6,6}$, which has genus 4; see [White 2001]. □

Corollary 13.2.21. *The Cayley graph* Cay$(A \times R_r, C \times R_r)$ *is planar if and only if $A \cong \mathbb{Z}_n$, $n \leq 3$ and $r = 2$.*

Proof. If Cay$(A \times R_r, C \times R_r)$ is planar, then $\underline{\text{Cay}}(A, C)$ does not contain a $K_{2,2}$ subdivision, by Lemma 13.2.20. Thus $A \cong \mathbb{Z}_n$, $n \leq 3$. In Example 13.2.25, we see that $\underline{\text{Cay}}(\mathbb{Z}_3 \times R_3, \{1\} \times R_3)$ is not planar. Therefore, $r = 2$, not looking at trivial cases. For the converse observe that $\underline{\text{Cay}}(\mathbb{Z}_3 \times R_2, \{1\} \times R_2) \cong C_4 + \overline{K}_2$, and $\underline{\text{Cay}}(\mathbb{Z}_2 \times R_2, \{1\} \times R_2) \cong C_4$, up to loops. □

Lemma 13.2.22. *If $r \geq 5$ and A a nontrivial group, then* Cay$(A \times R_r, C \times R_r)$ *cannot be embedded on the torus.*

Proof. The resulting graph $\underline{\text{Cay}}(A \times R_r, C \times R_r)$ contains $K_{5,5}$, which has genus 3; see [White 2001]. □

Hence, for the rest of the subsection we have to investigate Cay$(A \times R_r, C \times R_r)$ for all planar groups A and $1 \leq r \leq 4$.

Lemma 13.2.23. *If a planar Cayley graph* $\underline{\text{Cay}}(A, C)$ *is at least 3-regular, then* Cay$(A \times R_2, C \times R_2)$ *cannot be embedded on the torus.*

Proof. Since $\underline{\text{Cay}}(A, C)$ is at least 3-regular, $\underline{\text{Cay}}(A \times R_2, C \times R_2)$ is at least 6-regular.

Assume that Cay$(A \times R_2, C \times R_2)$ is embedded on the torus; then the Euler–Poincaré formula (Theorem 1.8.4) tells us that all faces are triangular. This implies that every edge of $\underline{\text{Cay}}(A \times R_2, C \times R_2)$ lies in at least two triangles, and hence every edge of $\underline{\text{Cay}}(A, C)$ lies in at least one triangle.

Let $c_1, c_2, c_3 \in C$ be the generators corresponding to a triangle a_1, a_2, a_3. Then $c_1^{\pm 1} c_2^{\pm 1} c_3^{\pm 1} = e_A$ for some signing, where e_A is the identity in A. If any two of the c_i are distinct, then one of the two is redundant; hence C was not inclusion minimal. Thus every $c \in C$ must be of order 3. Since A is not cyclic, we obtain that $\underline{\mathrm{Cay}}(A, C)$ is at least 4-regular. Then $\underline{\mathrm{Cay}}(A\times R_2, C\times R_2)$ is at least 8-regular, and the Euler–Poincaré formula yields that it cannot be embedded on the torus. □

Theorem 13.2.24. *Let $A \times R_r$ be a finite right group with $r \geq 2$. The minimal genus of* $\mathrm{Cay}(A \times R_r, C \times R_r)$ *among all generating sets $C \subseteq A$ of A is 1 if and only if $A \times R_r$ is isomorphic to one of the following right groups:*
- $\mathbb{Z}_n \times R_r$ *with* $(n, r) \in \{(2, 3), (2, 4), (3, 3), (i, 2)\}$ *for* $i \geq 4$;
- $D_n \times R_2$ *for all* $n \geq 2$.

Note that this list includes $\mathbb{Z}_2 \times D_n \times R_2 \cong D_{2n} \times R_2$ *and* $\mathbb{Z}_2 \times \mathbb{Z}_n \times R_2 \cong \mathbb{Z}_{2n} \times R_2$ *for odd* $n \geq 3$.

Proof. By Lemma 13.2.23, the group A has to be generated either by one element or by two elements of order 2 to be embeddable on the torus. This necessary condition is equivalent to (A, C) being $(\mathbb{Z}_n, \{1\})$ or $(D_n, \{a, b\})$, where $a^2 = b^2 = (ab)^n = 1_{D_n}$.

First, we consider the cyclic case. For $n = 2$, we have $\mathrm{Cay}(\mathbb{Z}_2 \times R_r, C \times R_r) = K_{r,r}$ which exactly for $r \in \{3, 4\}$ has genus 1.

Take $n = 3$. If $r = 2$, we obtain the planar graph $\mathrm{Cay}(\mathbb{Z}_3 \times R_2, \{1\} \times R_2)$ shown in the first figure of Example 13.2.25. If $r = 3$, the resulting graph $\underline{\mathrm{Cay}}(\mathbb{Z}_3 \times R_3, \{1\} \times R_3)$ contains $K_{3,3}$, so it cannot be planar. In Example 13.2.25, there is an embedding as a triangular grid on the torus. If $r = 4$, we have the complete tripartite graph $K_{4,4,4}$. Delete the entire set of 16 edges between two of the three partitioning sets. The remaining (nonplanar) graph has 12 vertices, 32 edges, and, assuming a toroidal embedding, 20 faces. A simple count shows that this cannot be realized without triangular faces. So for $r \geq 4$ the graph $\mathrm{Cay}(\mathbb{Z}_3 \times R_r, C \times R_r)$ is not toroidal. Take $n \geq 4$. Then the graph $\underline{\mathrm{Cay}}(\mathbb{Z}_n, \{1\})$ contains a $C_4 = K_{2,2}$ subdivision. If $r \geq 3$, then $\mathrm{Cay}(\mathbb{Z}_n \times R_r, \{1\} \times R_r)$ is not toroidal by Lemma 13.2.20. If $r = 2$, an embedding of $\underline{\mathrm{Cay}}(\mathbb{Z}_4 \times R_2, \{1\} \times R_2)$ as a square grid on the torus is shown in the right most figure of Example 13.2.25. This is instructive for the cases $n \geq 5$. Moreover, we see from this figure that the vertices $\{0_1, 0_2, 2_1\}$ and $\{1_1, 1_2, 3_1\}$ induce a $K_{3,3}$ subgraph of $\underline{\mathrm{Cay}}(\mathbb{Z}_4 \times R_2, \{1\} \times R_2)$. Generally, for $n \geq 4$ we have that $\underline{\mathrm{Cay}}(\mathbb{Z}_n \times R_2, \{1\} \times R_2)$ contains a $K_{3,3}$ subdivision, so it is not planar.

Second, if A is a dihedral group and C consists of two generators a, b of order 2, the graph $\mathrm{Cay}(D_n, C)$ is isomorphic to $\underline{\mathrm{Cay}}(\mathbb{Z}_{2n}, \{1\})$. Thus $\mathrm{Cay}(D_n \times R_2, \{a, b\} \times R_r)$ has genus 1 if and only if $r = 2$, by the cyclic case. Any different generating system C for D_n would have a generator of order greater than 2, and hence would yield $\mathrm{Cay}(D_n \times R_2, C \times R_2)$ with genus greater than 1, by Lemma 13.2.23. □

Example 13.2.25. Here, we draw some of the graphs from the theorem.

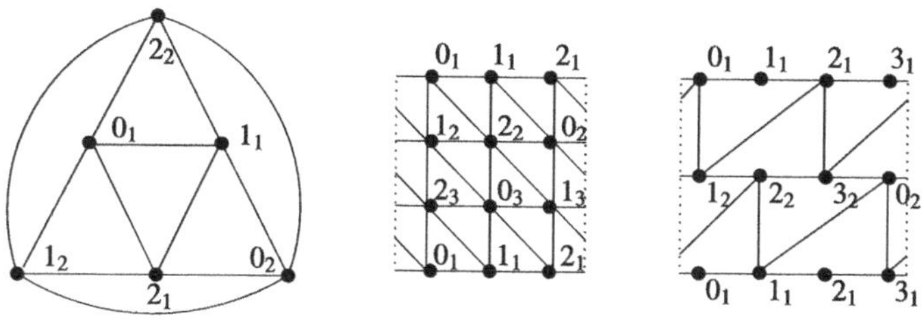

In the left most graph $\mathrm{Cay}(\mathbb{Z}_3 \times R_2, \{1\} \times R_2)$ (planar), the inner and the outer triangle are directed clockwise, the other arcs are directed counter clockwise; the second graph is $\underline{\mathrm{Cay}}(\mathbb{Z}_3 \times R_3, \{1\} \times R_3)$, and the third graph is $\underline{\mathrm{Cay}}(\mathbb{Z}_4 \times R_2, \{1\} \times R_2) \cong K_{4,4}$, both are toroidal. In all cases, x_1 stands for (x, r_1), x_2 for (x, r_2) and x_3 for (x, r_3), $x \in \mathbb{Z}_n$.

Remark 13.2.26. For $r = 1$, we have $A \times R_r \cong A$. Hence the characterization of toroidal groups due to V. K. Proulx [74], is the above theorem for $r = 1$.

We have seen in Theorem 13.2.16 that all right groups from Theorem 13.2.24 are planar, with the exception of $\mathbb{Z}_2 \times R_4$.

In the above proofs, we make strong use of Lemma 13.2.23, which tells us that 3-regular planar Cayley graphs will not be embeddable on the torus after taking the Cartesian product with R_2. The following small example from the next theorem shows that this operation can increase the genus from 0 to 3.

Theorem 13.2.27. *The genus of* $\underline{\mathrm{Cay}}(\mathbb{Z}_6 \times R_2, \{2,3\} \times R_2)$ *is* 3.

Proof. We observe that $\underline{\mathrm{Cay}}(\mathbb{Z}_6 \times R_2, \{2,3\} \times R_2)$ consist of two disjoint copies $(C_3 \square K_2)$ of $\underline{\mathrm{Cay}}(\mathbb{Z}_6, \{2,3\})$, say $(C_3 \square K_2)_1$ and $(C_3 \square K_2)_2$ with vertex sets $\{0_1, 1_1, 2_1, 3_1, 4_1, 5_1\}$ and $\{0_2, 1_2, 2_2, 3_2, 4_2, 5_2\}$, respectively. Every vertex v_1 of $(C_3 \square K_2)_1$ is adjacent to every neighbor of its copy v_2 in $(C_3 \square K_2)_2$. Figure 13.20 shows an embedding of $\underline{\mathrm{Cay}}(\mathbb{Z}_6 \times R_2, \{2,3\} \times R_2)$ into the orientable surface of genus 3, the *triple torus*.

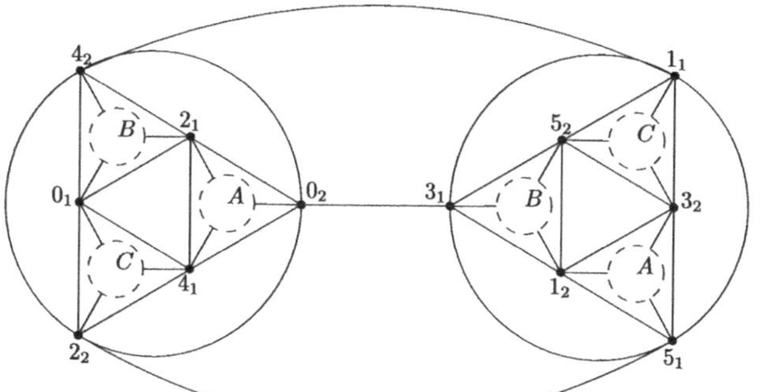

Figure 13.20: $\underline{\mathrm{Cay}}(\mathbb{Z}_6 \times R_2, \{2,3\} \times R_2)$ on the triple torus with handles A, B, C.

This graph is 6-regular with 12 vertices, so it has 36 edges.

Using Lemma 13.2.23, we will show that $\underline{\mathrm{Cay}}(\mathbb{Z}_6 \times R_2, \{2,3\} \times R_2)$ cannot be embedded on the double torus.

Assume that $\underline{\mathrm{Cay}}(\mathbb{Z}_6 \times R_2, \{2,3\} \times R_2)$ is 2-cell-embedded on the double torus. Delete the four edges connecting 1_1 and 1_2 with 5_1 and 5_2 and also the four edges connecting 0_1 and 0_2 with 4_1 and 4_2. The resulting graph H has 28 edges. It consists of two graphs X and Y which are copies of $K_{4,4}$, where X has the bipartition $(\{0_1, 0_2, 5_1, 5_2\}, \{2_1, 2_2, 3_1, 3_2\})$ and Y has the bipartition $(\{0_1, 0_2, 1_1, 1_2\}, \{3_1, 3_2, 4_1, 4_2\})$. They are glued at the four vertices with the same numbers, and the corresponding four edges are identified. Although H is no longer bipartite, it still is triangle-free. By the assumption it is 2-cell-embedded on the double torus.

By the Euler–Poincaré formula, this gives 14 faces, all of which are quadrangular. So the edges between $1_1, 1_2$ and $5_1, 5_2$ and between $0_1, 0_2$ and $4_1, 4_2$, which we have to put back in, have to be diagonals of these quadrangular faces. But then $\{2_2, 4_1, 2_1, 0_1\}$ and $\{2_2, 4_1, 2_1, 0_2\}$ are the only 4-cycles in H which contain, respectively, the vertices $4_1, 0_1$ and $4_1, 0_2$; they form faces of H. Since they have the common edges $\{2_2, 4_1\}$ and $\{2_1, 4_1\}$, we obtain a $K_{2,3}$ with bipartition $(\{2_1, 2_2\}, \{0_1, 0_2, 4_1\})$. We know from Theorem 1.8.3 that $K_{2,3}$ is not outer planar. Thus the region consisting of the glued 4-cycles $\{2_2, 4_1, 2_1, 0_1\}$ and $\{2_2, 4_1, 2_1, 0_2\}$ must contain one of the vertices $0_1, 0_2$ or 4_1 in its interior. Hence this vertex has only degree 2, which is a contradiction. □

13.3 On planar Clifford semigroups

Now we consider the following question: When does the Cay functor take a Clifford semigroup S with connection set C to a planar graph.

Lemma 13.3.1. *Let $S = \bigcup_{\xi \in Y} A_\xi$ be a Clifford semigroup and A a subgroup of S, then for the genus we have $\gamma(Y), \gamma(A) \leq \gamma(S)$.*

Proof. Note that a subgroup of A of S must be a subgroup of A_ξ for some $\xi \in A$. Thus by Proposition 11.3.14, we have $\gamma(A) \leq \gamma(A_\xi)$. Corollary 11.3.19 gives that Cayley graphs of Y and A_ξ are minors of $\mathrm{Cay}(S, C)$ for any generating system C. Thus, Lemma 1.8.1 yields the result. □

As Clifford semigroups are strong semilattices of groups, because of Lemma 13.3.1, a first step is to study planar semilattices, i. e., Clifford semigroups where all groups are one-element. It will turn out, that this is a big class, and a simple description by a finite list and a couple of infinite families or in purely algebraic terms seems unlikely.

In a second step, we will reduce our attention to two-component Clifford semigroup, i. e., $S = A_\beta \bigcup A_\alpha$ where the semilattice $Y = \{\alpha, \beta\}$, $\beta > \alpha$, is a two-element chain.

Planar semilattices

In this subsection, we characterize planar semilattices, i. e., planar Clifford semigroups in which every group is trivial. Since semilattices are the main players in this part, we use Latin letters instead of the small Greeks for their elements. If the poset Y is a meet semilattice, then it is well known that the unique minimal generating system of Y is given by the set of meet irreducibles, i. e.,

$$M(Y) = \{m \in Y \mid y_1 \wedge \ldots \wedge y_k = m \implies m \in \{y_1, \ldots, y_k\}\}.$$

In the Hasse diagram of Y, the set $M(Y)$ corresponds to those elements with at most one upward edge. We say that Y is planar if $\underline{\mathrm{Cay}}(Y, M(Y))$ is planar. We usually refer to the minimum element of Y by 0. For $x \in Y$, its *height* $h(x)$ is the number of elements of a longest chain from 0 to x. The *height* $h(P)$ of a poset P is the number of elements in one of its longest chains. See Figure 13.21 for an illustration.

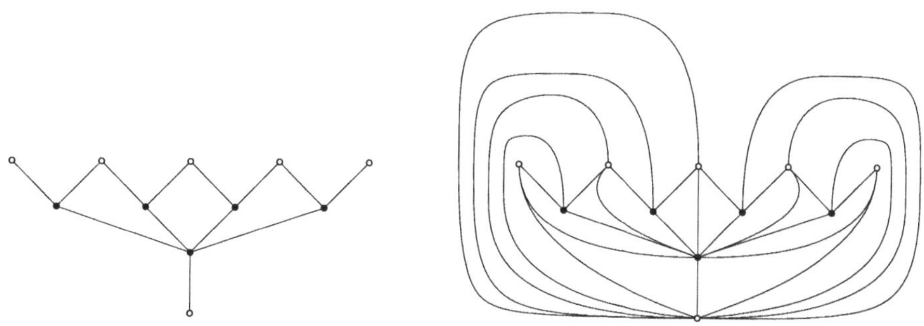

Figure 13.21: The Hasse diagram of a height 4 semilattice Y, with $M(Y)$ as unfilled vertices and a plane embedding of the Cayley graph $\underline{\mathrm{Cay}}(Y, M(Y))$. The subgraph $\underline{\mathrm{Cay}}(Y, M(Y)) \setminus \{0\}$ is outer planar.

Lemma 13.3.2. *If Y is a meet semilattice, then $\underline{\mathrm{Cay}}(Y, M(Y))$ contains a $K_{h(Y)}$ minor.*

Proof. Let $C = (y_1, \ldots, y_k)$ be a chain in Y with $k = h(Y)$ elements. By Remark 11.3.16, for each $1 \leq i \leq k$ there is an m_i such that $y_i \leq m_i$ but $y_j \not\leq m_i$ for all $i \leq j \leq k$. By Lemma 11.3.17, there is an arc (m_i, y_ℓ) in $\underline{\mathrm{Cay}}(Y, M(Y))$ for all $1 \leq \ell \leq i$. Thus, contracting the arcs of the form (m_i, y_i) for all $1 \leq i \leq k$, yields a complete subgraph on k vertices. □

Recall that a is an ***atom*** in a meet semilattice if $0 \prec a$.. We say that a semilattice is a decorated X, if it can be obtained from the lattice in Figure 13.22, page 286, by putting for any pair of an atom a and a maximum m of X at most one element between a and m. Moreover, for any atom a we can put a height 2 poset whose cover graph is a disjoint union of paths above a. See Figure 13.23, page 286, for an example.

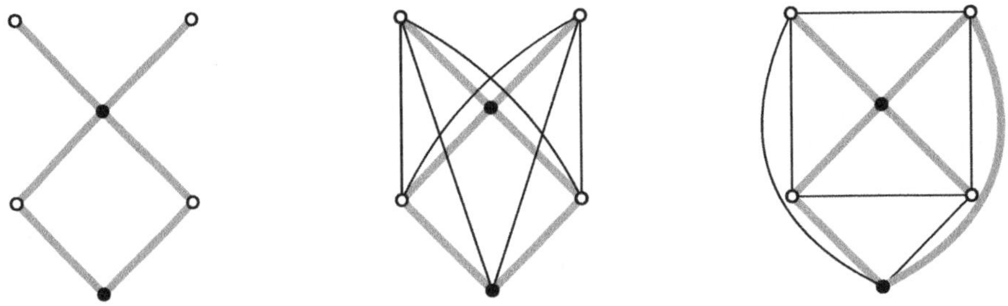

Figure 13.22: An X semilattice, with meet-irreducibles as unfilled vertices and its cover relations in bold gray. Form the left to right, the Hasse diagram, the Cayley graph, and a plane embedding of the latter (right).

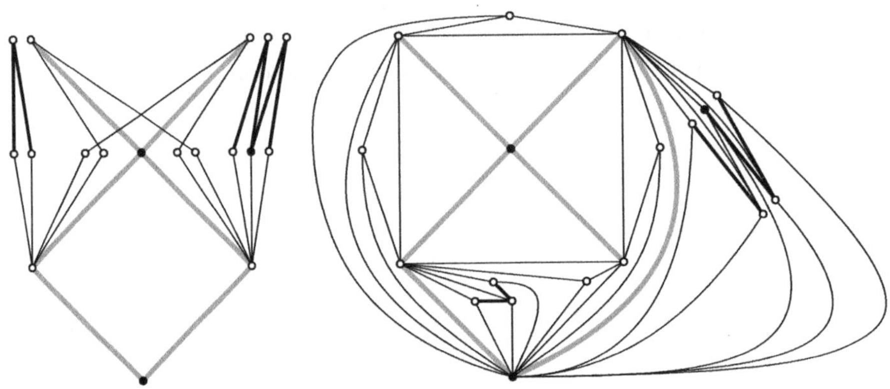

Figure 13.23: Left: A decorated X meet semilattice Y (on the left), with $M(Y)$ as unfilled vertices, the height 2 posets above atoms are bold black. The underlying X is drawn as bold gray. Right: a plane embedding of $\underline{\mathrm{Cay}}(Y, M(Y))$. Y is planar, but $Y \setminus \{0\}$ is not outer planar.

Lemma 13.3.3. *If the semilattice Y is planar, while* $\underline{\mathrm{Cay}}(Y, M(Y)) \setminus \{0\}$ *is not outer planar, then Y is a decorated X.*

Proof. Suppose, that $\underline{\mathrm{Cay}}(Y, M(Y)) \setminus \{0\}$ is not outer planar, but Y is planar. First, observe, that all meet irreducibles and atoms of Y lie on the outer face of a planar drawing of $\underline{\mathrm{Cay}}(Y, M(Y)) \setminus \{0\}$. Otherwise $\underline{\mathrm{Cay}}(Y, M(Y))$ would not be planar. Moreover, there are at least two atoms a_1, a_2, since otherwise $0 \in M(Y)$, all elements are connected to 0, and Y is not planar. Let $x \in Y$ be not on the outer face. Thus, x has height 3, because it is not an atom nor a maximum and Y is of height 4 by Lemma 13.3.2. Moreover, x is not a meet irreducible. Hence, x is covered by at least two maxima m_1, m_2 and covers at least one atom a_1. Suppose that a_2 is not covered by x. By Remark 11.3.16, there must be a meet irreducible $m \in \uparrow a_2 \setminus \uparrow x$. Therefore, we have $x \parallel m$ and $m \wedge x = 0$ and x must lie on the outer face – a contradiction. Thus, all atoms of Y lie below x.

With this information we identify the decorated X. It is easy to see, that if x lies above three atoms or below three maxima, then there is a $K_{3,3}$ subdivision in $\underline{\mathrm{Cay}}(Y, M(Y))$. Thus, $m_1, m_2, x, a_1, a_2, 0$ induce an X-poset. It is now easy to see, that in a planar Y there cannot be another x' with the same properties as x. Consult Figure 13.22, page 286 for the (unique) plane embedding of the so-far identified graph. Since, a_1, a_2 are the only atoms of Y, all further elements must lie above either a_1 or a_2. If there are two incomparable elements $a_1 \leq y \parallel z \leq m_1$, then both are meet irreducibles, i. e., have to be connected to 0. This contradicts planarity, compare Figure 13.22, page 286. If there is a poset P sitting above a_1 (but not below m_1 or m_2), then it has to be of height at most 2 since $h(Y) \leq 4$ by Lemma 13.3.2. If the diagram of P contains a cycle C, then its maxima have to be connected to 0 and its minima to a_1, but there is an edge $(a_1, 0)$—contradicting planarity. If P contains a minimum y covered by three elements z_1, z_2, z_3, then the latter are one partitioning set of a $K_{3,3}$, while the other one is given by $0, a_1, y$. If in P, there is a maximum z covering three elements y_1, y_2, y_3, then the meet of any of the latter with a_2 is 0 and they all are connected to 0. Therefore, y_1, y_2, y_3 form one partioning set of $K_{3,3}$, the other one is given by $0, a_1, z$. Thus, the diagram of P has maximum degree 2. Hence, Y is a decorated X. □

Theorem 13.3.4. *A semilattice Y is planar if and only if Y is a decorated X or $h(Y) \leq 4$ and the comparability graph of $Y \setminus \{0\}$ is outer planar.*

Proof. Let Y be a planar semilattice. Then $h(Y) \leq 4$ by Lemma 13.3.2, since otherwise a $\underline{\mathrm{Cay}}(Y, M(Y))$ has a K_5-minor and Y cannot be planar. Observe now that $\underline{\mathrm{Cay}}(Y, M(Y)) \setminus \{0\}$ is the comparability graph of $Y \setminus \{0\}$ since $h(Y) \leq 4$. Now, Lemma 13.3.3 implies that the comparability graph of $Y \setminus \{0\}$ is outer planar or Y is a decorated X.

Conversely, for a decorated X Figure 13.23, page 286, suggests how to find a plane embedding. For an outer planar $\underline{\mathrm{Cay}}(Y, M(Y)) \setminus \{0\}$, the graph $\underline{\mathrm{Cay}}(Y, M(Y))$ is planar, since the element 0 can be connected to all elements of $\underline{\mathrm{Cay}}(Y, M(Y)) \setminus \{0\}$. □

Theorem 13.3.4 calls for some characterization of posets of height ≤ 3 and outer planar comparability graph. There could be different kinds of answers to such a question, e. g., an algebraic description by specifying some laws of $\wedge$, or a constructive description by specifying operations to build such a poset starting from the single element lattice. Another way would be to employ the characterization of outer planar graph by Chartrand and Harary; see Theorem 1.8.3.

Planar Clifford semigroups with two groups

Parts of the following are taken from Xia Zhang [92]. However, the above paper misses a case in the main theorem, that is completed here; see Theorem 13.3.5.

We investigate Clifford semigroups $S = (A_\beta \bigcup A_\alpha; f_{\beta,\alpha})$. That is, S is a strong semilattice $Y = \{\beta > \alpha\}$ of groups $B := A_\beta$, $A := A_\alpha$ with structure homomorphism $f := f_{\beta,\alpha}$. By the way, the semigroup S is a monoid with the identity $e_S = e_B$. We use the notation $B \xrightarrow{f} A$ for this type of Clifford semigroup, or $B \xrightarrow{\text{inj}} A$ and $B \xrightarrow{\neq c_e} A$ instead of $B \xrightarrow{f\ \text{inj}} A$ and $B \xrightarrow{f \neq c_e} A$. We also refer to A and B as *lower* and *upper group*, respectively. If A, B contain elements with the same canonical name, we sometimes use x' for the elements of the lower group if the corresponding element of the upper group is called x.

For $D \cup C$ with $D \subseteq B$ and $C \subseteq A$, we write $B^D \xrightarrow{f} A^C$ instead of $(S, C \cup D)$. In particular, we write $\operatorname{Cay}(B^D \xrightarrow{f} A^C)$ instead of $\operatorname{Cay}(S, C \cup D)$. Recall, that for the purpose of this section we can assume that $D \cup C$ is an inclusion-minimal generating system of S.

Our characterization is the following.

Theorem 13.3.5. *The Clifford semigroup $B \xrightarrow{f} A$ is planar if and only if*
- $f = c_0$, $B = \mathbb{Z}_n$, $A = \mathbb{Z}_m$,
- $f = c_0$, $B \in \{\mathbb{Z}_1, \mathbb{Z}_2\}$, $A \in \{\mathbb{Z}_1, D_m, A_4, S_4, A_5, \mathbb{Z}_2 \times \mathbb{Z}_m\}$,
- $f = c_0$, $B = D_n$, $A = \mathbb{Z}_m$,
- $f \neq c_0$ *and noninjective*, $B = D_2$, $A \in \{\mathbb{Z}_2, Z_4, \mathbb{Z}_2 \times A_4, \mathbb{Z}_2 \times \mathbb{Z}_n\}$,
- f *injective*, $B = \mathbb{Z}_n$, $A \in \{D_n, \mathbb{Z}_n, \mathbb{Z}_{2n}, \mathbb{Z}_2 \times \mathbb{Z}_n\}$,
- f *injective*, $B = \mathbb{Z}_2$, $A \in \{D_m, \mathbb{Z}_2 \times A_4, \mathbb{Z}_2 \times \mathbb{Z}_m\}$,
- f *injective*, $B = D_n$, $A \in \{D_n, D_{2n}, \mathbb{Z}_2 \times D_n\}$.

The proof occupies the rest of this section. It consists of a relatively long case distinction, which will furthermore produce different planar generating systems in some cases. We will outline this result in more detail at the end of this section. We start with some general observations.

First, we reformulate Construction 11.3.21 and Corollary 11.3.19 for the present situation $\operatorname{Cay}(B^D \xrightarrow{f} A^C)$.

Construction 13.3.6. Let $D \cup C$ be an inclusion-minimal generating system of $S = B \xrightarrow{f} A$. The Cayley graph $\operatorname{Cay}(B^D \xrightarrow{f} A^C)$ consists of:
- the upper graph $\operatorname{Cay}(B, D)$,
- the lower graph $\operatorname{Cay}(A, C \cup f(D))$,
- for every $c \in C$ we have an arc from every $b \in \operatorname{Cay}(B, D)$ to $f(b)c \in \operatorname{Cay}(A, C \cup f(D))$.
- the structure homomorphism f induces a graph homomorphism $f : \operatorname{Cay}(B, D) \xrightarrow{f} \operatorname{Cay}(A, C \cup f(D))$.

Observe that, since $D \cup C$ generates S, D generates B, $C \cup f(D)$ generates A, and $|C| \geq 1$. In particular, we see that C is not necessarily a generating system of A. Note moreover that here, opposed to Construction 11.3.20, we have to be sure that $C \neq \emptyset$. Only if there is an element in $c \in C$, this can operate on elements of $b \in B$, i. e., $b * c = f(b)c \in A$.

In other words, elements $b \in B$ have arrows to elements of A in the respective Cayley graph, i. e., we get arrows $(b, f(b)c)$.

Again, we work on the basis of Maschke's theorem (Theorem 13.1.5) using planar groups A and B. Similar to the case of right-groups the direction of arcs of the planar Cayley graphs of groups turn out essential for the investigation of $\mathrm{Cay}(B \xrightarrow{f} A)$.

Using Construction 13.3.6, we prove the following basic lemma. Keep in mind, that C is not necessarily a generating system of A. It is only determined by the requirement that $D \cup C$ is a generating system of $B^D \xrightarrow{f} A^C$, i. e., D generates B, $C \cup f(D)$ generates A, and $C \neq \emptyset$.

Lemma 13.3.7. *Let* $\mathrm{Cay}(B^D \xrightarrow{f} A^C)$ *be a planar Cayley graph. Then* $\mathrm{Cay}(B, D)$ *is outer planar,* $\mathrm{Cay}(A, C \cup f(D))$ *has a plane representation such that the vertices* $f(B)C$ *lie on a common face, and* $1 \leq |C| \leq 2$.

Proof. As commented above the lemma, from Construction 13.3.6 follows that $1 \leq |C|$, as $D \cup C$ is a generating system of the Clifford semigroup.

We have to draw $\mathrm{Cay}(A, C \cup f(D))$ and $\mathrm{Cay}(B, D)$ simultaneously without crossings. Thus, $\mathrm{Cay}(B, D)$ is drawn inside one face F of $\mathrm{Cay}(A, C \cup f(D))$. Every vertex $b \in \mathrm{Cay}(B, D)$ has to be connected to $f(b)C$. Thus, the vertices $f(B)C$ all have to lie on F. Moreover, all vertices of $\mathrm{Cay}(B, D)$ have to lie on its outer face since otherwise they could not be connected without crossing to F.

If now $|C| \geq 3$, then at least three elements $e_A \neq c_i \in C$, $i = 1, 2, 3$, do not lie on the same face of $\mathrm{Cay}(A, C \cup f(D))$ – examination of the plane Cayley graphs of the planar groups shows it. But since left-multiplication with group elements gives a graph isomorphism, neither do $f(b)c_i \in C$, $i = 1, 2, 3$ lie on the same face for any $b \in B$. This contradicts the properties shown above and concludes the proof. □

The only outer planar groups are $\mathbb{Z}_n$ and D_n; cf. Remark 13.1.6. This shows that the above lemma already tells quite a bit about planar Clifford semigroups. In particular we get the following.

Corollary 13.3.8. *The Clifford semigroup* $B \rightarrow \mathbb{Z}_1$ *is planar if and only if* $B \in \{\mathbb{Z}_n, D_n\}$.

So from now on, we will always assume that $A \neq \mathbb{Z}_1$.

The case $S = B \xrightarrow{c_e} A$

Here, we restrict ourselves to the case where the structure homomorphism f is the constant mapping c_e, that sends all elements of B to the identity element of A. In this case C in $S = B^D \xrightarrow{c_e} A^C$ will always be a (planar) generating system of A, compare Construction 13.3.6. Also we immediately get that the generators in $D \subseteq B$ generate loops at all vertices of $\mathrm{Cay}(A)$ via $f(d) = e$ for all $d \in D$.

For the next result, compare Lemma 13.3.7.

Lemma 13.3.9. *The Cayley graph* $\mathrm{Cay}(B^D \xrightarrow{c_e} A^C)$ *is planar if and only if* $\mathrm{Cay}(A, C)$ *is planar,* $\mathrm{Cay}(B, D)$ *is outer planar, and* $|C| = 1$ *or* $C = \{c, c'\}, c \neq c'$, *c and* c' *lie on one face of* $\mathrm{Cay}(A, \{c, c'\})$, *and* $B \in \{\mathbb{Z}_1, \mathbb{Z}_2\}$.

Proof. Sufficiency. Suppose that $|C| = 1$, i. e., $A = \mathbb{Z}_m^{\{1\}}$, $m \in \mathbb{N}$. Take $B^D \in \{\mathbb{Z}_n^{\{1\}}, D_n^{\{a,b\}} | n \in \mathbb{N}\}$ with $a, b \in D_n$ of order 2, i. e., the upper and lower graphs are outer planar. As $f = c_0$ is the constant mapping, each point of $\mathbb{Z}_m$ gets one or two loops. Each point from the upper group, whose Cayley graph is a cycle, gets an arc to the generating element of $\mathbb{Z}_m$. Compare again Remark 13.1.6. The resulting graph is planar.

Suppose now that $C = \{c, c'\}, c \neq c'$. Then $\mathrm{Cay}(B^D \xrightarrow{c_e} A^C)$ is the union of $\mathrm{Cay}(B, D)$ and $\mathrm{Cay}(A, C)$ with arcs from every vertex of $\mathrm{Cay}(B, D)$ to the vertices c and c' in $\mathrm{Cay}(A, C)$. Since c and c' lie on the same face of $\mathrm{Cay}(A, C)$, the entire graph is planar, if $\mathrm{Cay}(B, D)$ is a path. See Figure 13.26, right, page 293, of $\mathrm{Cay}(\mathbb{Z}_2^1 \xrightarrow{c_e} A_4^{\{c,c'\}})$ for an example.

Necessity follows mostly from Lemma 13.3.7. If $|C| = 1$, i. e., $A = \mathbb{Z}_m^{\{1\}}$, $m \in \mathbb{N}$ we get from Lemma 13.3.7 that B is outer planar, i. e., $B \in \{\mathbb{Z}_n, D_n\}$, cf. Remark 13.1.6. For $C = \{c, c'\}, c \neq c'$, it can moreover be seen from Figure 13.26, right, page 293, that $\mathrm{Cay}(B^D \xrightarrow{c_e} A^C)$ is not planar if $\mathrm{Cay}(B, D)$ is not a path, i. e., if $\underline{\mathrm{Cay}}(B, D)$ contains a minor K_3. This implies $B \in \{\mathbb{Z}_1, \mathbb{Z}_2\}$. □

Theorem 13.3.10. *The Clifford semigroup* $S = B \xrightarrow{c_e} A$ *is planar if and only if* $A \in \{D_n, A_4, S_4, A_5, \mathbb{Z}_2 \times \mathbb{Z}_{2n}, \mathbb{Z}_2 \times A_4\}$ *and* $B \in \{\mathbb{Z}_1, \mathbb{Z}_2\}$, *or* $A = \mathbb{Z}_m$ *and* $B \in \{\mathbb{Z}_n, D_n\}$, *where* $m, n \in \mathbb{N}$.

Proof. By Lemma 13.3.7, we get that $|C| < 3$. So we have to examine the planar group graphs $\mathrm{Cay}(A, \{c, c'\})$. We find that the exactly the groups $D_n, A_4, S_4, A_5, \mathbb{Z}_2 \times \mathbb{Z}_{2n}, \mathbb{Z}_2 \times A_4$, have minimal generating systems of size two yielding planar Cayley graphs. More precisely, they have a set of generators $C = \{c, c'\}$, where c is of order 2. In particular, c and c' lie on one face of the respective Cayley graph. Planar groups with one generator are of the form $A = \mathbb{Z}_m$. Now Lemma 13.3.9 completes the proof. □

In the preceding result, we exclude only the planar groups $\mathbb{Z}_2 \times D_n, \mathbb{Z}_2 \times S_4, \mathbb{Z}_2 \times A_5$ since they need 3 planar generators. Indeed, we get all planar groups with 2 planar generators. They all have one planar generator which is of order 2. Because of this, c and c' are always on one face.

Note however, that the "one face condition" of Lemma 13.3.9 is not necessarily fulfilled if $|C| = 2$ and none of the generators is of order 2. As example consider Figure 13.2, page 263. Here, $\mathrm{Cay}(A_4, \{(123), (234)\})$ contains K_4. The new semipath between (123) and (234) generates a minor K_5. The same holds for $\mathrm{Cay}(S_4, \{(123), (1234)\})$ and $\mathrm{Cay}(A_5, \{(124), (12345)\})$ in Figures 13.2 and 13.6, pages 263 and 265.

The case $S = B \xrightarrow{\text{non-inj } \neq c_e} A$

In the next example, we discuss two Clifford semigroups with non-injective structure homomorphism $f \neq c_0$. They turn out to be non-planar.

Example 13.3.11. In the left of Figure 13.24, we have $\mathrm{Cay}(\mathbb{Z}_4^{\{1\}} \xrightarrow{f(1)=1'} \mathbb{Z}_2^{\{1'\}})$ and, in the middle, $\mathrm{Cay}(\mathbb{Z}_4^{\{1\}} \xrightarrow{f(1)=1'} \mathbb{Z}_2^{\{0'\}})$. Here, $K_{3,3}$ is underlying, and thus it is not planar.

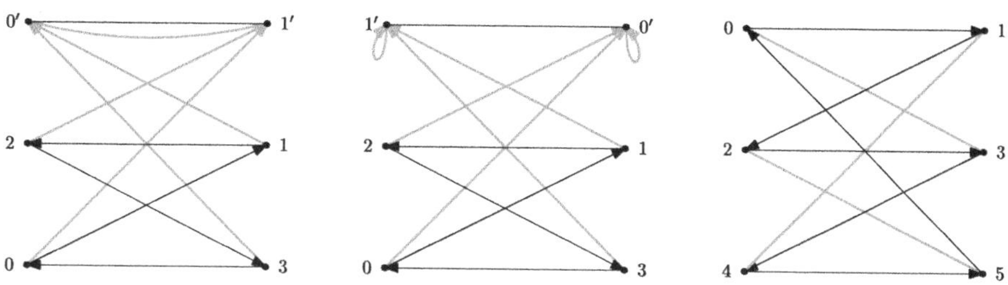

Figure 13.24: $\mathrm{Cay}(\mathbb{Z}_4^{\{1\}} \xrightarrow{f(1)=1'} \mathbb{Z}_2^{\{1'\}})$, $\mathrm{Cay}(\mathbb{Z}_4^{\{1\}} \xrightarrow{f(1)=1'} \mathbb{Z}_2^{\{0'\}})$, and $\mathrm{Cay}(\mathbb{Z}_6, \{1, 3\})$.

Now consider $\mathbb{Z}_4 \xrightarrow{f} \mathbb{Z}_2 \leq \mathbb{Z}_4$ with $f(\mathbb{Z}_4) = \{0', 2'\} \subseteq \mathbb{Z}_4$. So we get $\mathrm{Cay}(\mathbb{Z}_4^{\{1\}} \xrightarrow{f(1)=2'} \mathbb{Z}_4^{\{1'\}})$. Here again, we already have $K_{3,3}$ underlying as before.

The next result gives the only group B for which the Clifford semigroup $B \xrightarrow{\text{non-inj}\neq c_e} A$ is planar.

Lemma 13.3.12. *Let* $\mathrm{Cay}(B^D \xrightarrow{\text{non-inj}\neq c_e} A^C)$ *be planar, then* $B = \mathbb{Z}_2 \times \mathbb{Z}_2 \cong D_2$, $D = \{(0,1), (1,0)\}$, *and* $f^{-1}(e_A) = \{(0,0), (1,0)\}$, *or* $f^{-1}(e_A) = \{(0,0), (0,1)\}$.

Proof. As $f \neq c_e$ is not injective, we know that $f^{-1}(e_A) = N$ is a nontrivial normal divisor of B. Then all elements of N have an arc to $f(N)c = c$ for $c \in C$. Take the coset $Ng \subseteq B$ of N with $g \in B \setminus N$. Then all elements of Ng have an arc to $f(g)c \neq c$. Furthermore, since the graph of $\mathrm{Cay}(A, f(D) \cup C)$ is connected, it contains a path P from c to $f(g)c$. Since the entire Cayley graph is planar, then by Lemma 13.3.7 we have that $B \in \{\mathbb{Z}_n, D_n\}$ and the graph of B with respect to D is a cycle or a path. However, if B is a path, i. e., $B \in \{\mathbb{Z}_1, \mathbb{Z}_2\}$, then f cannot be noninjective different from c_e. Thus we can contract edges such that $N \cup Ng$ form a cycle X. It is easy to see that the subgraph consisting of P, X, and the arcs from $N \cup Ng$ to c and $f(g)c$ is planar if and only if N and Ng appear as two disjoint intervals on X. (If they interlace a $K_{3,3}$ subdivision can be found.) This has to hold for any choice of g, thus even without contracting, any two cosets of N are forbidden to interlace on the graph of B. In particular, N has to appear consecutively on the cycle graph corresponding to B.

It is easy to see that no proper subgroup of $\mathbb{Z}_n$ appears consecutively on the cycle, thus $B \neq \mathbb{Z}_n$. The only proper subgroups of D_n that appear consecutively on the cycle are of size 2. It is well known that they are not normal unless $n = 2$.

Since $D_2 \cong \mathbb{Z}_2 \times \mathbb{Z}_2$, this proves that $B = \mathbb{Z}_2 \times \mathbb{Z}_2$ and $D = \{(0,1),(1,0)\}$. Moreover, up to isomorphism we have $N = \{(0,0),(1,0)\}$ and $g = (0,1)$. This concludes the proof. □

Conversely, we can prove a positive result.

Lemma 13.3.13. *Let* $\mathbb{Z}_2 \times \mathbb{Z}_2 \xrightarrow{\text{non-inj}\neq c_e} A$ *be a Clifford semigroup such that* $f^{-1}(e_A) = \{(0,0),(1,0)\}$, *or* $f^{-1}(e_A) = \{(0,0),(0,1)\}$. *The graph* $\mathrm{Cay}(\mathbb{Z}_2 \times \mathbb{Z}_2^{\{(0,1),(1,0)\}} \xrightarrow{\text{non-inj}\neq c_e} A^C)$ *is planar if* $A^C \in \{\mathbb{Z}_2^{\{0\}}, \mathbb{Z}_2^{\{1\}}, \mathbb{Z}_4^{\{1\}}, D_n^{\{b\}}, (\mathbb{Z}_2 \times \mathbb{Z}_n)^{\{(0,1)\}}, (\mathbb{Z}_2 \times A_4)^{\{(0,(123))\}}\}$. *In these cases* $f(D) \setminus \{e_A\} = \{1\}, \{1\}, \{2\}, \{a\}, \{(1,0)\}, \{(1,(12)(34))\}$, *respectively.*

Proof. The Cayley graphs of $\mathbb{Z}_2 \times \mathbb{Z}_2^{\{(0,1),(1,0)\}} \xrightarrow{f(0,1)=a, f(1,0)=e} D_3^{\{b\}}$ and $\mathbb{Z}_2 \times \mathbb{Z}_2^{\{(0,1),(1,0)\}} \xrightarrow{f(0,1)=2, f(1,0)=0} Z_4^{\{1\}}$ in Figure 13.25, are instructive to find the embeddings for the first cases. For the other cases of A^C, compare the Cayley graphs of $(\mathbb{Z}_2 \times A_4)^{\{(1,(12)(34)),(0,(123))\}}$, in Figure 13.1, page 262, and of $(\mathbb{Z}_2 \times \mathbb{Z}_3)^{\{(1,0),(0,1)\}}$ as an example for $(\mathbb{Z}_2 \times \mathbb{Z}_n)^{\{(1,0),(0,1)\}}$, Figure 13.26, left, page 293. It is easy to see that the preconditions of Lemma 13.3.7 are satisfied and embeddings can be found. □

The above list turns out to be complete, see Corollary 13.3.20. The following lemma is the basis of the proof.

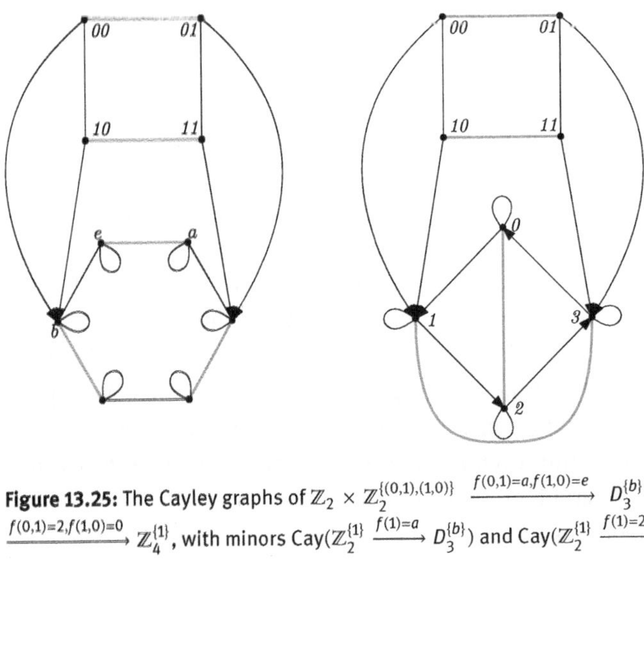

Figure 13.25: The Cayley graphs of $\mathbb{Z}_2 \times \mathbb{Z}_2^{\{(0,1),(1,0)\}} \xrightarrow{f(0,1)=a, f(1,0)=e} D_3^{\{b\}}$ and $\mathbb{Z}_2 \times \mathbb{Z}_2^{\{(0,1),(1,0)\}} \xrightarrow{f(0,1)=2, f(1,0)=0} \mathbb{Z}_4^{\{1\}}$, with minors $\mathrm{Cay}(\mathbb{Z}_2^{\{1\}} \xrightarrow{f(1)=a} D_3^{\{b\}})$ and $\mathrm{Cay}(\mathbb{Z}_2^{\{1\}} \xrightarrow{f(1)=2} \mathbb{Z}_4^{\{1\}})$, respectively.

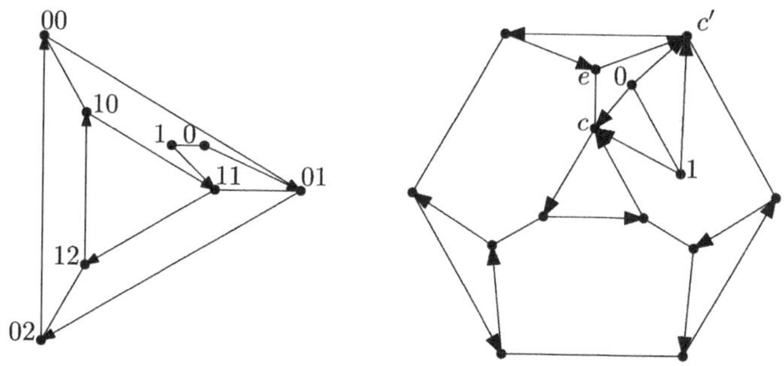

Figure 13.26: The plane Cayley graphs of $\mathbb{Z}_2^{\{1\}} \xrightarrow{f(1)=(1,0)} \mathbb{Z}_2\times\mathbb{Z}_3^{\{(0,1)\}}$ and the graph $\underline{\mathrm{Cay}}(\mathbb{Z}_2^{\{1\}} \xrightarrow{c_e} A_4^{\{c,c'\}})$.

Lemma 13.3.14. *Any Cayley graph of the Clifford semigroup* $(\mathbb{Z}_2\times\mathbb{Z}_2)^{\{(1,0),(0,1)\}} \xrightarrow{\text{non-inj}\neq c_e} A^C$ *with* $f^{-1}(e_A) = \{(0,0),(1,0)\}$ *or* $f^{-1}(e_A) = \{(0,0),(0,1)\}$ *has* $\mathrm{Cay}(\mathbb{Z}_2^{\{1\}} \xrightarrow{inj} A^C)$ *as a minor.*

Proof. In the graph of $(\mathbb{Z}_2\times\mathbb{Z}_2)^{\{(1,0),(0,1)\}}$ contract the edge $\{(0,0),(1,0)\}$ and the edge $\{(0,1),(1,1)\}$ in case $f^{-1}(e_A) = \{(0,0),(1,0)\}$, analogously, in the other case. □

The Cayley graphs in Figure 13.25, page 292, of two cases with noninjective $f \neq c_e$ will in particular exemplify this situation.

The case $S = \mathbb{Z}_2 \xrightarrow{\text{inj}} A$

Consider $\mathrm{Cay}(\mathbb{Z}_2^{\{1\}} \xrightarrow{f(1)=2'} \mathbb{Z}_4^{\{1'\}})$. This graph is planar, which can be seen from the second Cayley graph in Figure 13.25, page 292. More generally, we have the following.

Lemma 13.3.15. *Take* $A = \mathbb{Z}_n$ *for* $n > 1$. *The Cayley graph* $\mathrm{Cay}(\mathbb{Z}_2^{\{1\}} \xrightarrow{f(1)\neq e_A} A^C)$ *is planar if and only if* $A^C \in \{\mathbb{Z}_2^{\{1\}}, \mathbb{Z}_2^{\{0\}}, \mathbb{Z}_4^{\{1'\}}\}$, *where* $f(1) = 1$ *in the first two cases, and* $f(1) = 2'$ *in the third case.*

Proof. Necessity. Since f is a group homomorphism from $\mathbb{Z}_2$ to A, we get $A \in \{\mathbb{Z}_2, \mathbb{Z}_{2n}\}$. If now $n = 3$, say, we consider $\mathrm{Cay}(\mathbb{Z}_6, \{1', 3'\})$, i. e., the lower part of $\mathrm{Cay}(\mathbb{Z}_2^{\{1\}} \xrightarrow{f(1)=3'} \mathbb{Z}_6^{\{1'\}})$; see Figure 13.24, page 291, in Example 13.3.19. This graph is already not planar. We have the same situation for all $n > 2$. Thus, $A \in \{\mathbb{Z}_2, \mathbb{Z}_4\}$. In the first case, we have $f(1) = 1$ and C is either $\{0\}$ or $\{1\}$. If $A = \mathbb{Z}_4$, then $f(1) = 2'$ and by Lemma 13.3.7 and minimality of the generating system we get $C = \{1'\}$.

Sufficiency follows as we have minors, Lemma 13.3.14, of planar graphs, Lemma 13.3.13. □

In direct analogy with the proof of Lemma 13.3.15, we get the following.

Corollary 13.3.16. *Take* $A = \mathbb{Z}_n$ *for* $n > 1$. *The Cayley graph* $\mathrm{Cay}(\mathbb{Z}_m^{\{1\}} \xrightarrow{f(1)\neq e_A} A^C)$ *is planar if and only if* $A^C \in \{\mathbb{Z}_m^{\{1\}}, \mathbb{Z}_m^{\{0\}}, \mathbb{Z}_{2m}^{\{1'\}}\}$, *where* $f(1) = 1$ *in the first two cases, and* $f(1) = 2'$ *in the third case.*

An example for $n = 3$ is Figure 13.27, right, page 297, in Example 13.3.23.

Lemma 13.3.17. *The Cayley graph* $\mathrm{Cay}(\mathbb{Z}_2^{\{1\}} \xrightarrow{f(1)\neq e_A} A^C)$ *is not planar if* $A^C \in \{D_n^{\{(1\ldots n)\}}, A_4^{\{(123)\}}, S_4^{\{(123)\}}, S_4^{\{(1234)\}}, A_5^{\{(124)\}}, A_5^{\{(12345)\}}\}$, *where* $f(1) \in \{a, (12)(34), (34), (12), (23)(45)\}$, *respectively, and* $a \in D_n$ *denotes a generator of order 2.*

Proof. Consider $\mathrm{Cay}(\mathbb{Z}_2^{\{1\}} \xrightarrow{f(1)=(12)(34)} A_4^{\{(123)\}})$. Here, all $\vec{C_3}$ are oriented counterclockwise. The graph is not planar by Lemma 13.3.7, as $f(0)(123) = (123)$ and $f(1)(123) = (12)(34)(123)$ lie on different faces. In the other cases the same argument applies correspondingly. Compare the respective graphs in Figure 13.3, page 263 in Example 13.1.3, where we have $D_n^{\{a,(1\ldots n)\}}$ for $n = 3, 4, 5$. □

For further planar cases, compare the figures of the Cayley graphs of $(\mathbb{Z}_2 \times A_4)^{\{(1,(12)(34)),(0,(123)\}}$, in Figure 13.1, page 262, and of $(\mathbb{Z}_2 \times \mathbb{Z}_3)^{\{(1,0),(0,1)\}}$ as an example for $(\mathbb{Z}_2 \times \mathbb{Z}_n)^{\{(1,0),(0,1)\}}$, Figure 13.26, left, page 293.

We conclude the following.

Theorem 13.3.18. *The Clifford semigroup* $\mathbb{Z}_2^{\{1\}} \xrightarrow{f(1)\neq e_A} A^C$ *is planar if and only if* $A^C \in \{\mathbb{Z}_2^{\{1\}}, \mathbb{Z}_2^{\{0\}}, \mathbb{Z}_4^{\{1'\}}, D_n^{\{b\}}, (\mathbb{Z}_2 \times \mathbb{Z}_n)^{\{(0,1)\}}, (\mathbb{Z}_2 \times A_4)^{\{(0,(123))\}}\}$, *where* $n > 1$, $a, b \in D_n$ *denote generators of order 2 and* $f(1) \in \{1, 1, 2', a, (1,0), (1,(12)(34))\}$, *respectively.*

Proof. Suppose that the Clifford semigroup $\mathbb{Z}_2^{\{1\}} \xrightarrow{f(1)\neq e_A} A^C$ is planar. By Lemma 13.3.7, we have $|C| \leq 2$. But if $|C| = 2$, then $\{f(1)\} \cap C = \emptyset$. Thus, $\mathrm{Cay}(A, C \cup \{f(1)\})$ is a planar Cayley graph with three generators. By examination, we find that no such graph satisfies the one face condition from Lemma 13.3.7.

The case $\{f(1)\} \cap C \neq \emptyset$ implies $A = \mathbb{Z}_2$ by Lemma 13.3.15.

The remaining case concerns $C = |1|$ and $\{f(1)\} \cap C = \emptyset$. By Lemma 13.3.15, this implies $A \in \{\mathbb{Z}_2, \mathbb{Z}_4\}$ if A is cyclic. Otherwise, $C \cup \{f(1)\}$ is a minimal generating system of A, where $f(1) = d$ is a generator of order 2. By examination, we find that $A^{\{c,d\}}$ is one of $D_n^{\{a,b\}}$, $D_n^{\{(1\ldots n),b\}}$, $A_4^{\{(123),(12)(34)\}}$, $S_4^{\{(123),(34)\}}$, $S_4^{\{(1234),(12)\}}$, $A_5^{\{(124),(23)(45)\}}$, $A_5^{\{(12345),(23)(45)\}}$, $(\mathbb{Z}_2 \times \mathbb{Z}_n)^{\{(0,1),(1,0)\}}$, $(\mathbb{Z}_2 \times A_4)^{\{(0,(123)),(1,(12)(34))\}}$. But by Lemma 13.3.17 we can exclude $D_n^{\{(1\ldots n),b\}}$, $A_4^{\{(123),(12)(34)\}}$, $S_4^{\{(123),(34)\}}$, $S_4^{\{(1234),(12)\}}$, $A_5^{\{(124),(23)(45)\}}$, $A_5^{\{(12345),(23)(45)\}}$. For the positive statement, we use that here we have planar minors by Lemma 13.3.14 of planar graphs from Lemma 13.3.13. □

Example 13.3.19 (Different embeddings of $\mathbb{Z}_2$). We have seen in Theorem 13.3.18 that $\mathbb{Z}_2 \xrightarrow{\mathrm{inj}} \mathbb{Z}_2 \times \mathbb{Z}_n$ is planar. So in particular, $\mathbb{Z}_2^{\{1\}} \xrightarrow{f(1)=(1,0)} (\mathbb{Z}_2 \times \mathbb{Z}_3)^{\{(1,0),(0,1)\}}$ is planar. Despite of $\mathbb{Z}_2 \times \mathbb{Z}_3 \cong \mathbb{Z}_6$, the Cayley graph of $\mathbb{Z}_2^{\{1\}} \xrightarrow{f(1)=3} \mathbb{Z}_6^{\{1\}}$ is not planar. Here, already the lower graph $\mathrm{Cay}(\mathbb{Z}_6, \{1, 3\})$ contains $K_{3,3}$; see Figure 13.24 right, page 291, in Example 13.3.11. Compare also the proof of Lemma 13.3.15. A closer analysis shows that $\mathrm{Cay}(\mathbb{Z}_2 \times \mathbb{Z}_3, \{(1,0), (0,1)\}) \cong \mathrm{Cay}(\mathbb{Z}_6, \{2, 3\})$. With this, we get that $\mathbb{Z}_2^{\{1\}} \xrightarrow{f(1)=3} \mathbb{Z}_6^{\{2,3\}} \cong \mathbb{Z}_2^{\{1\}} \xrightarrow{f(1)=(1,0)} (\mathbb{Z}_2 \times \mathbb{Z}_3)^{\{(1,0),(0,1)\}}$ is planar.

But $\mathbb{Z}_2 \xrightarrow{f(1)=4} \mathbb{Z}_8$, e. g., is not planar. Since $\mathbb{Z}_8 \not\cong \mathbb{Z}_2 \times \mathbb{Z}_4$, a different embedding is not possible. Here, planarity is destroyed by the edges generated by $4 = f(1)$ in the lower graph $\mathrm{Cay}(\mathbb{Z}_8, \{1,4\})$. We get a situation analogous to Figure 13.24, page 291, where the lower graph is the Cayley graph of $\mathbb{Z}_6^{\{1,3\}}$.

Using Lemma 13.3.14 and the fact that graphs with nonplanar minors are nonplanar from Theorem 13.3.18, we obtain the following.

Corollary 13.3.20. *The Clifford semigroup $\mathbb{Z}_2 \times \mathbb{Z}_2 \cong D_2 \xrightarrow{non\text{-}inj \neq c_0} A$ is planar if and only if $A^C \in \{\mathbb{Z}_2^{\{1\}}, \mathbb{Z}_2^{\{0\}}, \mathbb{Z}_4^{\{1\}}, D_n^{\{a,b\}}, \mathbb{Z}_2 \times \mathbb{Z}_n, (\mathbb{Z}_2 \times A_4)^{\{(0,(123)),(1,(23)(45))\}}\}$, where $n > 1$.*

The case $S = B \xrightarrow{\text{inj}} A$

The only remaining case is $B \xrightarrow{\text{inj}} A$ with $B \neq \mathbb{Z}_2$, since $B = \mathbb{Z}_2$ we studied in the previous part.

As f is an injective homomorphism from B to A we get that $B < A$ is a subgroup relation with $B \in \{\mathbb{Z}_n, D_n\}$. Moreover, $\mathrm{Cay}(B, D)$ is outer planar and $\mathrm{Cay}(A, C \cup f(D))$ planar by Lemma 13.3.7. We can thus explore pairs $B < A$ where A has a planar generating system containing generators of the form $f(D)$. Moreover, we have that f induces a graph isomorphism.

The following subgroup-group relations $B < A$ from Table 13.1, page 267 with outer planar subgroups B have to be considered:

- $\mathbb{Z}_3 < A_4, S_4, A_5, \mathbb{Z}_2 \times A_4, \mathbb{Z}_2 \times S_4, \mathbb{Z}_2 \times A_5$,
- $\mathbb{Z}_4 < S_4, \mathbb{Z}_2 \times S_4$,
- $\mathbb{Z}_5 < A_5, \mathbb{Z}_2 \times A_5$,
- $\mathbb{Z}_6 < \mathbb{Z}_2 \times A_4, \mathbb{Z}_2 \times S_4, \mathbb{Z}_2 \times A_5$,
- $\mathbb{Z}_{10} < \mathbb{Z}_2 \times A_5$,
- $D_2 < A_4, S_4, A_5, \mathbb{Z}_2 \times A_4, \mathbb{Z}_2 \times S_4, \mathbb{Z}_2 \times A_5$,
- $D_3 < S_4, A_5, \mathbb{Z}_2 \times S_4, \mathbb{Z}_2 \times A_5$,
- $D_4 < S_4, \mathbb{Z}_2 \times S_4$,
- $D_5 < A_5, \mathbb{Z}_2 \times A_5$,
- $D_6 < \mathbb{Z}_2 \times S_4, \mathbb{Z}_2 \times A_5$,
- $D_{10} < \mathbb{Z}_2 \times A_5$.

Indeed, the Cayley graph of the corresponding $A^{C \cup f(D)}$ either is not planar or does not fulfill the "one face condition" of Lemma 13.3.7. This is proved in Example 13.3.21 for $\mathbb{Z}_3 < A_4$.

Example 13.3.21 (Not on one face). Consider $\mathrm{Cay}(\mathbb{Z}_3^{\{1\}} \xrightarrow{f(1)=(123)} A_4^{\{(12)(34)\}})$. Then f maps the $\vec{C_3} = \{0,1,2\}$ of the graph of $\mathbb{Z}_3^{\{1\}}$ onto the directed triangle $\vec{C_3}$ consisting of the points $e, e(123), e(123)^2$ in the graph of $A_4^{\{(123),(12)(34)\}}$. So, $0 * (12)(34) = e(12)(34) = (12)(34)$, $1 * e(123)(12)(34) = (123)(12)(34)$ and so on. Consequently, the generator

(12)(34) connects the points 0, 1, 2 of $\vec{C_3}$ to $e(12)(34), e(123)(12)(34), e(123)^2(12)(34)$ of $\mathrm{Cay}(A_4^{\{(123),(12)(34)\}})$. The result is not planar as these three points are not on one face of $\mathrm{Cay}(A_4^{\{(123),(12)(34)\}})$, as can be seen from Example 13.1.3.

That also the other cases from the above list cannot be planar, can be seen by inspection of the respective figures. Consequently only the following subgroup-group relations remain:

- $\mathbb{Z}_n < \mathbb{Z}_{kn}, D_{kn}, \mathbb{Z}_2 \times \mathbb{Z}_{kn}, \mathbb{Z}_2 \times D_{kn}$,
- $D_n < D_{kn}, \mathbb{Z}_2 \times D_{kn}$.

We will first concentrate on the case $B = \mathbb{Z}_n$. By Corollary 13.3.16, we know that $\mathbb{Z}_n \overset{\text{inj}}{\longrightarrow} \mathbb{Z}_{kn}$ and similarly $\mathbb{Z}_n \overset{\text{inj}}{\longrightarrow} \mathbb{Z}_2 \times \mathbb{Z}_{kn}$ for $k \geq 3$ are nonplanar. Moreover, one can see that $\mathbb{Z}_n \overset{\text{inj}}{\longrightarrow} \mathbb{Z}_2 \times \mathbb{Z}_{2n}$ is nonplanar, since the only possible lower graph $\mathrm{Cay}(\mathbb{Z}_2 \times \mathbb{Z}_{2n}, \{(1,0), (0,1), (0,2)\})$ is not planar. Similarly, for $\mathbb{Z}_n \overset{\text{inj}}{\longrightarrow} D_{kn}$ and $k \geq 2$ one can see that the planar generating system of D_{kn} consisting of one order two and one order kn generator cannot be extended by a generator of order n and remain planar. If alternatively, one generates D_{kn} with two elements of order two and adds one element of order n, then the resulting lower graph is planar but violates the "one face condition" of Lemma 13.3.7. Finally, note that $\mathbb{Z}_n, n > 2$, is not a subgroup of the planar generated $\mathbb{Z}_2 \times D_{kn}$, since a plane representation of $\mathbb{Z}_2 \times D_{kn}$ needs three generators of order 2, compare Table 13.1, page 267. Hence, also this case is nonplanar.

Let us now show that the remaining cases are planar: For $\mathbb{Z}_n \overset{\text{inj}}{\longrightarrow} \mathbb{Z}_n, \mathbb{Z}_n \overset{\text{inj}}{\longrightarrow} \mathbb{Z}_{2n}, \mathbb{Z}_n \overset{\text{inj}}{\longrightarrow} \mathbb{Z}_2 \times \mathbb{Z}_n$; see Corollary 13.3.16 and Figure 13.27, page 297 for $n = 3$. Planarity for the case $\mathbb{Z}_n \overset{\text{inj}}{\longrightarrow} D_n$ and the case $\mathbb{Z}_n \overset{\text{inj}}{\longrightarrow} \mathbb{Z}_2 \times \mathbb{Z}_n$ illustrated on the left of Figure 13.27, p. 297, for the latter only the middle cycle has to be reversed.

We collect the results in the following theorem.

Theorem 13.3.22. *For $n > 2$, the Clifford semigroup $\mathbb{Z}_n \overset{inj}{\longrightarrow} A$ is planar if and only if $A \in \{\mathbb{Z}_n, \mathbb{Z}_{2n}, \mathbb{Z}_2 \times \mathbb{Z}_n, D_n\}$. Moreover, $\mathrm{Cay}(\mathbb{Z}_n^{\{1\}} \overset{f(1)=1}{\longrightarrow} \mathbb{Z}_n^{\{0\}})$, as well as $\mathrm{Cay}(\mathbb{Z}_n^{\{1\}} \overset{f(1)=1}{\longrightarrow} \mathbb{Z}_n^{\{1\}})$ are $\vec{C_n}$-prisms, $\mathrm{Cay}(\mathbb{Z}_n^{\{1\}} \overset{f(1)=(0,1)}{\longrightarrow} (\mathbb{Z}_2 \times \mathbb{Z}_n)^{\{(1,0)\}})$ and $\mathrm{Cay}(\mathbb{Z}_n^{\{1\}} \overset{f(1)=(1\ldots n)}{\longrightarrow} D_n^{\{a\}}), a^2 = e$, are double $\vec{C_n}$-prisms, and $\mathrm{Cay}(\mathbb{Z}_n^{\{1\}} \overset{f(1)=2'}{\longrightarrow} \mathbb{Z}_{2n}^{\{1'\}})$ is a $\vec{C_n}$-prism over a $\vec{C_n}$-antiprism.*

Example 13.3.23 (Double $\vec{C_3}$-prism, $\vec{C_3}$-prism over $\vec{C_3}$-antiprism). The lower graph of the Cayley graph of $\mathbb{Z}_3^{\{1\}} \overset{f(1)=(0,1)}{\longrightarrow} (\mathbb{Z}_2 \times \mathbb{Z}_3)^{\{(1,0)\}}$, is a $\vec{C_3}$-prism. The entire graph is planar, Figure 13.27, page 297.

Now we turn our attention to the case where $B = D_n$. In all cases, $B = D_n$ is generated by two degree two elements $D = \{a, b\}$. First, consider $A = \mathbb{Z} \times D_{kn}$ with $k \geq 2$, where $C \cup f(D)$ has to consist of three generators of order two. However, since f is a homomorphism this implies that $|C \cup f(D)| > 3$, which contradicts planarity. Now consider $A = D_{kn}$ for $k \geq 2$ and suppose that C contains an additional element w

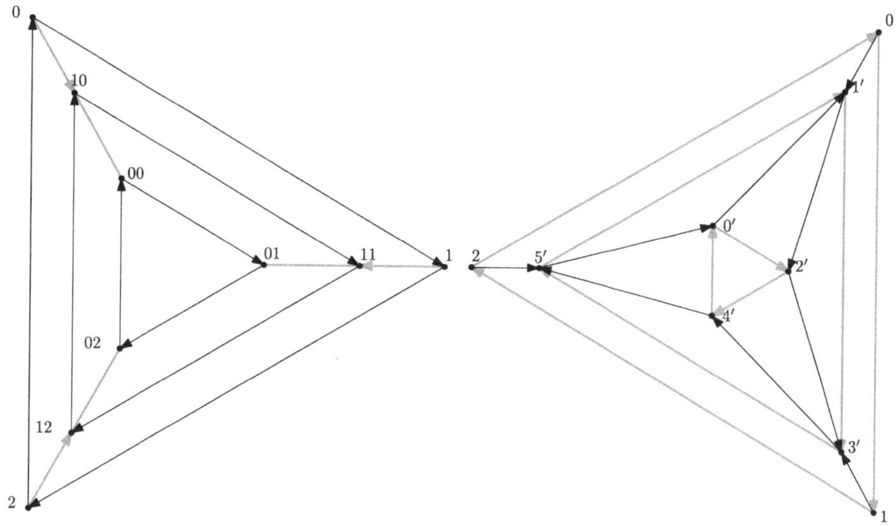

Figure 13.27: A plane representation of $\mathrm{Cay}(\mathbb{Z}_3^{\{1\}} \xrightarrow{f(1)=(0,1)} (\mathbb{Z}_2 \times \mathbb{Z}_3)^{\{(1,0)\}})$ and $\mathrm{Cay}(\mathbb{Z}_3^{\{1\}} \xrightarrow{f(1)=2'} \mathbb{Z}_6^{\{1'\}})$.

of order kn. The only way to choose a, b, w such that $\mathrm{Cay}(D_{kn}, C \cup f(D))$ is planar is such that $f(a) = wa'$ and $f(b) = b'$. The resulting graph is a prism with diagonals in some four-faces. However, one can check that for any $k \geq 1$, the "one face condition" of Lemma 13.3.7 is violated. Suppose now, that C contains an additional element a' of order two. Since f is a homomorphism, one can set $f(b) = b'$ of order two but $f(a) = (a'b')^{k-1}a' = w$ is necessary for respecting the homomorphism, i. e., w has to be of order two and wb' of order n. One finds that these long edges corresponding to w destroy planarity of $\mathrm{Cay}(D_{kn}, C \cup f(D))$ if $k \geq 3$. For instance, observe that in $\mathrm{Cay}(D_2^{\{a,b\}} \xrightarrow{f(a)=a'b'a'b'a', f(b)=b'} D_6^{\{a'\}})$, already $\mathrm{Cay}(D_6, \{a', b', a'b'a'b'a'\})$ is not planar. The positive cases, i. e., $A \in \{D_n, \mathbb{Z}_2 \times D_n, D_{2n}\}$ are illustrated in Figure 13.28, page 298, in Example 13.3.25.

We resume the preceding discussion, where a, b, a', b' stand for order two generators of the corresponding dihedral groups.

Theorem 13.3.24. *The Clifford semigroup* $D_n \xrightarrow{inj} A$ *is planar if and only if* $A \in \{D_n, \mathbb{Z}_2 \times D_n, D_{2n}\}$. *Moreover,* $\mathrm{Cay}(D_n^{\{a,b\}} \xrightarrow{f(a)=a, f(b)=b} D_n^{\{e\}})$ *is a 2n-prism and* $\mathrm{Cay}(D_n^{\{a,b\}} \xrightarrow{f(a)=(0,a), f(b)=(0,b)} (\mathbb{Z}_2 \times D_n)^{\{(1,e)\}})$ *as well as* $D_n^{\{a,b\}} \xrightarrow{f(a)==a'b'a'b'a', f(b)=b'} D_{2n}^{\{a'\}}$ *are double 2n-prisms.*

Example 13.3.25 (Several dihedral prisms). In Figure 13.28, we exhibit several dihedral prisms. Observe that in $\mathrm{Cay}(D_2^{\{a,b\}} \xrightarrow{f(a)=a'b'a'b'a', f(b)=b'} D_6^{\{a'\}})$, already $\mathrm{Cay}(D_6, \{a',$

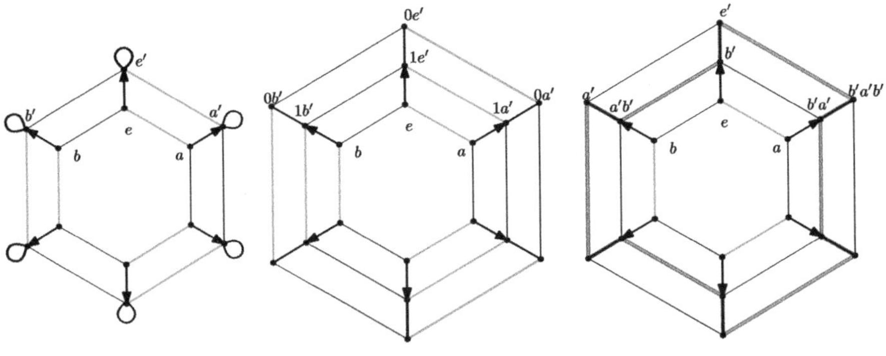

Figure 13.28: Several dihedral prisms: $\mathrm{Cay}(D_3^{\{a,b\}} \xrightarrow{f(a)=a',f(b)=(b')} D_3^{\{e'\}})$, $\mathrm{Cay}(D_3^{\{a,b\}} \xrightarrow{f(a)=(0,a'),f(b)=(0,b')} (\mathbb{Z}_2 \times D_3)^{\{(1,e')\}})$, $\mathrm{Cay}(D_3^{\{a,b\}} \xrightarrow{f(a)=a'b'a',f(b)=b'} D_6^{\{a'\}})$, $(a'b'a'b')^2 = e' = e_{D_6}$.

$b', a'b'a'b'a'\})$ is not planar. Observe that, considering $\mathrm{Cay}(D_3^{\{a,b\}} \xrightarrow{f(a)=a',f(b)=(b')} D_3^{\{a'\}})$ instead of $\mathrm{Cay}(D_3^{\{a,b\}} \xrightarrow{f(a)=a',f(b)=(b')} D_3^{\{e'\}})$ in the left most figure, gives a minor $K_{3,3}$.

The main result of this section, Theorem 13.3.5 in detail

The following list collects all planar Clifford semigroups $B \xrightarrow{f} A$, it is exhaustive. We also give planar generating systems and cite the corresponding results and figures. By $a, b, a', b' \in D_m$ we denote generators of order 2. As usual, $m, n \geq 1$.

Case 1. $B^D \xrightarrow{c_{e_A}} A^C$: (Theorem 13.3.10)

(a) $\mathbb{Z}_1$ or $\mathbb{Z}_2 \xrightarrow{c_e} D_n, A_4, S_4, A_5, \mathbb{Z}_2 \times \mathbb{Z}_{2n}$, $|C| = 2$, (Figure 13.26, page 293);

(b) $\mathbb{Z}_n \xrightarrow{c_0} \mathbb{Z}_m^{\{1\}}$ or $\mathbb{Z}_1^{\{0\}}$;

(c) $D_n^{\{a,b\}} \xrightarrow{c_0} \mathbb{Z}_m^{\{1\}}$ or $\mathbb{Z}_1^{\{0\}}$.

Case 2. $B^D \xrightarrow{\text{non-inj} \neq c_e} A^C$ implies $B = D_2 (\cong \mathbb{Z}_2 \times \mathbb{Z}_2)$: (Corollary 13.3.20)

Here, $a \in D_2$ corresponds to $(1,0) \in \mathbb{Z}_2 \times \mathbb{Z}_2$ and $b \in D_2$ to $(0,1) \in \mathbb{Z}_2 \times \mathbb{Z}_2$.

(a) $D_2 \xrightarrow{f(a)=1,f(b)=0} \mathbb{Z}_2^{\{0\}}$ or $\mathbb{Z}_2^{\{1\}}$;

(b) $D_2 \xrightarrow{f(a)=2,f(b)=0} \mathbb{Z}_4^{\{1\}}$, (Figure 13.25, page 292);

(c) $D_2 \xrightarrow{f(a)=a',f(b)=e'} D_n^{\{b'\}}$, (Figure 13.25, page 292);

(d) $D_2 \xrightarrow{f(a)=(1,(12)(34)),f(b)=(0,e)} \mathbb{Z}_2 \times A_4^{\{(0,(123))\}}$.

(e) $D_2 \xrightarrow{f(a)=(1,0),f(b)=(0,0)} \mathbb{Z}_2 \times \mathbb{Z}_n^{\{(0,1)\}}$.

Case 3*a*. $B = \mathbb{Z}_2 \xrightarrow{\text{inj}}$: (Theorem 13.3.18)

(a) $\mathbb{Z}_2 \xrightarrow{f(1)=1} \mathbb{Z}_2^{\{0\}}$ or $\mathbb{Z}_2^{\{1\}}$;

(b) $\mathbb{Z}_2 \xrightarrow{f(1)=2'} \mathbb{Z}_4^{\{1'\}}$;

(c) $\mathbb{Z}_2 \xrightarrow{f(1)=a} D_n^{\{b\}}$;

(d) $\mathbb{Z}_2 \xrightarrow{f(1)=(1,0)} \mathbb{Z}_2 \times \mathbb{Z}_n^{\{(0,1)\}}$, (Figure 13.26, page 293);

(e) $\mathbb{Z}_2 \xrightarrow{f(1)=(1,(12)(34))} \mathbb{Z}_2 \times A_4^{\{(0,(123))\}}$.

Case 3*b*. $B = \mathbb{Z}_n \xrightarrow{\text{inj}}$, $n > 2$: (Theorem 13.3.22)

(a) $\mathbb{Z}_n \xrightarrow{\text{inj}} \mathbb{Z}_n^{\{0\}}$ or $\mathbb{Z}_n^{\{1\}}$;

(b) $\mathbb{Z}_n \xrightarrow{f(1)=(0,1)} \mathbb{Z}_2 \times \mathbb{Z}_n^{\{(1,0)\}}$, (Figure 13.27, page 297);

(c) $\mathbb{Z}_n \xrightarrow{f(1)=2'} \mathbb{Z}_{2n}^{\{1'\}}$, (Figure 13.27, page 297);

(d) $\mathbb{Z}_n \xrightarrow{f(1)=(1\dots n)} D_n^{\{a\}}$.

Case 4. $B = D_n \xrightarrow{\text{inj}}$: (Theorem 13.3.24, Figure 13.28, page 298)

(a) $D_n \xrightarrow{f(a)=a, f(b)=b} D_n^{\{e\}}$;

(b) $D_n \xrightarrow{f(a)=(0,a), f(b)=(0,b)} \mathbb{Z}_2 \times D_n^{\{(1,e)\}}$;

(c) $D_n \xrightarrow{f(a)=w, f(b)=b'} D_{2n}^{\{a'\}}, (wb')^2 = e_{D_{2n}}$.

Some generalizations

Project 13.3.26.

(1) Take the strong semilattice $S = (B \xrightarrow{c_e} A)$ with a two-generated planar group A, where B is a Clifford semigroup. Then $\text{Cay}(S, C \cup D)$ is planar if and only if $\text{Cay}(B, D)$ is a path, i. e., $B \in \{\mathbb{Z}_1 \xrightarrow{c_0} \mathbb{Z}_2, \mathbb{Z}_1 \xrightarrow{c_0} \mathbb{Z}_1, \mathbb{Z}_2, \mathbb{Z}_1\}$. Compare Remark 13.1.6. For this Clifford semigroup, we write $\mathbb{Z}_1^{\{0\}} \xrightarrow{c_0} \mathbb{Z}_2^{\{1\}} \xrightarrow{c_e} A^C$.
If $A = \mathbb{Z}_m$, we can take for B any outer planar semigroup. So, e. g., we get that $\text{Cay}((\mathbb{Z}_n \times R_2)^{\{(1,r_1),(0,r_2)\}} \xrightarrow{c_0} \mathbb{Z}_m^{\{1\}})$ is planar but not Clifford. See Figure 13.10, page 270, for $n = 6$. For B, we can also take any disjoint union of connected outer planar semigroups.
If $A \cong D_n$, we can take instead of B two outer planar (semi)groups with structure homomorphism $c_{e_{D_n}}$ and get a planar (Clifford) semigroup $B_1 \cup B_2 \xrightarrow{c_{e_{D_n}}, c_{e_{D_n}}} D_n$.
Other Cayley graphs of Clifford semigroups are in Diagram (b)⊕(d) of Example 11.3.22 (planar). There $\text{Cay}(S, \{1_\beta\}) \oplus \text{Cay}(S, \{1_\gamma\})$ comes from $S_\beta^{\{1_\beta\}} \cup S_\gamma^{\{1_\gamma\}} \xrightarrow{f_{\beta,\alpha}(1_\beta)=0_\alpha, f_{\gamma,\alpha}(1\gamma)=1_\alpha} S_\alpha^{\emptyset}$.
In Example 11.3.25, we have the nonplanar Cayley graph of $S_\beta^{\{1_\beta\}} \cup S_\gamma^{\{1_\gamma\}} \xrightarrow{f_{\beta,\alpha}(1_\beta)=0_\alpha, f_{\gamma,\alpha}(1\gamma)=2_\alpha} S_\alpha^{\{1_\alpha\}}$.

(2) Faces of a 2-cell embedded graph are called ***(vertex) spanning faces*** if they contain all vertices of the graph. Denote by $\text{sf}(G)$ the minimal number of spanning faces of a graph G. Observe that $\text{Cay}(A_4, \{(12)(34), (123)\})$ (see Figure 13.1, p. 262) has 3 spanning hexagons $\overrightarrow{C_6}$ (clockwise) or 4 spanning triangles $\overrightarrow{C_3}$ (counterclockwise). That is $\text{sf}(A_4) = 3$. Then we claim, that $\text{Cay}\,(A_4^{\{(12)(34),(123)\}} \xrightarrow{f} \mathbb{Z}_1)$ has genus $2 = \text{sf}(A_4) - 1$.

The corresponding graph is $\mathrm{Cay}(A_4, \{(12)(34), (123)\}) + K_1$. To show the claim, we proceed as follows. We put one point, i. e., K_1, in one spanning face, here a hexagon, and connect it to all points on the surrounding circle. Then we build sf − 1 bridges to the remaining (two) spanning faces and connect this point over the bridges to all points on the respective surrounding circle.
Similarly, $\mathrm{Cay}(\mathbb{Z}_2 \times A_4)$ and $\mathrm{Cay}(S_4)$ in Figure 13.1, page 262, have 8 spanning triangles $\overrightarrow{C_3}$, (which in the first case are all counterclockwise directed, in the second case 4 are counterclockwise and 4 are clockwise directed,) or 4 spanning octagons (which in the first case are $\overrightarrow{C_8}$ clockwise, in the second case undirected). So $\mathrm{sf}(\mathbb{Z}_2 \times A_4) = \mathrm{sf}(S_4) = 4$ and the genus of $\mathbb{Z}_2 \times A_4 \xrightarrow{f} \mathbb{Z}_1$ and of $S_4 \xrightarrow{f} \mathbb{Z}_1$ would be 3.
In $\mathrm{Cay}(A_5, \{(124), (23)(45)\})$ we have 10 spanning 10-gons ($\overrightarrow{C_{10}}$, clockwise) or 20 spanning triangles ($\overrightarrow{C_3}$, counterclockwise). So $\mathrm{sf}(A_5) = 10$ and the genus of $A_5 \xrightarrow{f} \mathbb{Z}_1$ would be 9.

(3) If we take a planar semilattice like those in Theorem 13.3.4, we can replace the minimum 0 by $\mathbb{Z}_m^{\{1\}}$ and get a planar Clifford semigroup. Generally, take a decorated X from Theorem 13.3.4 and replace points, i. e., groups $\mathbb{Z}_1$, by larger groups.
(4) In the situation of Theorem 13.3.18 and Corollary 13.3.20, the Clifford semigroups have genus 1, if the planar group $A^C \notin \{\mathbb{Z}_2^{\{1\}}, \mathbb{Z}_2^{\{0\}}, \mathbb{Z}_4^{\{1\}}, D_n^{\{a,b\}}, (\mathbb{Z}_2 \times A_4)^{\{(0,(123)),(1,(23)(45))\}}\}$, i. e., if $A^C \in \{D_n^{\{a,(1\dots n)\}}, A_4^{\{(12)(34),(123)\}}, S_4^{\{(34),(123)\}}, S_4^{\{(12),(1234)\}}, A_5^{\{(23)(45),(124)\}}, A_5^{\{(23)(45),(12345)\}}\}$.
(5) Find toroidal Clifford semigroups (which are not groups) consisting of toroidal groups.

13.4 Comments

We see the possibility to catalogue all planar Clifford semigroups containing at least one nontrivial group, which is not the minimum of the underlying semilattice.

Moreover, besides the genus of strong semilattices of groups and the genus of right and left groups, one might want to consider the genus of strong semilattices of right or left groups. Here, the results and examples of Section 11.3 will be quite useful.

The question of planar semigroups which are direct products of cyclic semigroups, has been brought up by D. V. Solomatin, e. g., in [86].

Recall also Remark 11.2.4: A study of semigroups which are subdirect products, as presented in [Petrich/Reilly 1999], will lead to many interesting questions concerning the interaction between semigroups and graphs, among them the questions of their genus.

The book by A. K. Zvonkin and S. K. Lando [Zvonkin/Lando 2010], is related to the subject of this chapter, but goes far beyond of what we have discussed here. We just mention quantum field theory and Galois theory in connection with Grothendieck's program, precise references can be found in the book. The authors of this book cite Grothendieck with the words "the objects are so simple that a child will discover them when playing," we suppose that in the first line planar graphs are meant.

This leads to another book (in Russian), which cares about this aspect and starts with planar graphs: Larisa Ju. Berezina [Berezina 2009].

List of cited papers, theses etc.

[1] S. Antohe and E. Olaru, *On homomorphisms and congruences of graphs*, Bull. Univ. Galat 11 (1978) 15–23 (in German). 27

[2] S. Arworn, *An algorithm for the numbers of endomorphisms on paths*, Discrete Math. 309 (2009) 94–103. 190

[3] Sr. Arworn, U. Knauer, and N. N. Chiangmai, *Characterization of digraphs of right (left) zero unions of groups*, Thai J. Math. 1 (2003) 131–140. 235

[4] Sr. Arworn, U. Knauer, and S. Leeratanavalee, *Locally strong enomorphisms of paths*, Discrete Math. 308 (2008) 2525–2532. 190

[5] L. Babai, *Groups of graphs on given surfaces*, Acta Math. Acad. Sci. Hung. 24 (1973) 215–221. 161

[6] L. Babai, *On the minimum order of graphs with groups*, Can. Bull. Math. 17 (1974) 467–470. 149, 150

[7] L. Babai, *Automorphism group and category of cospectral graphs*, Acta Math. Acad. Sci. Hung. 31 (1978) 295–306. 46

[8] L. Babai and J. Nešetřil, *High chromatic rigid graphs I*, Colloq. Math. Soc. János Bolyai 18 (1976) 53–60. 154

[9] L. W. Beineke, *Characterization of derived graphs*, J. Comb. Theory 9 (1970) 129–135. 100

[10] T. Bıyıkoğlu, J. Leydold, and P. F. Stadler, *Laplacian Eigenvectors of Graphs: Perron–Frobenius and Faber–Krahn type theorems* (Lecture Notes in Mathematics 1915), Springer 2007. 47

[11] M. Böttcher and U. Knauer, *Endomorphism spectra of graphs*, Discrete Math. 109 (1992) 45–57. 21, 142

[12] M. Böttcher and U. Knauer, *Postscript "Endomorphism spectra of graphs"*, Discrete Math. 270 (2003) 329–331. 21

[13] R. L. Brooks, C. A. Smith, A. H. Stone, W. T. Tutte, *The dissection of rectangles into squares*, Duke Math. J. 7 (1940), 312–340. 129

[14] L. Collatz and U. Singowitz, *Spektren endlicher Graphen*, Abh. Math. Semin. Univ. Hamb. 21 (1957) 63–77. 173

[15] K. Culik, *On the theory of graphs*, Casopis Pest. Mat. 83 (1958) (in German). 27

[16] A. Espinoza-Valdeza, R. Femata, F. C. Ordaz-Salazarb, *A model for renal arterial branching based on graph theory*, Math. Biosci. 225 (2010) 36–43. 135

[17] S. Fan, *Graphical strong representations of monoids*, Int. Conf. on Semigroups and Related Topics, Kunming 1995, Springer Singapur, pp. 130–139. 149

[18] S. Fan, *On End-regular graphs*, Discrete Math. 159 (1996) 95–102. 217

[19] S. Fan and Y. Zeng, *On Cayley graphs of bands*, Semigroup Forum 74 (2007) 99–105. 225, 260

[20] G. Foldes, G. Sabidussi, *On semigroups of graph endomorphisms*, Discrete Math. 30 (1986) 117–120. 154

[21] R. Frucht, *Herstellung von Graphen mit vorgegebener abstrakter Gruppe*, Compos. Math. 6 (1936) 239–250. 148

[22] R. Frucht, *Graphs of degree 3 with a given abstract group*, Can. J. Math. (1949) 365–378. 151

[23] M. Frenzel, *Starke Endomorphismen und Komposition von Graphen*, Diplomarbeit, Oldenburg 1986. 218, 223

[24] I. García-Marco, K. Knauer and G. Mercui-Voyant, *Cayley posets*, (in preparation). 245

[25] R. Goebel, *$\mathbb{Z}_n$-critical digraphs*, Elektr. Informationsverarb. und Kommunik. 22 (1986) 25–30. 154

[26] A. Gerwitz, L. V. Quintas, *Connected extremal graphs having symmetric automorphism group with given semigroups*, in [Tutte 1966], see also [Halin 1980] II, p. 122. 150

https://doi.org/10.1515/9783110617368-014

[27] P. Grillet, *Computing finite commutative semigroups*, Semigroup Forum 53 (1996) 140–154. 186
[28] P. Grillet, *Computing finite commutative semigroups: Part III*, Semigroup Forum 67 (2003) 185–204. 186
[29] G. Hahn, C. Tardiff, *Graph homomorphism: structure and symmetry*, in [Hahn/Sabidussi 1997]. 84
[30] Yifei Hao, Xing Gao, Yanfeng Luo, *On the Cayley graphs of Brandt semigroups*, Commun. Algebra 39 (2011) 2874–2883. 269
[31] F. Harary, C. King, A. Mowshowitz, *Cospectral graphs and digraphs*, Bull. Lond. Math. Soc. 3 (1971) 321–328. 170
[32] Z. Hedrlín, A. Pultr, *Symmetric relations (undirected graphs) with given semigroups*, Monatshefte Math. 69 (1965) 318–322. 139, 148
[33] Z. Hedrlín, A. Pultr, *On rigid undirected graphs*, Can. J. Math. 18 (1966) 1237–1242. 139
[34] P. Heidemann, *Automorphismengruppen und Endomorphismenmonoide von Box- und Kreuzprodukten endlicher Graphen*, Diplomarbeit, Oldenburg 1993. 223
[35] E. Heilbronner, *Some comments on cospectral graphs*, Match 5 (1979) 105–133. 169
[36] A. Heinze, *Construction of commuting graphs*, in K. Denecke and H.-J. Vogel (eds.), General Algebra and Discrete Mathematics: Proceedings of the Conference on General Algebra and Discrete Mathematics in Potsdam 1998. Shaker Verlag 1999, pp. 113–120. 37
[37] P. Hell, *Subdirect products of bipartite graphs* (Coll. Math. Janos Bolyai 10, Infinite and finite sets, 1973, Vol II), pp. 857–866, North Holland, Amsterdam 1975. 27
[38] P. Hell, *On some independence results in graph theory* in Proc. Conf. Algebraic Aspects of Combinatorics, Univ. Toronto, pp. 89–122, Univ. Toronto, Winnipeg 1975. 153
[39] J. Hoffman, *On the exceptional case in the characterization of the arcs of a complete graph*, IBM J. Res. Dev. 4 (1960) 487–496. 101
[40] W. Imrich, *Graphs with transitive Abelian automorphism group*, in: Combinatorial Theory and Its Applications II, North-Holland, Amsterdam 1970, pp. 651–656. 179
[41] H. Izbicki, *Reguläre Graphen beliebigen Grades mit vorgegebenen Eigenschaften*, Monatshefte Math. 61 (1957) 42–50. 152
[42] Z. Jiang, *An answer to a question of Kelarev and Praeger on Cayley graphs of semigroups*, Semigroup Forum, 69 (2004) 457–461. 260
[43] R. Kaschek, *Über das Endomorphismenmonoid des lexikographischen Produktes endlicher Graphen*, Dissertation, Oldenburg 1990. 212, 213
[44] A. V. Kelarev and C. E. Praeger, *On transitive Cayley graphs of groups and semigroups*, Eur. J. Comb. 24 (2003) 59–72. 254, 257, 260
[45] B. Khosravi, *On Cayley graphs of completely simple semigroups*, Bull. Malays. Math. Sci. Soc. 41 (2018) 741–749. 225
[46] B. Khosravi and B. Khosravi, *On Cayley graphs of semilattices of semigroups*, Semigroup Forum 86 (2013) 114–132. 225
[47] M. Kilp, U. Knauer, *Graph operations and categorical constructions*, Acta Comment. Univ. Tartu Math. 5 (2001) 43–57. 84
[48] K. Knauer and U. Knauer, *On toroidal right groups*, Thai J. Math. 8 (2010) 483–490. 279
[49] K. Knauer and U. Knauer, *On planar right groups*, Semigroup Forum 92(1) (2016) 142–157. 270
[50] U. Knauer, *Unretractive and S unretractive joins and lexicographic products of graphs*, J. Graph Theory 11 (1987) 429–440. 211, 219
[51] U. Knauer, *Endomorphisms of graphs II. Various unretractive graphs*, Arch. Math. 55 (1990) 193–203. 140, 211, 219
[52] U. Knauer, *Endomorphism types of trees*, in M. Ito (ed.), Words, Languages and Combinatorics, World Scientific, Singapore 1992, pp. 273–287. 24

[53] U. Knauer, *Endomorphism types of bipartite graphs*, in M. Ito, H. Jürgensen (eds.), Words, Languages and Combinatorics II, World Scientific, Singapore 1994, pp. 234–251. 24

[54] U. Knauer, M. Michels, *The congruence classes of paths and cycles*, Discrete Math. 309 (2009) 5352–5359. 190

[55] U. Knauer and A. Wanichsombat, *Completely regular endomorphisms of split graphs*, Ars Comb. 115 (2014) 367–384. 208

[56] U. Knauer, Y. Wang, and X. Zhang, *Functorial properties of Cayley constructions*, Acta Comment. Univ. Tartu Math. 10 (2006) 17–29. 225

[57] J. Koetters, Almost all graphs are rigid – revisited, in [Kaschek/Knauer 2009], pp. 5420–5424. 144

[58] A. D. Korschunov, *Basic properties of stochastic graphs*, Usp. Mat. Nauk 40 (1985) 107–173 (in Russian, with English translation). 144

[59] H. Maschke, *The representation of finite groups, especially of the rotation groups of the regular bodies of three- and four-dimensional space, by Cayley's color diagrams*, Am. J. Math. 18 (Apr 1896) 156–194. 263, 267

[60] J. Meksawang, S. Panma, and U. Knauer, *Characterization of completely simple semigroup digraphs*, Algebra Discrete Math. 12 (2011) 53–69. 225

[61] E. Mendelsohn, *On a technique for representing semigroups as endomorphism semigroups of graphs with given properties*, Semigroup Forum 4 (1972) 238–294. 154

[62] B. Mohar, *Some applications of Laplacian eigenvalues of graphs*, in [Hahn/Sabidussi 1997], pp. 227–275. 47

[63] A. Mowshowitz, *Graphs, groups and matrices*, Proc. Canad. Math. Congr. (1971) 509–522. 176

[64] A. Mowshowitz, *The characteristic polynomial of a graph*, J. Comb. Theory B 12 (1972) 177–193. 45

[65] A. Mowshowitz, *The adjacency matrix and the group of a graph*, in [Harary/Palmer 1973], pp. 129–148. 176

[66] R. Novakovski and I. Rival, *Fixed-edge theorem for graphs without loops*, J. Graph Theory 3 (1979) 339–350. 188

[67] R. Novakovski and I. Rival, *Retract rigid Cartesian products of graphs*, Discrete Math. 70 (1988) 169–184. 23

[68] U. Nummert, *On the monoid of strong endomorphisms of wreath products of graphs*, Mat. Zametki 41 (1987) 844–853, the journal is translated into English under the title "Mathematical Notes". 214

[69] Sayan Panma, *On transitive Cayley graphs of strong semilattices of some completely regular semigroups*, PhD thesis, Chiang Mai 2007. 247

[70] S. Panma, U. Knauer, and S. Arworn, *On transitive Cayley graphs of right (left) groups and of Clifford semigroups*, Thai J. Math. 2 (2004) 183–195. 247

[71] S. Panma, U. Knauer, and S. Arworn, *On transitive Cayley graphs of strong semilattices of right (left) groups*, Discrete Math. 309 (2009) 5393–5403. 247

[72] S. Panma, N. Na Chiangmai, U. Knauer, and S. Arworn, *Characterization of Clifford semigroup digraphs* Discrete Math. 306 (2006) 1247–1252. 237

[73] M. Petersdorf and H. Sachs, *Spectrum und Automorphismengruppe von Graphen*, in: Combinatorial Theory and Its Applications III, North-Holland, Amsterdam 1969, pp. 891–907. 177

[74] Viera K. Proulx, *Classification of toroidal groups*, J. Graph Theory 2 (1978) 269–279. 268, 283

[75] G. Sabidussi, *Graph multiplication*, Math. Z. 72 (1956) 446–457. 153, 223

[76] G. Sabidussi, *Graphs with given groups and given graph theoretical properties*, Can. J. Math. (1957) 515–525. 152

[77] G. Sabidussi, *The composition of graphs*, Duke Math. J. 26 (1959) 693–696. 212, 215

[78] G. Sabidussi, *Vertex-transitive graphs*, Monatshefte Math. 68 (1964) 426–438. 160
[79] G. Sabidussi, *Graphs with given infinite groups*, Monatshefte Math. 64 (1969) 64–67. 148
[80] G. Sabidussi, *Subdirect representations of graphs* (Coll. Math. Janos Bolyai 10, Infinite and finite sets, 1973, Vol. III), pp. 1199–1226, North-Holland, Amsterdam 1975. 27
[81] H. Sachs, *Beziehungen zwischen den in einem Graphen enthaltenen Kreisen und seinem charakteristischen Polynom*, Publ. Math. (Debr.), 11 (1964) 119–134. 39
[82] H. Sachs and M. Stiebitz, *Simple eigenvalues of transitive graphs*, Studia Sci. Math. Hung. 17 (1982) 77–90. 179
[83] A. J. Schwenk and R. J. Wilson, *On the eigenvalues of a graph*, in [Beineke/Wilson 1978]. 45
[84] J. J. Seidel, *Strongly regular graphs with* $(-1, 1, 0)$ *adjacency matrix having eigenvalue* 3, Linear Algebra Appl. 1 (1968) 281–298. 166
[85] S. S. Shrikhande, *The uniqueness of the* L_2 *association scheme*, Ann. Math. Stat. 30 (1959) 791–798. 101
[86] D. V. Solomatin, *Direct products of cyclic semigroups admitting a planar Cayley graph*, Siber. Electron. Math. Rep. 3 (2006) 238–252, http://semr.math.nsc.ru (in Russian). 300
[87] M. Stamer, *Endomorphismen von Koprodukten endlicher Graphen*, Diplomarbeit, Oldenburg 1993. 219
[88] S. Wang and Y. Li, *On Cayley graphs of completely 0-simple semigroups*, Open Math. 11 (2013) 924–930. 260
[89] A. Wanichsombat, *Algebraic Structure of Endomorphism Monoids of Finite Graphs*, PhD Thesis, Oldenburg 2011. 217
[90] H. Whitney, *Congruent graphs and the connectivity of graphs*, Am. J. Math. 54 (1932) 150–168. 138, 140
[91] E. Wilkeit, *Graphs with a regular endomorphism monoid*, Arch. Math. 66 (1996) 344–352. 187
[92] X. Zhang, *Clifford semigroups with genus zero*, in V. Laan, S. Bulman-Fleming and R. Kaschek (eds.), *Semigroups, Acts and Categories with Applications to Graphs* (Proceedings of the Conference 2007), Estonian Mathematical Society, Tartu 2008, pp. 151–160. 287

List of books

[Ahuja et al. 1993] R. K. Ahuja, T. L. Magnanti, J. Orlin, *Network Flows*, Prentice Hall, Englewood Cliffs, NJ 1993.

[Aigner 1976] M. Aigner, *Kombinatorik, II. Matroide und Transversaltheorie*, Springer 1976.

[Aigner 1984] M. Aigner, *Graphentheorie: Eine Entwicklung aus dem 4-Farben-Problem*, Teubner, Stuttgart 1984 (English translation 1987).

[Aigner 1987] M. Aigner (transl. L. Boron, C. Christenson, B. Smith), *Graph Theory: A Development from the 4-Color Problem*, BGS Associates, Moscow, ID 1987.

[Aldous/Wilson 2000] J. M. Aldous, R. J. Wilson, *[Graphs and Applications. An Introductory Approach]*, Springer, New York 2000.

[Anderson 1970] S. S. Anderson, *Graph Theory and Finite Combinatorics*, Markham, Chicago 1970.

[Andrafai 1977] B. Andrafai, *Introductory Graph Theory*, Akadémiai Kiado, Budapest 1977.

[Appel/Haken 1989] K. Appel, W. Haken, *Every Planar Map is Four Colorable*, American Mathematical Society, Providence, RI 1989.

[Arlinghaus et al. 2002] S. Arlinghaus, W. C. Arlinghaus, F. Harary, *Graph Theory and Geography: An Interactive View E-Book*, John Wiley & Sons, New York 2002.

[Arlinghaus 1985] W. C. Arlinghaus, *The Classification of Minimal Graphs with Given Abelian Automorphism Group* (Memoirs Amer. Math. Soc. 330), American Mathematical Society, Providence, RI 1985. 150

[Asratian et al. 1998] A. S. Asratian, Ý. M. Denley, R. Haegkvist, *Bipartite Graphs and Their Applications*, Cambridge University Press, Cambridge 1998.

[Avondo-Bodino 1962] G. Avondo-Bodino, *Economic Applications of the Theory of Graphs*, Gordon and Breach, London 1962.

[Balaban 1976] A. T. Balaban (ed.), *Chemical Applications of Graph Theory*, Academic Press, New York 1976.

[Balakrishnan/Ranganathan 2000] R. Balakrishnan, K. Ranganathan, *A Textbook of Graph Theory*, Springer, New York 2000.

[Balakrishnan 1997] V. K. Balakrishnan, *Graph Theory* (Schaum's Outline), McGraw-Hill, Boston, MA 1997.

[Bang-Jensen/Gutin 2002] J. Bang-Jensen, G. Gutin, *Digraphs*, Springer, New York 2002.

[Bapat 2011] R. B. Bapat, *Graphs and Matrices*, Springer, London 2011. 47

[Barbut 1972] M. Barbut (ed.), *Combinatoire, Graphs et Algébre*, Gauthier-Villars, Paris 1972.

[Barnette 1983] D. Barnette, *Map Colorings, Polyhedra and the Four-Color Problem*, Mathematical Association of America, New York 1983.

[Behzad/Chartrand 1971] M. Behzad, G. Chartrand, *Introduction to the Theory of Graphs*, Allyn and Bacon, Boston 1971.

[Behzad et al. 1979] M. Behzad, G. Chartrand, L. Lesniak-Foster, *Graphs & Digraphs*, Prindle, Weber & Schmidt, Boston 1979. 39, 100, 138, 140, 151, 172

[Beineke/Wilson 1978] L. W. Beineke, R. J. Wilson (eds.), *Selected Topics in Graph Theory*, Academic Press, London 1978. 41, 44, 306

[Beineke/Wilson 1983] L. W. Beineke, R. J. Wilson (ed.), *Selected Topics in Graph Theory 2*, Academic Press, London 1983.

[Beineke/Wilson 1988] L. W. Beineke, R. J. Wilson, *Selected Topics in Graph Theory 3*, Academic Press, London 1988.

[Beineke/Wilson 1997] L. W. Beineke, R. J. Wilson (eds.), *Graph Connections: Relationships Between Graph Theory and Other Areas of Mathematics*, Oxford University Press, Oxford 1997. 148, 156

https://doi.org/10.1515/9783110617368-015

[Beineke et al. 2004] L. W. Beineke, R. J. Wilson, P. J. Cameron (Eds.) *Topics in Algebraic Graph Theory*, Cambridge Univ. Press 2004.

[Bellmann et al. 1970] R. Bellmann, K. L. Cooke, J. A. Lockett, *Algorithms, Graphs and Computers*, Academic Press, New York 1970.

[Berge 1962] C. Berge, *The Theory of Graphs and its Applications*, Methuen, London 1962 (original French edition published in 1958).

[Berezina 2009] Larisa Ju. Berezina, *Graphs and their Applications, a popular book for pupils and teachers (in Russian)*, Moskva 2009. 301

[Berge 1973] C. Berge, *Graphs and Hypergraphs*, North-Holland, London 1973.

[Berge 1982] C. Berge, *The Theory of Graphs and Its Applications*, Greenwood Publishing Group, Westport, CT 1982.

[Berge 1985] C. Berge, *Graphs* (2nd revised edn.), North-Holland, New York 1985.

[Berge 1987] C. Berge, *Hypergraphes*, Gauthier-Villars, Paris 1987.

[Berge 1989] C. Berge, *Hypergraphs: Combinatorics of Finite Sets*, Elsevier Science and Technology, San Diego 1989.

[Berge 2001] C. Berge, *The Theory of Graphs*, Dover Publications, Mineola, NY 2001.

[Berman 1978] G. Berman, *Applied Graph Theory Bibliography*, University of Waterloo, Waterloo, ON 1978.

[Biggs 1974] N. Biggs, *Algebraic Graph Theory*, Cambridge University Press, Cambridge 1974.

[Biggs 1993] N. Biggs, *Algebraic Graph Theory* (2nd edn.), Cambridge University Press, Cambridge 1993.

[Biggs 1996] N. Biggs, *Algebraic Graph Theory*, Cambridge University Press, Cambridge 1996. 39, 42, 43, 100, 101, 121, 158, 159, 160

[Biggs et al. 1976] N. L. Biggs, E. K. Lloyd, R. J. Wilson, *Graph Theory 1736–1936*, Oxford University Press, London 1976.

[Boffey 1982] T. B. Boffey, *Graph Theory in Operations Research*, MacMillan Press, London 1982.

[Bollabas 1978] B. Bollabas, *Extremal Graph Theory*, Academic Press, London 1978.

[Bollobás 1979] B. Bollobás, *Graph Theory: An Introductory Course*, Springer, New York 1979.

[Bollobás 1985] B. Bollobás, *Random Graphs*, Academic Press, London 1985.

[Bollobás 2001] B. Bollobás, *Random Graphs* (2nd edn.), Cambridge University Press, Cambridge 2001.

[Bollobás 2002] B. Bollobás, *Modern Graph Theory*, Springer, New York 2002.

[Bondyet al. 2006] A. Bondy, J. Fonlupt, J.-L. Fouquet, J.-C. Fournier, J. L. Ramírez Alfonsín, *Graph Theory in Paris: Proceedings of a Conference in Memory of Claude Berge*, Birkhäuser 2006.

[Bondy/Murty 2007] A. Bondy, U. S. R. Murty, *Graph Theory: An Advanced Course*, Springer 2007.

[Bondy/Murty 1976] J. A. Bondy, U. S. R. Murty, *Graph Theory with Applications*, North-Holland, New York 1976.

[Bonnington/Little 1995] C. P. Bonnington, C. H. C. Little *The Foundations of Topological Graph Theory*, Springer, New York 1995.

[Borgelt/Kruse 2002] C. Borgelt, R. Kruse, *Graphical Models: Mehods for Data Analysis and Mining*, Wiley, New York 2002.

[Bornholdt/Schuster 2003] S. Bornholdt, H. G. Schuster (eds.), *Handbook of Graphs and Networks: From the Genome to the Internet*, Wiley, New York 2003.

[Bosak 1990] J. Bosak, *Decompositions of Graphs*, Kluwer Academic, Boston 1990.

[Brandstädt 1994] A. Brandstädt, *Graphen und Algorithmen*, Teubner, Stuttgart 1994.

[Brieskorn 1985] E. Brieskorn, *Lineare Algebra und analytische Geometrie (Band 1)*, Vieweg Verlag 1985. 42, 43

[Brouwer et al. 89] A. E. Brouwer, A. M. Cohen, A. Neumaier, *Distance Regular Graphs*, Springer, Berlin 1989.

[Bryant/Perfect 1980] V. Bryant, H. Perfect, *Independence Theory in Combinatorics*, Chapman and Hall, London 1980.

[Brylawski/Kelly 1980] T. Brylawski, D. Kelly, *Matroids and Combinatorial Geometry*, University of North Carolina, Chapel Hill 1980.

[Buckley/Harary 1990] F. Buckley, F. Harary, *Distance in Graphs*, Addison-Wesley, Redwood 1990.

[Buckley/Lewinter 2003] F. Buckley, M. Lewinter, *A Friendly Introduction to Graph Theory*, Prentice Hall, Upper Saddle River, NJ 2003.

[Busacker/Saaty 1965] R. G. Busacker, T. L. Saaty, *Finite Graphs and Networks: An Introduction With Applications*, McGraw-Hill, New York 1965.

[Cameron/Lint 1980] P. J. Cameron, J. H. Lint, *Graphs, Codes and Designs*, Cambridge University Press, Cambridge 1980.

[Cameron/van Lint 1975] P. J. Cameron, J. H. van Lint, *Graph Theory, Coding theory and Block designs*, Cambridge University Press, Cambridge 1975.

[Capobianco/Molluzzo 1977] M. Capobianco, J. C. Molluzzo, *Examples and Computerexamples in Graph Theory*, Springer, New York 1977.

[Capobianco/Molluzzo 1978] M. Capobianco and J. C. Molluzzo, *Examples and Counterexamples in Graph Theory*, North-Holland, New York 1978.

[Carrie 1979] B. Carrie, *Graphs and Networks*, Clarendon Press, Oxford 1979.

[Chachra et al. 1979] V. Chachra, P. M. Ghare, J. M. Moore, *Applications of Graph Theory Algorithmus*, North-Holland, New York 1979.

[Chartrand 1977] G. Chartrand, *Graphs as Mathematical Models*, Prindle, Weber & Schmidt, Boston 1977.

[Chartrand 1984] G. Chartrand, *Introductory Graph Theory*, Dover 1984.

[Chartrand 1985] G. Chartrand, *Introductory Graph Theory*, Dover Publications, Mineola, NY 1985.

[Chartrand/Jacobson 1996] G. Chartrand, M. S. Jacobson (eds.), *Surveys in Graph Theory*, Congressus Numerantium, Winnipeg 1996.

[Chartrand/Lesniak 1986] G. Chartrand, L. Lesniak, *Graphs & Digraphs* (2nd edn.), Wadsworth & Brooks/Cole, Menlo Park, CA 1986.

[Chartrand/Lesniak 1996] G. Chartrand, L. Lesniak, *Graphs & Digraphs* (3rd edn.), Chapman & Hall, London 1996.

[Chartrand/Lesniak 2004] G. Chartrand, L. Lesniak, *Graphs & Digraphs* (4th edn.), CRC Press, Boca Raton, FL 2004. 144

[Chartrand/Oellermann 1993] G. Chartrand, O. R. Oellermann, *Applied and Algorithmic Graph Theory*, McGraw-Hill, New York 1993.

[Chen 1971] W.-K. Chen, *Applied Graph Theory*, North-Holland, Amsterdam 1971.

[Christofides 1975] N. Christofides, *Graph Theory, An Algorithmic Approach*, Academic Press, London 1975.

[Chung 1997] F. R. K. Chung, *Spectral Graph Theory*, American Mathematical Society, Providence, RI 1997.

[Chung/Graham 1998] F. R. K. Chung, R. L. Graham, *Erdős on Graphs*, A. K. Peters, Wellesley, MA 1998.

[Clark/Holton 1991] J. Clark, D. A. Holton, *A First Look at Graph Theory*, World Scientific, New Jersey 1991.

[Cori 1975] R. Cori, *Un code pour les graphes planaires et ses applications*, Société Mathématique de France, Paris 1975.

[Coxeter et al. 1981] H. S. M. Coxeter, R. Frucht, D. L. Powers, *Zero-Symmetric Graphs*, Academic Press, New York 1981.

[Cverkovic et al. 2004] D. Cverkovic, P. Rowlinson, S. Simic, *Spectral Generalizations of Line Graphs*, Cambridge University Press, Cambridge 2004.

[Cvetcovic et al. 1997] D. Cvetcovic, P. Rowlinson, S. Simic, *Eigenspaces of Graphs*, Cambridge University Press, Cambridge 1997.

[Cvetković et al. 1988] D. Cvetković, M. Doob, I. Gutman, A. Torgasev, *Recent Results in the Theory of Graph Spectra*, North-Holland, Amsterdam 1988.

[Cvetković et al. 1979] D. Cvetković, M. Doob, H. Sachs, *Spectra of Graphs*, Academic Press, New York 1979. 37, 38, 42, 44, 46, 76, 81, 102, 176, 177

[Cvetkovich et al. 1980] D. M. Cvetkovich, M. Doob, H. Sachs, *Spectra of Graphs*, Deutscher Verlag der Wissenschaften, Berlin 1980 (Russian edition published in 1984).

[Battista et al. 1999] G. Di Battista, P. Eades, R. Tamassia, I. G. Tollis, *Graph Drawing: Algorithms for the Visualization of Graphs*, Prentice Hall 1999.

[Diestel 1990] R. Diestel *Graph Decompositions*, Clarendon Press, Oxford 1990.

[Diestel 2017] R. Diestel, *Graph Theory*, Springer, New York 2017, and earlier editions.

[Dörfler/Mühlbacher 1973] W. Dörfler, J. Mühlbacher, *Graphentheorie für Informatiker*, de Gruyter, Berlin 1973.

[Ebert 1985] J. Ebert, *Effiziente Graphenalgorithmen*, Akademische Verlagsgesellschaft, Wiesbaden 1985. 135

[Egle 1977] K. Egle, *Graphen und Präordnungen*, Bibliographisches Institut Mannheim 1977.

[Ehrig et al. 2006] H. Ehrig, K. Ehrig, U. Prange, G. Taentzer, *Fundamentals of Algebraic Graph Transformation*, (Monographs in Theoretical Computer Science. An EATCS Series) Springer 2006.

[Elmaghraby 1970] S. E. Elmaghraby, *Some Network Models in Management Science*, Springer, Berlin 1970.

[Even 1979] S. Even, *Graph Algorithmus*, Computer Science Press, Potomac, MD 1979.

[Faradzev et al. 1993] I. A. Faradzev, A. A. Ivanov, M. Klin, A. J. Woldar (Eds.), *Investigations in Algebraic Theory of Combinatorial Objects*, Springer 1993.

[Felsner 2004] S. Felsner, *Geometric Graphs and Arrangements*, Springer Vieweg 2004.

[Fiorini/Wilson 1977] S. Fiorini, R. J. Wilson, *Edge-Colourings of Graphs*, Pitman, London 1977.

[Fishburn 1985] P. C. Fishburn, *Interval Orders and Interval Graphs*, John Wiley & Sons, New York 1985.

[Flament 1963] C. Flament, *Applications of Graph Theory to Group Structure*, Prentice Hall, Engelwood Cliffs, NJ 1963 (original French edition published in 1962).

[Flament 1968] C. Flament, *Théorie des graphes et structures sociales*, Gauthier-Villar, Paris 1968.

[Fleischner 1990] H. Fleischner, *Eulerian Graphs and Related Topics* (Part 1, Vol. 1, Annals of Discrete Mathematics 45), North-Holland, Amsterdam 1990.

[Fleischner 1991] H. Fleischner, *Eulerian Graphs and Related Topics* (Part 1, Vol. 2, Annals of Discrete Mathematics 50), North-Holland, Amsterdam 1991.

[Ford/Fulkerson 1962] L. R. Ford, D. R. Fulkerson, *Flows in Networks*, Princeton University Press, Princeton 1962.

[Foulds 1992] L. R. Foulds, *Graph Theory Applications*, Springer, New York 1992.

[Framer/Stanford 1995] D. W. Framer, T. B. Stanford, *Knots and Surfaces*, American Mathematical Society, Providence, RI 1995.

[Fritsch/Fritsch 1994] R. Fritsch, G. Fritsch, *Der Vierfarbensatz*, Bibliographisches Institut, Mannheim 1994.

[Fulkerson 1975] D. R. Fulkerson (ed.), *Studies in Graph Theory, Parts 1 and 2*, Mathematical Association of America, New York 1975.

[Geogakopoulos 2017] A. Georgakopoulos, *The planar cubic Cayley graphs*, American Mathematical Soc., Providence, Rhode Island 2017. 269

[Gera et al. 2018] R. Gera, T. W. Haynes, S. Hedetniemi (Eds.) *Graph Theory. Favorite Conjectures and Open Problems - 2*, Springer 2018.

[Gera et al. 2016] R. Gera, S. Hedetniemi, C. Larson (Eds.) *Graph Theory. Favorite Conjectures and Open Problems - 1*, Springer 2016.

[Gibbons 1985] A. Gibbons, *Algorithmic Graph Theory*, Cambridge University Press, Cambridge 1985.

[Giblin 1977] P. J. Giblin, *Graphs, Surfaces and Homology*, Chapman and Hall, London 1977. 135

[Godehardt 1988] E. Godehardt, *Graphs as Structural Models*, Vieweg, Braunschweig 1988.

[Godsil/Royle 2001] Ch. Godsil, G. Royle, *Algebraic Graph Theory*, Springer, New York 2001. 91, 144, 176

[Golumbic/Hartmann 2005] M. Golumbic, I. B.-A. Hartmann (Eds.), *Graph Theory, Combinatorics and Algorithms*, Springer 2005.

[Golumbic/Trenk 2004] M. Golumbic, A. Trenk, *Tolerance Graphs*, Cambridge University Press, Cambridge 2004.

[Golumbic 1980] M. C. Golumbic, *Algorithmic Graph Theory and Perfect Graphs*, Academic Press, New York 1980.

[Gondran/Minoux 1984] M. Gondran, M. Minoux, *Graphs and Algorithms*, Wiley-Interscience, New York 1984 (French edition published by Eyrolles, Paris in 1979).

[Gould 1988] R. Gould, *Graph Theory*, Benjamin Cummings, Menlo Park, CA 1988.

[Graham et al. 1980] R. L. Graham, B. L. Rothschild, J. H. Spencer, *Ramsey Theory*, Wiley-Interscience, New York 1980.

[Graham et al. 1990] R. L. Graham, B. L. Rothschild, J. H. Spencer, *Ramsey Theory* (2nd edn.), Wiley-Interscience, New York 1990.

[Graver/Watkins 1977] J. E. Graver, M. E. Watkins, *Combinatorics With Emphasis on the Theory of Graphs*, Springer, New York 1977.

[Gross/Yellen 1999] J. Gross, J. Yellen, *Graph Theory and Its Applications*, CRC Press, Boca Raton, FL 1999.

[Gross/Yellen 2004] J. Gross, J. Yellen (eds.), *Handbook of Graph Theory*, CRC Press, Boca Raton, FL 2004.

[Gross/Tucker 1987] J. L. Gross, T. W. Tucker, *Topological Graph Theory*, Wiley-Interscience, New York 1987. 263, 267, 280

[Grossmann/Magnus 1964] I. Grossmann, W. Magnus, *Groups and Their Graphs*, Random House, New York 1964. 268

[Großmann/Magnus 1971] I. Großmann, W. Magnus, *Gruppen und ihre Graphen*, Klett, Stuttgart 1971.

[Haggard 1980] G. Haggard, *Excursions in Graph Theory*, University of Maine, Orono 1980.

[Hahn/Sabidussi 1997] G. Hahn, G. Sabidussi (eds.), *Graph Symmetry*, Kluwer, Dordrecht 1997. 232, 304, 305

[Halin 1980] R. Halin, *Graphentheorie I, II*, Wissenschaftliche Buchgesellschaft, Darmstadt 1980 (one-volume 2nd edn. published in 1981). 267, 303

[Hansen/Marcotte 1999] P. Hansen, O. Marcotte, *Graph Colouring and Applications*, (Crm Proceedings and Lecture Notes, Band 23), American Mathematical Soc., Providence, Rhode Island 1999.

[Harary 1969] F. Harary, *Graph Theory*, Addison-Wesley, Reading, MA 1969 (German edition published in 1974). 138, 148, 218

[Harary/Norman 1953] F. Harary, R. Z. Norman, *Graph Theory as a Mathematical Model in Social Science*, University of Michigan 1953.

[Harary et al. 1965] F. Harary, R. Z. Norman, D. Cartwright, *Structural Models: An Introduction to the Theory of Directed Graphs*, John Wiley & Sons, New York 1965.

[Harary/Palmer 1973] F. Harary, E. M. Palmer, *Graphical Enumeration*, Academic Press, New York 1973. 305

[Harris 1970] B. Harris (ed.), *Graph Theory and Its Applications*, Academic Press, New York 1970.

[Harris et al. 2008] J. Harris, J. L. Hirst, M. Mossinghoff, *[Adventures in Graph Theory]*, Springer 2008.

[Hartsfield/Ringel 2003] N. Hartsfield, G. Ringel, *Pearls in Graph Theory. A Comprehensive Introduction*, Dover 2003.

[Hartsfield/Ringel 1990] N. Hartsfield, G. Ringel, *Pearls in Graph Theory: A Comprehensive Introduction*, Academic Press, Boston 1990.

[Hartsfield/Ringel 1994] N. Hartsfield, G. Ringel, *Pearls in Graph Theory: A Comprehensive Introduction* (revised and augmented edition), Academic Press, Boston 1994.

[Haynes et al. 1998a] T. W. Haynes, S. T. Hedetniemi, P. J. Slater, *Fundamentals of Domination in Graphs*, Marcel Dekker, New York 1998.

[Haynes et al. 1998b] T. W. Haynes, S. T. Hedetniemi, P. J. Slater, *Domination in Graphs: Advanced Topics*, Marcel Dekker, New York 1998.

[Heesch 1970] H. Heesch, *Untersuchungen zum Vierfarbenproblem*, Bibliographisches Institut, Mannheim 1970.

[Heim 1977] O. Heim, *Graphentheorie für Anwender*, Bibliographisches Institut, Mannheim 1977.

[Hell 1974] P. Hell, *Absolute Retracts in Graphs* (Lecture Notes in Mathematics 406), Springer, Berlin 1974. 142

[Hell/Nešetřil 2004] P. Hell, J. Nešetřil, *Graphs and Homomorphisms*, Oxford University Press, Oxford 2004. 144

[Henning/Yeo 2013] M. Henning, A. Yeo *[Total Domination in Graphs]*, Springer 2013.

[Herrlich/Strecker 1973] H. Herrlich, G. Strecker, *Category Theory*, Allyn & Bacon, Boston 1973. 49

[Holton/Sheehan 1993] D. A. Holton, J. Sheehan, *The Petersen Graph*, Cambridge University Press, Cambridge 1993. 93

[Hässig 1979] K. Hässig, *Graphentheoretische Methoden des Operations Research*, Teubner, Stuttgart 1979.

[Imrich/Klavžar 2000] W. Imrich, S. Klavžar, *Product Graphs*, John Wiley & Sons, New York 2000. 27, 179, 215, 280

[Janson et al. 2000] S. Janson, T. Luczak, A. Rucinski, *Random Graphs*, John Wiley & Sons, New York 2000.

[Jemelitschev et al. 1990] V. A. Jemelitschev, O. I. Melnikov, V. I. Sarvanov, R. I. Tyschkevitsch, *Vorlesungen über Graphentheorie* (originally in Russian), Nauka, Moskow 1990.

[Jensen/Toft 1995] T. R. Jensen, B. Toft, *Graph Coloring Problems*, John Wiley & Sons, New York 1995.

[Johnson/Johnson 1972] D. E. Johnson, J. R. Johnson, *Graph Theory With Engineering Applications*, Ronald Press, New York 1972.

[Joyner/Melles 2017] D. Joyner, C. G. Melles, *[Adventures in Graph Theory]*, Springer 2017.

[Jungnickel 1982] D. Jungnickel, *Transversaltheorie*, Akademische Verlagsgesellschaft, Leipzig 1982.

[Jungnickel 1987] D. Jungnickel, *Graphen, Netzwerke und Algorithmen*, Bibliographisches Institut, Mannheim 1987 (2nd edn. 1990).

[Jungnickel 2013] D. Jungnickel *[Graphs, Networks and Algorithms]*, Springer 2013.

[Karpinski/Rytter 1998] M. Karpinski, W. Rytter, *Fast Algorithms for Graph Matching Problems*, Clarendon Press, London 1998.

[Kaschek/Knauer 2009] R. Kaschek, U. Knauer (eds.), *Graph Asymmetries*, Discrete Mathematics 309 (special issue) (2009) 5349–5424. 305

[Kaufmann 1967] A. Kaufmann, *Graphs, Dynamic Programming and Finite Games*, Academic Press, New York 1967.

[Kaufmann 1971] A. Kaufmann, *Einführung in die Graphentheorie*, Oldenburg, München 1971 (original French edition published in 1968).

[Kaufmann/Wagner 2000] M. Kaufmann, D. Wagner (eds.), *Drawing Graphs: Methods and Models*, Teubner, Stuttgart 2000.

[Kemeny/Snell 1962] J. Kemeny, J. Snell, *Mathematical Models in Social Sciences*, Blaisdell Publishing Company 1962.

[Kilp et al. 2000] M. Kilp, U. Knauer, A. V. Mikhalev, *Monoids, Acts and Categories*, De Gruyter, Berlin 2000. 49, 154, 197, 205, 227

[Klin et al. 1988] M. Klin, R. Pöschel, K. Rosenbaum, *Angewandte Algebra*, Vieweg, Braunschweig 1988. 145

[Knödel 1969] W. Knödel, *Graphentheoretische Methoden und ihre Anwendungen*, Springer, Berlin 1969. 135

[Kocay/Kreher 2005] W. Kocay, D. L. Kreher, *Graphs, Algorithms, and Optimization*, Chapman & Hall, Boca Raton, FL 2005. 135

[Kolchin 1999] V. F. Kolchin, *Random Graphs*, Cambridge University Press, Cambridge 1999.

[König 1936] D. König, *Theorie der endlichen und unendlichen Graphen*, Teubner, Leipzig 1936.

[König 1990] D. König (transl. R. McCoart), *Theory of Finite and Infinite Graphs*, Birkhäuser, Boston 1990.

[Korte et al. 91] B. Korte, L. Lovász, R. Schrader, *Greedoids*, Springer, Berlin 1991.

[Kreweras 1972] G. Kreweras, *Graphes, chaines de Markov et quelques applications économiques*, Dalloz, Paris 1972.

[Kroschunov 1985] A. D. Kroschunov, *Grundlegende Eigenschaften zufälliger Graphen mit vielen Ecken und Kanten*, UMN 1985, vol. 40, no. 1, pp. 107–173.

[Kummer 1979] B. Kummer, *Spiele auf Graphen*, Deutscher Verlag der Wissenschaften, Berlin 1979.

[Lando/Zvonkin 2004] S. K. Lando, A. K. Zvonkin, *Graphs on Surfaces and Their Applications*, Springer, New York 2004.

[Lau 1989] H. T. Lau, *Algorithms on Graphs*, Tab Books, Blue Ridge Summit, PA 1989.

[Laue 1971] R. Laue, *Graphentheorie und ihre Anwendung in den biologischen Wissenschaften*, Vieweg, Braunschweig 1971. 135

[Lauri/Scapellato 2003] J. Lauri, R. Scapellato, *Topics in Graph Automorphisms and Reconstruction*, Cambridge University Press, Cambridge 2003.

[Liebling 1970] T. N. Liebling, *Graphentheorie in Planungs- und Tourenproblemen*, Springer, Berlin 1970.

[Liu 1995] Y. P. Liu, *Embeddability in Graphs*, Kluwer, Dordrecht 1995. 27

[Liu 2017a] Y. P. Liu, *Topological theory of graphs*, De Gruyter, Berlin/Boston 2017.

[Liu 2017b] Y. P. Liu, *Algebraic elements of graphs*, De Gruyter, Berlin/Boston 2017.

[Lovász et al. 1995] L. Lovász, R. L. Graham, M. Grötschel (eds.), *Handbook of Combinatorics*, Elsevier Science, Amsterdam 1995.

[Lovász/Plummer 1986] L. Lovász, M. D. Plummer, *Matching Theory*, Akademiai Kiado, Budapest 1986.

[Läuchli 1991] P. Läuchli, *Algorithmische Graphentheorie*, Birkhäuser, Basel 1991.

[Mahadev/Peled 1995] N. V. R. Mahadev, U. N. Peled, *Threshold Graphs and Related Topics* (Annals of Discrete Mathematics 56), North-Holland, Amsterdam 1995.

[Malkevitsch/Meyer 1974] J. Malkevitsch, N. Meyer, *Graphs, Models and Finite Mathematics*, Prentice Hall, Engelwood Cliffs, NJ 1974.

[Marchette 2004] D. R. Marchette, *Random Graphs for Statistical Pattern Recognition*, Wiley, New York 2004.

[Marshall 1971] C. W. Marshall, *Applied Graph Theory*, Wiley-Interscience, New York 1971. 135

[Matousek/Nesetril 1998] J. Matousek, J. Nesetril, *Invitation to Discrete Mathematics*, Clarendon Press, Oxford 1998.

[Maxwell/Reed 1971] L. M. Maxwell, M. B. Reed, *The Theory of Graphs: A Basis for Network Theory*, Pergamon Press, New York 1971.

[Mayeda 1972] W. Mayeda, *Graph Theory*, Wiley-Interscience, New York 1972.

[McHugh 1990] J. A. McHugh, *Algorithmic Graph Theory*, Prentice Hall, Englewood Cliffs, NJ 1990.

[McKee/McMorris 1999] T. A. McKee, F. R. McMorris, *Topics in Intersection Graph Theory*, SIAM, Philadelphia 1999.

[Mehlhorn 1984] K. Mehlhorn, *Graph Algorithmus and NP-Completeness*, Springer, Berlin 1984.

[Merris 2000] R. Merris, *Graph Theory*, John Wiley & Sons, New York 2000.

[Minieka 1978] E. Minieka, *Optimization Algorithms for Networks and Graphs*, Marcel Dekker, New York 1978.

[Mirkin/Rodin 1984] B. G. Mirkin, S. N. Rodin (transl. H. L. Beus), *Graphs and Genes*, Springer, Berlin 1984.

[Mohar/Thomassen 2001] B. Mohar, C. Thomassen, *Graphs on Surfaces*, Johns Hopkins University Press, Baltimore 2001.

[Molloy/Reed 2002] M. Molloy, B. Reed, *Graph Colouring and the Probabilistic Method*, Springer, New York 2002.

[Moon 1968] J. W. Moon, *Topics on Tournaments*, Holt, Rinehart and Winston, New York 1968.

[Moon 1970] J. W. Moon, *Counting Labelled Trees*, Canadian Mathematical Congress, Montreal 1970.

[Nakanishi 1971] N. Nakanishi, *Graph Theory and Feynman Integrals*, Gordon and Breach, New York 1971.

[Nebeský 1969] L. Nebeský, *Algebraic Properties of Trees*, Universita Karlova, Praha 1969.

[Netschepurenko et al. 1990] M. I. Netschepurenko, V. K. Popkov, S. M. Mainagaschev, S. B. Kaul, V. A. Prosurjakov, V. A. Kochov, A. B. Gryzunov, *Algorithmen und Programme zur Lösung von Problemen für Graphen und Netze*, Novosibirsk 1990 (in Russian).

[Neumann 1985] K. Neumann, *Graphentheorie und verwandte Gebiete, Graphen und Netzwerke*, Fernuniversität Hagen, Hagen 1985.

[Nishizeki/Chiba 1988] T. Nishizeki, N. Chiba, *Planar Graphs: Theory and Algorithms*, North-Holland, Amsterdam 1988.

[Noltemeier 1976] H. Noltemeier, *Graphentheorie mit Algorithmen und Anwendungen*, de Gruyter, Berlin 1976.

[Novak/Gibbons 1999] L. Novak, A. Gibbons, *Hybrid Graph Theory and Network Analysis*, Cambridge University Press, Cambridge 1999.

[Nägler/Stopp 1996] G. Nägler, F. Stopp, *Graphen und Anwendungen*, Teubner, Stuttgart 1996.

[Ore 1962] O. Ore, *Theory of Graphs*, American Mathematical Society, Providence, RI 1962.

[Ore 1963] O. Ore, *Graphs and Their Uses*, Random House, New York 1963 (German edition published by Klett, Stuttgart in 1973).

[Ore 1967] O. Ore, *The Four-Color Problem*, Academic Press, New York 1967.

[Ore/Wilson 1990] O. Ore, R. J. Wilson, *Graphs and their Uses* (updated), Mathematical Association of America, New York 1990.

[Palmer 1985] E. M. Palmer, *Graphical Evolution*, John Wiley & Sons, New York 1985.

[Pautsch 1991] J. Pautsch, *Subtour – Eliminations-Verfahren zur Lösung asymmetrischer Routing-Probleme*, Vandenhoek & Ruprecht, Göttingen 1991.

[Perl 1981] J. Perl, *Graphentheorie, Grundlagen und Anwendungen*, Akademische Verlagsgesellschaft, Wiesbaden 1981.

[Pesch 1988] E. Pesch, *Retracts of Graphs*, Athenaeum, Frankfurt 1988.

[Petrich/Reilly 1999] M. Petrich, N. Reilly, *Completely Regular Semigroups*, Wiley, New York 1999. 183, 231, 245, 260, 300

[Picard 1972] C. F. Picard, *Graphes et questionnaires (two volumes)*, Gauthier-Villars, Paris 1972.

[Picard 1973] C. F. Picard, *Theorie der Fragebogen*, Akademie-Verlag, Berlin 1973 (original French edition published in 1965).

[Price 1971] W. L. Price, *Graphs and Networks: An Introduction*, Auerbach, New York 1971.

[Prisner 1995] E. Prisner, *Graph Dynamics*, Longman, Essex 1995.

[Pultr/Trnkova 1980] A. Pultr, V. Trnková, *Combinatorial, algebraic and topological representations of groups, semigroups and categories*, North-Holland Mathematical Library, Vol. 22. Amsterdam, New York, Oxford: North-Holland Publishing Company, 1980. 27

[Ramirez-Alfonsin/Reed 2001] J. L. Ramirez-Alfonsin, B. A. Reed, *Perfect Graphs*, Wiley-Interscience, New York 2001.

[Read 1972] R. C. Read (ed.), *Graph Theory and Computing*, Academic Press, New York 1972.

[Read/Wilson 1998] R. C. Read, R. J. Wilson, *An Atlas of Graphs*, Oxford University Press, New York 1998.

[Recski 1987] A. Recski, *Matroid Theory and its Applications*, Springer, Berlin 1987.

[Recski 1989] A. Recski, *Matroid Theory and its Applications*, Springer, Berlin 1989.

[Ringel 1959] G. Ringel, *Färbungsprobleme auf Flächen und Graphen*, VEB Deutscher Verlag der Wissenschaften, Berlin 1959.

[Ringel 1974] G. Ringel, *Map Color Theorem*, Springer, Berlin 1974.

[Rival 1985] I. Rival (Ed.), *Graphs and Order: The Role of Graphs in the Theory of Ordered Sets and Its Applications*, Nato Science Series, Springer 1985.

[Roberts 1978] F. S. Roberts, *Graph Theory and Its Applications to Problems of Society*, SIAM, Philadelphia 1978.

[Robinson/Foulds 1980] D. F. Robinson, L. R. Foulds, *Digraphs: Theory and Techniques*, Gordon, New York 1980.

[Roman 1978] T. Roman, *Reguläre und halbreguläre Körper*, Deutsch Verlag, Frankfurt 1978. 268

[Roy 1970] B. Roy, *Sous-ensembles de sommets remarquables d'un graph*, Dunod, Paris 1970.

[Roy/Horps 1969] B. Roy, M. Horps, *Algébre moderne et théorie des graphes orientées vers les sciences économique et sociales*, Dunod, Paris 1969.

[Saaty/Kainen 1977] T. L. Saaty, P. C. Kainen, *The Four-Color Problem: Assaults and Conquests*, McGraw-Hill, New York 1977.

[Sache 1974] A. Sache, *La théorie des graphes*, Presses Universitaire de France, Paris 1974.

[Sachs 1972] H. Sachs, *Einführung in die Theorie der endlichen Graphen II*, Teubner, Leipzig 1972. 166

[Scheinerman/Ullman 1997] E. R. Scheinerman, D. H. Ullman, *Fractional Graph Theory: A Rational Approach to the Theory of Graphs*, John Wiley & Sons, New York 1997.

[Schmidt/Stöcklein 1989] G. Schmidt, Th. Stöcklein, *Relationen und Graphen*, Springer, Berlin 1989.

[Sedlacek 1964] J. C. Sedlacek, *Einführung in die Graphentheorie*, Teubner, Leipzig 1968 (original Czech edition published in 1964).

[Seshu/Reed 1961] S. Seshu, M. B. Reed, *Linear Graphs and Electrical Networks*, Addison-Wesley, Reading, MA 1961.

[Simon 1992] K. Simon, *Effiziente Algorithmen für perfekte Graphen*, Teubner, Stuttgart 1992.

[Steinbach 1990] P. Steinbach, *Field Guide to Simple Graphs*, Design Lab, Albuquerque 1990.

[Taylor 1970] H. F. Taylor, *Balance in Small Groups*, Van Nostrand Reinhold, London 1970.

[Teh/Shee 1976] H. H. Teh, S. C. Shee, *Algebraic Theory of Graphs*, Lee Kong Chien Institute of Mathematics and Computer Science, Singpore 1976.

[Temperley 1981] H. N. V. Temperley, *Graph Theory and Applications*, Halstead Press, New York 1981.

[Thulasiraman/Swarmy 1992] K. Thulasiraman, M. N. S. Swarmy, *Graphs: Theory and Applications*, John Wiley & Sons, New York 1992.

[Tinhofer 1976] G. Tinhofer, *Methoden der angewandten Graphentheorie*, Springer, Wien 1976.

[Tinhofer 1980] G. Tinhofer, *Zufallsgraphen*, Hanser, München 1980.

[Trotter 1993] W. T. Trotter (ed.), *Perfect Graphs*, American Mathematical Society, Providence, RI 1993.

[Trudeau 1976] R. J. Trudeau, *Dots and Lines*, Kent State University Press, Kent, OH 1976.

[Trudeau 1994] R. J. Trudeau, *Introduction to Graph Theory*, Dover Publications, Mineola, NY 1994.

[Tutte 1966] W. T. Tutte, *Connectivity in Graphs*, University of Toronto Press, London, ON 1966. 303

[Tutte 1984] W. T. Tutte, *Graph Theory*, Cambridge University Press, Cambridge 1984.

[Tutte 1998] W. T. Tutte, *Graph Theory As I Have Known It*, Clarendon Press, Oxford 1998. 47, 128, 129

[Tutte 2001] W. T. Tutte, *Graph Theory*, Cambridge University Press, Cambridge 2001.

[Uebe 1970] G. Uebe, *Optimale Fahrpläne*, Springer, Berlin 1970.

[Vago 1985] I. Vago, *Graph Theory: Application to the Calculation of Electrical Networks*, Elsevier, Amsterdam 1985.

[Volkmann 1991] L. Volkmann, *Graphen und Digraphen*, Springer, Vienna 1991.

[Volkmann 1996] L. Volkmann, *Fundamente der Graphentheorie*, Springer, Berlin 1996.

[Voloshin 2002] V. Voloshin, *Coloring Mixed Hypergraphs: Theory, Algorithms and Applications*, American Mathematical Society, Providence, RI 2002.

[Voss 1991] H.-J. Voss, *Cycles and Bridges in Graphs*, Kluwer Academic, Norweel, MA 1991.

[Wagner 1970] K. Wagner, *Graphentheorie*, Bibliographisches Institut, Mannheim 1970.

[Wagner/Bodendiek 1989] K. Wagner, R. Bodendiek, *Graphentheorie I, II, III*, Bibliographisches Institut, Mannheim 1989, 1990, 1992.

[Wallis 1997] W. D. Wallis, *One-Factorizations*, Kluwer, Dordrecht 1997.

[Wallis 2000] W. D. Wallis, *A Beginner's Guide to Graph Theory*, Birkhäuser, Boston 2000.

[Wallis 2007] W. D. Wallis, *A Beginners Guide to Graph Theory*, Birkhäuser 2007.

[Walter 1979] H. Walter, *Anwendungen der Graphentheorie*, Deutscher Verlag der Wissenschaften, Berlin 1979.

[Walter/Voss 1974] H. Walter, H. J. Voss, *Über Kreise in Graphen*, Deutscher Verlag der Wissenschaften, Berlin 1974.

[Walther 1984] H. Walther, *Ten Applications of Graph Theory*, Reidel, Dordrecht 1984.

[Walther/Nägler 1987] H. Walther, G. Nägler, *Graphen, Algorithmen, Programme*, Springer, Vienna 1987.

[Watkins 2004] J. J. Watkins, *Across the Board: The Mathematics of Chess Problems*, Princeton University Press, Princeton 2004.

[Weisfeiler 1976] B. Weisfeiler, *On Construction and Identification of Graphs*, Springer, Berlin 1976.

[Welsh 1976] D. J. A. Welsh, *Matroid Theory*, Academic Press, London 1976.

[West 1996] D. B. West, *Introduction to Graph Theory*, Prentice Hall, Upper Saddle River, NJ 1996.

[West 2001] D. B. West, *Introduction to Graph Theory* (2nd edn.), Prentice Hall, Upper Saddle River, NJ 2001. 144

[White 1973] A. T. White, *Graphs, Groups and Surfaces*, North-Holland, Amsterdam 1973.

[White 1984] A. T. White, *Graphs, Groups and Surfaces* (revised edn.), North-Holland, Amsterdam 1984.

[White 2001] A. T. White, *Graphs of Groups on Surfaces: Interactions and Models*, North-Holland, Amsterdam 2001. 27, 267, 268, 269, 280, 281

[White 1987a] N. White (ed.), *Theory of Matroids*, Cambridge University Press, Cambridge 1987.

[White 1987b] N. White, *Combinatorial Geometry*, Cambridge University Press. Cambridge 1987.

[Wilson 2003] R. Wilson, *Four Colors Suffice: How the Map Problem Was Solved*, Princeton University Press, Princeton 2003.

[Wilson 2002] R. A. Wilson, *Graph Colourings and the Four-Colour Theorem*, Oxford University Press, Oxford 2002.

[Wilson 1972] R. J. Wilson, *Introduction to Graph Theory*, Academic Press, New York 1972 (German edition published in 1976).

[Wilson 1996] R. J. Wilson, *Introduction to Graph Theory* (4th edn.), Addison-Wesley/Longman, Harlow 1996.

[Wilson/Beineke 1979] R. J. Wilson, L. W. Beineke (eds.), *Applications of Graph Theory*, Academic Press, London 1979.

[Wilson/Watkins 1990] R. J. Wilson, J. J. Watkins, *Graphs: An Introductory Approach: A First Course in Discrete Mathematics*, John Wiley & Sons, New York 1990.

[Yap 1986] H. P. Yap, *Some Topics in Graph Theory* (London Mathematical Society Lecture Notes 108), London Mathematical Society, London 1986.

[Yap 1996] H. P. Yap, *Total Colourings of Graphs* (Lecture Notes in Mathematics 1623), Springer 1996. 104

[Ziegler 1995] G. M. Ziegler, *Lectures on Polytopes*, Springer 1995.

[Zhang 1997] C.-Q. Zhang, *Integer Flows and Cycle Covers of Graphs*, Marcel Dekker, New York 1997.

[Zvonkin/Lando 2010] A. K. Zvonkin, S. K. Lando, *Graphs on Surfaces and their Embedding(in Russian)*, Moskva 2010. 300

[Zykov 1969] A. A. Zykov, *Teoriia Konechnykh Grafov*, Nauka, Novosibirsk 1969.

[Zykov 1987] A. A. Zykov, *Fundamentals of Graph Theory*, Nauka, Moscow 1987 (in Russian, English translation 1990).

[Zykov 1990] A. A. Zykov, *Fundamentals of Graph Theory*, BCS Associates, Moscow, ID 1990.

Index

Index of symbols

De Gruyter Studies in Mathematics

Volume 43
Mark M. Meerschaert, Alla Sikorskii
Stochastic Models for Fractional Calculus, 2nd Edition, 2019
ISBN 978-3-11-055907-1, e-ISBN 978-3-11-056024-4, e-ISBN (ePUB) 978-3-11-055914-9

Volume 52
Miroslav Pavlović
Function Classes on the Unit Disc, 2nd Edition, 2019
ISBN 978-3-11-062844-9, e-ISBN 978-3-11-063085-5, e-ISBN (ePUB) 978-3-11-062865-4

Volume 73
Carol Jacoby, Peter Loth
Abelian Groups: Structures and Classifications, 2019
ISBN 978-3-11-043211-4, e-ISBN 978-3-11-042768-4, e-ISBN (ePUB) 978-3-11-042786-8

Volume 72
Francesco Aldo Costabile
Modern Umbral Calculus: An Elementary Introduction with Applications to Linear Interpolation and Operator Approximation Theory, 2019
ISBN 978-3-11-064996-3, e-ISBN 978-3-11-065292-5, e-ISBN (ePUB) 978-3-11-065009-9

Volume 34/1
József Lörinczi, Fumio Hiroshima, Volker Betz
Feynman-Kac-Type Theorems and Gibbs Measures: Volume 1, 2019
ISBN 978-3-11-033004-5, e-ISBN 978-3-11-033039-7, e-ISBN (ePUB) 978-3-11-038993-7

Volume 34/2
József Lörinczi, Fumio Hiroshima
Applications in Rigorous Quantum Field Theory: Volume 2, 2019
ISBN 978-3-11-040350-3, e-ISBN 978-3-11-040354-1, e-ISBN (ePUB) 978-3-11-040360-2

Volume 71
Wolfgang Herfort, Karl H. Hofmann, Francesco G. Russo
Periodic Locally Compact Groups: A Study of a Class of Totally Disconnected Topological Groups, 2018
ISBN 978-3-11-059847-6, e-ISBN 978-3-11-059919-0, e-ISBN (ePUB) 978-3-11-059908-4

Volume 70
Christian Remling
Spectral Theory of Canonical Systems, 2018
ISBN 978-3-11-056202-6, e-ISBN 978-3-11-056323-8, e-ISBN (ePUB) 978-3-11-056228-6

Printed and bound by CPI Group (UK) Ltd, Croydon, CR0 4YY
13/09/2026
14961093-0002